Pharmaceutical drug delivery systems and vehicles

Pharmaceutical drug delivery systems and vehicles

Suryakant Swain

Chinam Niranjan Patra

M. E. Bhanoji Rao

WOODHEAD PUBLISHING INDIA PVT LTD

New Delhi

Published by Woodhead Publishing India Pvt. Ltd.
Woodhead Publishing India Pvt. Ltd.,
303, Vardaan House, 7/28, Ansari Road,
Daryaganj, New Delhi - 110002, India
www.woodheadpublishingindia.com

First published 2016, Woodhead Publishing India Pvt. Ltd.
© Woodhead Publishing India Pvt. Ltd., 2016

Woodhead Publishing India Pvt. Ltd. ISBN: 978-93-85059-00-1
Woodhead Publishing India Pvt. Ltd. e-ISBN: 978-93-85059-57-5

Typeset by Mind Box Solutions, New Delhi
Printed and bound by Replika Press Pvt. Ltd.

Contents

Dedication

We dedicate this book to our well-wishers who always encourage us in achieving higher goal and motivate us writing this type of book for Pharmacy education and research.

Preface

It is our prime intention to cover the chapters of this series as comprehensively as possible. Thus, we are very pleased to introduce this volume focusing on pharmaceutical drug delivery systems and vehicles. And, as with most technological developments, they have all encountered a vast array of difficulties, ranging from problems in the synthesis of the carriers and drug conjugates to unfavourable pharmacokinetics and toxicity. Furthermore, lack of knowledge on the anatomical and physiological barriers in the body hampered application. The present volume is in several respects unique. It provides a map of the body from the viewpoint of drug targeting/drug delivery systems. We are extremely grateful to all the contributors in this book, who have given up encouragement to write all the selected chapters of pharmaceutics and drug delivery systems in order to create imagination in the mind of M.Pharma students and PhD scholars. Preparing this book took longer than anticipated, and it contains more pages than expected. This present book should prove to be useful to pharma students studying in the pharmaceutics or pharmaceutical technology specialization and could help to develop their career in the field of pharma industries as Formulation R&D scientists, technicians and mangers. I would like to think that this book might fit any of the above description, depending on the reader's need. I heartily thankful to my beloved research guide and author Prof. M.E. Bhanoji Rao and my beloved teacher Prof. Chinam Niranjan Patra of their constant and keen involvement in compilation, edition as well as creation of flow charts and figures of few selected chapters during writing of this book. Finally, I would like to thank my loving wife Linarani Swain for her love, understanding and constant support during the time of preparation of this book especially pharma students. I would like to thank my loving son Priyans Swain for giving me some time to make this book.

Dr. Suryakanta Swain

Foreword

There are very few text books covering most of the advanced aspects of novel drug delivery systems and vehicles. A need for such a book written in a simple, direct, lucid and relevant language was appropriately realized. The combined efforts of Dr. Swain, Prof. Rao and Prof. Patro have resulted in a faithful publication of this book covering major area related to new formulation techniques, development of novel drug delivery systems and application of drug delivery vehicles in academic and industrial pharmaceutical research labs. The subject matter is written with adequate practical tips and theoretical background and presented in a simple manner for better understanding among the B.Pharm, M.Pharm and PhD research scholars. I hope that you as a reader, whether you are a student, teacher in India or abroad, researcher, academicians would be pleased for this book, which should certainly add to your current knowledge, understanding and insights of theoretical and practical concepts.

It is a matter of immense pleasure for me to write a foreword for this book. I wish the editor and co-editors and all the contributing authors a great success in this combined venture and hope that their contribution to pharmaceutics or pharmaceutical sciences literature will contribute endlessly.

The chief-editor of this text book, Dr. Suryakanta Swain, is a well-established scientist with extensive academic teaching and research knowledge in the field of pharmaceutics. Dr. Swain has been doing research in pharmaceutical drug delivery systems and vehicles for last 8 years, and has patented and published peer-review journals, articles of research & review, invited editorials, thematic issues, international books and book chapters.

Prof. (Dr.) Sitty Manohar Babu
M. Pharm, Ph.D.
Principal
Southern Institute of Medical Sciences (SIMS)
College of Pharmacy, SIMS Group of Institutions
Mangaladas Nagar, Vijayawada Road
Guntur-522 001

Acknowledgements

I acknowledge with grateful appreciation the major contributions of co-editors Prof. Patro and Prof. Rao in sustaining the vitality of this book. Their respective expertise in the fields of novel drug delivery systems and vehicles has allowed the integrated approach utilized in this text book. I extend my gratitude with appreciation to all the contributing authors and academic colleagues, industry friends and my well-wishers who have shared their thoughts with me. I especially thank to the publisher, acquisitions editor, managing editor, copy editor and production manager of Woodhead Publishing India, New Delhi, contributed so expertly to the planning, preparation and production of first edition of this text book. Finally I acknowledge the support that my wife Mrs Linarani Swain and my son Priyans have lend me during my entire assignment.

Dr. Suryakanta Swain
Editor

List of contributors

Dr. Suryakanta Swain
M. Pharm, Ph.D.,
Associate Professor,
Southern Institute of Medical Sciences,
College of Pharmacy,
SIMS Group of Institutions,
Mangaladas Nagar, Vijayawada Road,
Guntur-522 001, Andhra Pradesh, (INDIA)
swain_suryakant@yahoo.co.in

Dr. Muddana Eswara Bhanoji Rao
M. Pharm, Ph.D.,
Professor-cum-Principal
Department of Pharmaceutics,
Roland Institute of Pharmaceutical
Sciences,
P.O: Khodasingi, Berhampur (Ganjam),
Pin-760 010, Odisha (INDIA)
drmebrao@yahoo.co.in

Dr. Sarwar Beg
M. Pharm (Ph.D),
Meritorious Research Fellow in Science,
University Institute of Pharmaceutical
Sciences, Panjab University (PU),
Chandigarh, (INDIA)
sarwar.beg@gmail.com

Dr. Satya Prakash Singh
Ph.D., M. Pharm,
Associate Professor
Department of Pharmaceutics,
Integral University, Lucknow,
Pin- 226026, (INDIA)
singh.satyaprakash@rediffmail.com

Dr. Chinam Niranjan Patra
M. Pharm, Ph.D., F.I.C
Professor in Pharmaceutics,
Department of Pharmaceutics,
Roland Institute of Pharmaceutical Sciences,
P.O: Khodasingi, Berhampur (Ganjam),
Pin-760 010, Odisha (INDIA)
drchniranjanpatro@gmail.com

Dr. Jnyanaranjan Panda
M. Pharm, Ph.D.,
Associate Professor
Department of Pharmachemistry,
Roland Institute of Pharmaceutical Sciences,
P.O: Khodasingi, Berhampur (Ganjam),
Pin-760 010, Odisha (INDIA)
jrpanda77@gmail.com

Dr. Kahnu Charan Panigrahi
M. Pharm (Ph.D),
Assistant Professor
Department of Pharmaceutics,
Roland Institute of Pharmaceutical Sciences,
P.O: Khodasingi, Berhampur (Ganjam),
Pin-760 010, Odisha (INDIA)
kanhu.pharma@gmail.com

Dr. Himansu Bhusan Samal
M. Pharm (Ph.D)
Associate Professor
Department of Pharmaceutics,
Guru Nanak Institutions Technical
Campus-School of Pharmacy,
Ibrahimpatnam, Hyderabad-501506 (INDIA)
himansubhusansamal@gmail.com

1

Self-emulsifying drug delivery systems

Suryakanta Swain[1*], Chinam Niranjan Patra[2] and Muddana Eswara Rao[2]
[1]Southern Institute of Medical Sciences, College of Pharmacy, Department of Pharmaceutics, SIMS Group of Institutions, Mangaldas Nagar, Vijyawada Road, Guntur-522 001, Andhra Pradesh, India
[2]Roland Institute of Pharmaceutical Sciences, Department of Pharmaceutics, Khodasinghi, Berhampur-760 010, Ganjam, Odisha, India

1.1　　　Introduction

Oral route of drug delivery is the simplest and easiest way of administering drugs due to their greater stability, high patient compliance, smaller bulk, accurate dosage, cost effectiveness and easy production [1]. Therefore, most of the new chemical entities (NCE) under development are intended to be used as a solid dosage form that originate an effective and reproducible in vivo plasma concentration after oral administration [2, 3]. In fact, most NCEs are poorly water-soluble drugs, not well absorbed after oral administration, which can detract from the drug's inherent efficacy [4]. Because of their low oral bioavailability, these drugs are not completely released in the gastrointestinal tract. A drug with poor bioavailability is one with poor aqueous solubility and slow dissolution rate in biological fluids [5]. The low bioavailable drugs have poor stability, less permeable and extensive presystemic metabolism at physiological pH through biomembrane [2, 6]. Figure 1.1 depicts the mechanisms of the physiological pathways through which the bioavailability of a drug from the conventional formulations tends to get impeded. The tremendous pharmaceutical research in understanding the causes of low oral bioavailability has led to the development of novel technologies to address these challenges. Amongst these, oral lipid-based drug delivery systems (DDS) had proved their immense potential in improving the poor and inconsistent drug absorption of poorly water-soluble drugs, especially following their administration after meals. These include various types of lipid suspensions, solutions and emulsions [7–10]. Pouton et al. studied that self-emulsifying drug delivery systems (SEDDS) can improve the bioavailability of poorly water-soluble drugs (PWSD) using Miglyol 812 and Tween 85 as surfactant and cosurfactant [11]. Self-emulsifying drug delivery systems are relatively newer lipid-based technological approach for enhancing oral bioavailability of poorly absorbed drugs [12]. These formulations have shown to reduce the slow and incomplete dissolution of a drug, facilitate the formation of its solubilized phase, increase the extent of its transportation via intestinal lymphatic system, and bypass the P-gp efflux, thereby increasing drug absorption from the GI tract [13]. SEDDS formulations can be simple binary systems: lipophilic phase and drug, or lipophilic phase, surfactant and drug [14]. SEDDS or self-emulsifying oil formulations (SEOF) are defined as isotropic mixtures of natural or synthetic oils, solid or liquid surfactants or, alternatively, one or more hydrophilic solvents and co-solvents/ surfactants [15]. Upon mild agitation followed by dilution in aqueous media, such as gastrointestinal fluids, these systems can form fine oil-in-water (o/w) emulsions or micro-emulsions (SMEDDS). SEDDS is a broad term, typically producing emulsions with a droplet size ranging from a few nano meters to several microns. Self-micro emulsifying drug delivery systems (SMEDDS) indicate the formulations forming transparent micro emulsions with oil droplets ranging between 100 nm and 250 nm. Self-nano-emulsifying drug delivery systems are a recent term construing the globule size range less than 100 nm. Fine oil droplets would pass rapidly from the stomach and promote wide distribution of drug throughout the GI tract, thereby minimizing the irritation frequently encountered during extended contact between bulk drug substances and the gut wall [16]. SEDDS SEDDS over simple oily solutions is that they provide a large interfacial are physically stable formulations as compared to emulsions which are

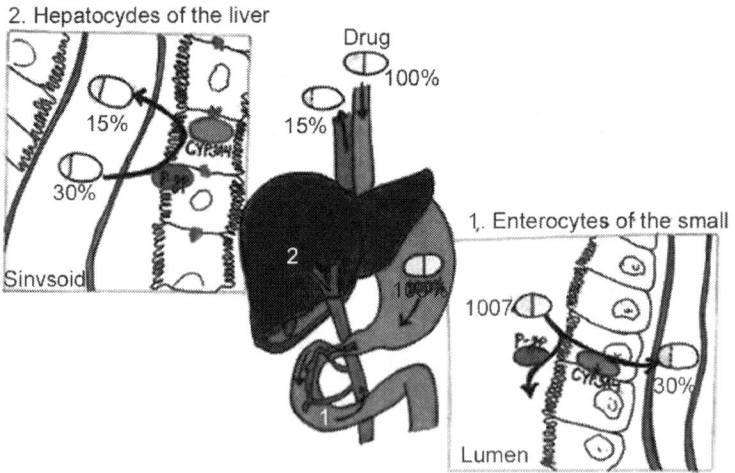

Figure 1.1 Physiological pathways leading to reduction in drug bioavailability through oral conventional dosage forms

sensitive and metastable dispersed forms. An additional advantage of area for partitioning of the drug between oil and water. Thus, SEDDS can be an efficient strategy for improving oral bioavailability of class II to IV molecules of biopharmaceutical classification system drugs [17].

1.1.1   Biopharmaceutical drug classification system

As per the Biopharmaceutical Classification System (BCS), a drug can be classified in one of the four possible categories, class I to IV on the basis of solubility and permeability characteristics [18]. The class I drugs, being highly soluble and permeable, do not normally pose any problem of rate and extent of bioavailability, with more than 90% of drug absorption [19]. Bioavailability of poorly soluble class II drugs, on the contrary, is dependent on their aqueous solubility/ dissolution rate [20]. Absorption of class III drugs is a distinct function of the permeability across GI barriers. Class IV drug compounds have neither sufficient solubility nor permeability for oral absorption to be complex [21]. For accomplishing better solubility or dissolution rate of class II drugs, use of techniques like micronization [22], cosolvents [23], micellar solubilization [24], solid dispersions [25] and complexation [26] have been employed. Diverse penetration enhancers have been used to enhance drug absorption of class III drugs effectively. The class IV drugs, on the other hand, are considered problematic for product development pharmacist, as it is difficult to improve the solubility as well as permeability of a drug using the conventional approaches. In this context self-emulsifying systems offer a unique feature of augmenting the solubility and permeability both of diverse medicinal agents. Figure 1.2 diagrammatically outlines the potential of the SEDDS formulations to overcome the problems of poor solubility and/or permeability of the BCS class II to IV drugs, leading eventually to their bioavailability enhancement.

1.2   Lipid Formulation Classification System

The Lipid Formulation Classification System (LFCS) was introduced as a working model in 2000, and an extra 'type' of formulation was added in 2006. The main purpose of the LFCS is to enable in vivo studies to be interpreted more readily and, subsequently, to facilitate the identification of the most appropriate formulations for specific drugs (i.e. with reference to their physicochemical properties) [27–32]. Table 1.1 shows the fundamental differences and their composition between types I, II, IIIA, IIIB and IV lipid-based formulations.

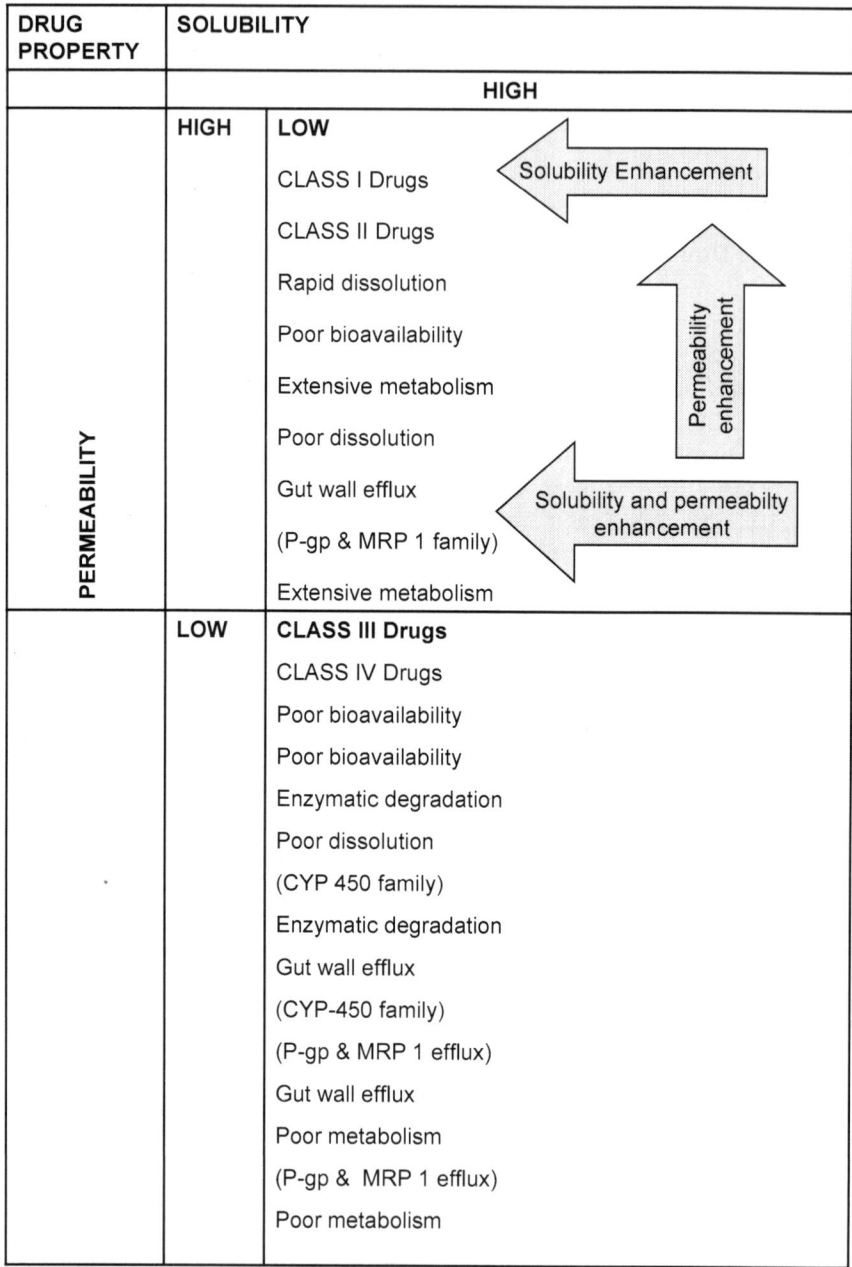

DRUG PROPERTY	SOLUBILITY	
	HIGH	
	HIGH	**LOW**
		CLASS I Drugs
		CLASS II Drugs
		Rapid dissolution
		Poor bioavailability
		Extensive metabolism
PERMEABILITY		Poor dissolution
		Gut wall efflux
		(P-gp & MRP 1 family)
		Extensive metabolism
	LOW	**CLASS III Drugs**
		CLASS IV Drugs
		Poor bioavailability
		Poor bioavailability
		Enzymatic degradation
		Poor dissolution
		(CYP 450 family)
		Enzymatic degradation
		Gut wall efflux
		(CYP-450 family)
		(P-gp & MRP 1 efflux)
		Gut wall efflux
		Poor metabolism
		(P-gp & MRP 1 efflux)
		Poor metabolism

Solubility Enhancement

Permeability enhancement

Solubility and permeabilty enhancement

Figure 1.2 Overcoming the problems of solubility and/ or intestinal permeability of Biopharmaceutical Classification System class II to IV drugs employing SEDDS

Table 1.1 Composition and salient features of various types of lipid formulations

Constituents/ Attributes	Type I	Type II	Type III a	Type III b	Type IV
Characteristics	Oils without surfactants Non-dispersing; requires digestion	SEDDS without water-soluble components	SEDDS/ SMEDDS with water-soluble components	SMEDDS with water-soluble components and low oil content	Oil-free formulation based on surfactants and cosolvents
Triglycerides or mixed surfactants	100%	40–80%	40–80%	<20%	–
Water insoluble surfactants	–	20–60%	–	–	0–20%
Water-soluble surfactants	–	–	20–40%	20–50%	30–80%
Hydrophilic cosolvents	–	–	0–40%	20–50%	0–50%
Advantages	GRAS status; simple; excellent capsule compatibility	Unlikely to lose solvent capacity on dispersion	Clear or almost clear dispersion; absorption without digestion	Clear dispersion; drug absorption without digestion	Good solvent capacity for many drugs; disperses to micellar solution
Disadvantages	Formulation has poor solvent capacity unless drug is highly lipophilic	Turbid o/w dispersion (particle size 0.25–2 µm)	Possible loss of solvent capacity on dispersion; less easily digested	Likely loss of solvent capacity on dispersion	Loss of solvent capacity on dispersion; may not be digestible

1.3   Need of SEDDS

SEDD system is a good option for the following cases: Oral delivery of poorly water-soluble compounds is to pre-dissolve the compound in a suitable solvent and fill the formulation into capsules. The main advantage of this approach is that pre-dissolving the compound overcomes the initial rate-limiting step of particulate dissolution in the aqueous environment within the GI tract. A potential problem is that the drug may precipitate out in the solution when the formulation disperses in the GI tract, particularly if a hydrophilic solvent is used (e.g. polyethylene glycol). If the drug can be dissolved in a lipid vehicle, there is less potential for precipitation on dilution in the GI tract [18]. Another strategy for poorly soluble drugs is to formulate a solid solution using a water-soluble polymer to aid the solubility of the drug compound. For example, poly-vinyl-pyrrolidone (PVP) and polyethylene glycol (PEG 6000) have been used for preparing solid solutions with poorly soluble drugs. One potential problem with this type of formulation is that the drug may favour a more thermodynamically stable state, which can result the compound crystallizing in the polymer matrix. Therefore the physical stability of such formulations needs to be assessed using techniques such as differential scanning calorimetry or X-ray crystallography [33].

1.4   Advantages and disadvantages of SEDDS

The potential advantages of self-emulsifying drug delivery systems can be summarized as follows: Enhanced oral bioavailability enabling reduction in dose, more consistent temporal profiles of drug absorption, it acts as substitute for traditional oral formulation of lipophilic drugs, predictable therapy due to reduced variability including food effects, protection of drug from hostile environment in gut, selective targeting of drug toward specific absorption window in GIT, control of delivery profiles, protection of sensitive drug substances, high drug payloads, the liquid can be used for liquid as well as solid dosage forms, ease of manufacturing and scale-up, require lower dose of drug with respect to conventional dosage forms and emulsion cannot be autoclaved as they have phase inversion temperature, while SEDDS can be autoclaved. The major hurdles in the development of self-emulsifying drug delivery systems are as follows: due to presence of high surfactant concentration, there may be chances of instabilities of drugs, sometimes irritation of git observe due to high content of surfactant in self-emulsifying formulation which may be avoided by utilizing optimum less amount of surfactants, sometime co-solvents remain into formulation and cause degradation of drugs; it may allow less drug loading; this system also lack good predictive in vitro models for assessment of the formulations, traditional dissolution methods do not work, because these formulations potentially are dependent on digestion prior to release of the drug, further development will be based on in vitro – in vivo correlations and therefore different proto-type lipid-based formulations need to be developed and tested in vivo in a suitable animal model [34, 35].

1.5   Mechanism of self-emulsification

Self-emulsification takes place when the entropy change favouring dispersion is greater than the energy required to increase the surface area of the dispersion [36, 37]. The free energy (ΔG) of a conventional emulsion is a direct function of the energy required to create a new surface between the oil and water phases and can be described by Equation (1.1) as follows:

$$r^2\sigma = \Sigma Ni4m^2\sigma \qquad ...(\text{Eq. 1.1})$$

Where ΔG is the free energy associated with the process (ignoring the free energy of mixing), N is the number of droplets of radius r and σ represents the interfacial energy [11]. The two phases of emulsion tend to separate with time to reduce the interfacial area, and subsequently, the emulsion is stabilized by emulsifying agents, which forms a monolayer of emulsion droplets, and hence reduce the interfacial energy, as well as providing a barrier to prevent coalescence. On the other hand, emulsification occurs spontaneously with SEDDS, as the free energy required to form the emulsion is low, whether positive or negative [38]. For emulsification to occur, it is necessary for the interfacial structure to have no resistance to surface shearing [39]. The addition of a binary mixture (oil/ non-ionic surfactant) to water results in interface formation between the oil and aqueous penetration through the interface; this tends to occur until the solubilization limit is close to the interface [40]. Following are the mechanisms responsible for enhanced drug absorption by SEDDS:

1.5.1   *In vivo* solubilisation of drug

The presence of lipids in the GIT stimulates an increase in the secretion of bile salts (BS) and endogenous biliary lipids including phospholipids (PL) and cholesterol (CH), leading to the formation of BS/PL/CH intestinal mixed micelles and an increase in the solubilization capacity of the GI mixed micelles and an increase in the solubilization capacity of the GIT. However, intercalation of administered (exogenous) lipids into these BS structures either directly or secondary to digestion, leads to swelling of the micellar structures either directly or secondary to digestion, leads to swelling of the micellar structures and a further increase in solubilization capacity [41].

1.5.2   Prolongation of gastric residence time

Lipids in the GI tract provoke delay in gastric emptying, i.e. gastric transit time is increased. As a result, the residence time of the co-administered lipophilic drug in the small intestine increases. This enables better dissolution of the drug at the absorptive site, and thereby improves absorption [36, 37].

1.5.3   Promotion of intestinal lymphatic transport

For highly lipophilic drugs, lipids may enhance the extent of lymphatic transport and increase bioavailability directly or indirectly via reduction in first pass metabolism [37].

1.5.4   Affecting intestinal permeability

A variety of lipids have been shown to change the physical barrier function of the gut wall, and hence, to enhance permeability [40].

1.5.5   Reduced metabolism and efflux activity

Recently, certain lipids and surfactants have been shown to reduce the activity of efflux transporters in the GI wall, and hence increase the fraction of drug absorbed. Because of the interplay between P-gp and CYP3A4 activity, this metabolism may reduce intra-enterocyte metabolism as well (Figure 1.3) [40].

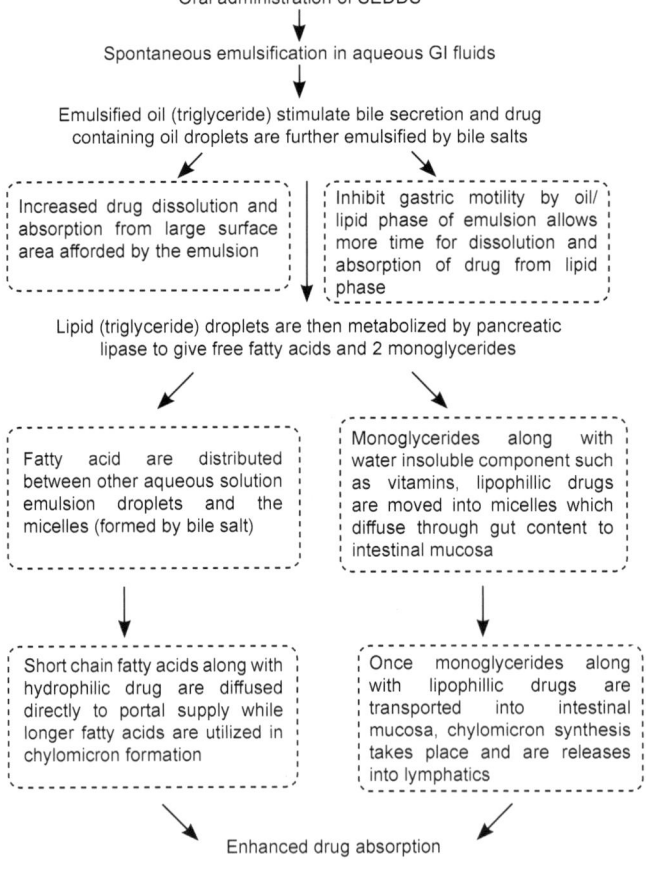

Figure 1.3 Fate of SEDDS following oral administration and mechanisms proposed for oral bioavailability enhancement of drugs

1.6   Factors affecting self-emulsification

1.6.1   Polarity of the lipophilic phase

The polarity of the lipid phase is one of the main factors that govern the drug release from the micro-emulsions. The polarity of the droplet is governed by the HLB, the chain length and degree of unsaturation of the fatty acid, the molecular weight of the hydrophilic portion and the concentration of the emulsifier. In fact, the polarity reflects the affinity of the drug for oil and/ or water, and the type of forces formed. The high polarity will promote a rapid rate of release of the drug into the aqueous phase. This is confirmed by the observations of Sang-Cheol Chi, who observed that the rate of release of idebenone from SEDDS is dependent upon the polarity of the oil phase used. The highest release was obtained with the formulation that had oil phase with highest polarity [42].

1.6.2   Nature and dose of the drug

Drugs which are administered at very high dose are not suitable for SEDDS unless they have extremely good solubility in at least one of the components of SEDDS, preferably lipophillic phase. The drugs which have limited or less solubility in water and lipids are most difficult to deliver by SEDDS. The ability of SEDDS to maintain the drug in solubilised form is greatly influenced by the solubility of the drug in oil phase. As mentioned above if surfactant or co-surfactant is contributing to the greater extent in drug solubilisation, then there could be a risk of precipitation, as dilution of SEDDS will lead to lowering of solvent capacity of the surfactant or co-surfactant. Equilibrium solubility measurements can be carried out to anticipate potential cases of precipitation in the gut [43]. However, crystallisation could be slow in the solubilising and colloidal stabilizing environment of the gut. Pouton's study reveal that such formulations can take up to 5 days to reach equilibrium and that the drug can remain in a super-saturated state for up to 24 hours after the initial emulsification event. It could thus be argued that such products are not likely to cause precipitation of the drug in the gut before the drug is absorbed, and indeed that super-saturation could actually enhance absorption by increasing the thermodynamic activity of the drug. There is a clear need for practical methods to predict the fate of drugs after the dispersion of lipid systems in the gastro-intestinal tract [42].

1.6.3   Concentration of surfactant or co-surfactant

If surfactant or co-surfactant is contributing to the greater extent in drug solubilization then there could be a risk of precipitation, as dilution of SMEDDS will lead to lowering or solvent capacity of the surfactant or co-surfactant [44].

1.7   Selection criteria of drugs for SEDDS

Upholding drug solubility within the gastro-intestinal tract and, in particular, maximizing drug solubility within the major absorptive site of the gut is one of the prime challenges to any formulation design system [45]. SEDDS offers a potential platform in enhancing the oral bioavailability of poorly aqueous soluble drugs especially those belonging to BCS class II and class IV, candidates of class II are poorly water soluble with high permeability but once dissolved they absorbed over the GIT membrane and class IV compounds are drugs with poor solubility and poor permeability, respectively [46]. Selection procedure of suitable drug candidate for SMEDDS depends majorly on under mentioned criteria.

1.7.1   Lipophilicity

The drug candidate should have sufficient solubility in pharmaceutically acceptable lipid excipients, i.e. in the United States, the Food and Drug Administration (FDA) has published listings in the Code of Federal Regulations for Generally Recommended as Safe (GRAS) substances that are generally recognized as safe its Packaging/Generally Recognized as Safe. Over the years, the Agency has also maintained a list entitled 'Inactive Ingredient Guide' for excipients that have been approved incorporated in marketed products. This guide is helpful in that it provides the database of allowed excipients with the maximum dosage level by route of administration or dosage form for each excipient [47, 48].

1.7.2   Food effects

Meals rich in fatty acids in stomach instead of empty stomach favor the absorption of drug from the lipid-based formulation because the absorption of lipophilic drug usually exhibit dissolution-rate limited. To explain the inclination for oral absorption, Lipinski's rule of five has been widely used as a qualitative predictive model. The rule of five explains poor absorption or poor permeation in terms of situation where more than five H-bond donors, more than ten H-bond acceptors, the molecular weight >500 and the calculated log p >5 are available. Both BCS and Lipinski's rule of five are useful, mainly at the primary screening stage but they have limitations. It is considered that the rule of five is only applicable to those compounds which are not substrates for active transporters, and with increasing evidence suggesting that for some efflux or uptake transporters, this limitation might be notable [49].

1.7.3   Log P value

This can be considered as the prime characteristic for lipodic system where higher log P (>4) values are desirous. For e.g. cinnarizine, atorvastatin, etc., a lipophilic drug, having log p values greater than 5 is strong candidate for SEDDS [50].

1.7.4   Melting point and dose

Low melting point and dose are desirable for development of lipoids systems, drugs having melting point and log P values (around 2) are not suitable candidates for SEDDS. Figure 1.4 illustrates major criteria for selecting suitable drug candidate for SEDDS [48].

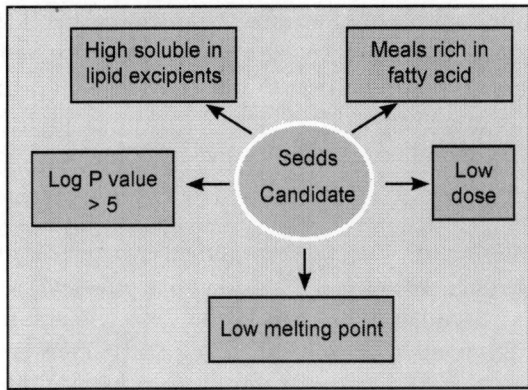

Figure 1.4 Criteria for suitable drug candidate selection for SEDDS

1.8   Excipients used in SEDDS

Self-emulsification has been shown to be specific to the nature of the oil-surfactant pair, surfactant concentration and oil/surfactant ratio, and the temperature and pH at which self-emulsification occurs. It has been demonstrated that only very specific pharmaceutical excipient combinations could lead to efficient self-emulsifying systems [27]. An ideal excipient should bear the following properties [51]. Be safe, inert and available at a purity level suitable for human use; it should not degrade during manufacturing or storage; be capable of solubilizing the drug dose in a volume not exceeding that of an oral capsule; preferably possess surface active properties to enable self-emulsification or complete dissolution of the drug dose; reliably and reproducibly enhance the oral bioavailability of the drug relative to a conventional formulation; be physically and chemically stable and compatible with a wide range of drugs and other excipients; be non-hygroscopic and inert to the capsule shell or other packaging components; and allow simple and efficient dosage form manufacture and permit ready scale-up from bench top to production-sized batches.

1.8.1   Factors affecting the choice of excipients for lipid-based formulations

There are some factors which affect the choice of excipients for lipid-based formulations as follows: regulatory issues such as irritancy, toxicity, knowledge and experience, solvent capacity of the lipid formulation on dispersion which could lead to precipitation of the drug, miscibility of the excipients that affect self-emulsification, morphology at room temperature (i.e. melting point of the formulation, self-dispersibility and role in promoting self-dispersion of the formulation, digestibility of the excipients and fate of digested products, capsule compatibility, purity of the lipid excipients and chemical stability, which could affect capsule compatibility and cost of goods [27, 52].

1.8.1.1   Lipids

Lipids represent one of the most important excipients in the SEDDS formulation not only because it can solubilise marked amounts of the lipophilic drug or facilitate self-emulsification but also and mainly because it can increase the fraction of lipophilic drug transported via the intestinal lymphatic system, thereby increasing absorption from the GI tract depending on the molecular nature of the triglyceride [53]. The oil phase used to prepare SEDDS can be formulated from various non-polar components. The formation, stability and properties of dispersions formed from SEDDS often depend on the bulk physicochemical characteristics of the oil phase, e.g. polarity, water-solubility, interfacial tension with the water phase, viscosity, density, phase behaviour and chemical stability [54]. Both long and medium chain triglyceride oils with different degrees of saturation have been used for the design of self-emulsifying formulations. Many lipids are used in oral lipid-based formulations as shown in Table 1.2. Furthermore, edible oils which could represent the logical and preferred lipid excipient choice for the development of SEDDS are not frequently selected due to their poor ability to dissolve large amounts of lipophilic drugs. Modified or hydrolyzed vegetable oils have been widely used since these excipients form good emulsification systems with a large number of surfactants approved for oral administration and exhibit better drug solubility properties. They offer formulative and physiological advantages and their degradation products resemble the natural end products of intestinal digestion. Novel semi-synthetic medium chain derivatives, which can be defined as amphiphilic compounds with surfactant properties, are progressively and effectively replacing the regular medium chain triglyceride oils in the SEOFs. Both unsaturated and saturated fatty acids have been widely used in the formulation of lipodic systems. However, the SEDDS, in particular, comprise of saturated fatty acids like, caproic, caprylic, capric, lauric and myristic acid [55]. Small developed a physicochemical system to classify lipids (including surfactants) into non-polar and polar lipids based on their interaction with bulk water and their behaviour at the water-air interface. This classification is presented in Table 1.3. Non-polar lipids do not spread to form a monolayer on water surface and are insoluble in bulk water (examples: alkanes, liquid paraffin, cholesterol esters, and fatty-acid esters, including waxes). Polar lipids are divided into four different classes and are described as insoluble non-swelling, insoluble swelling, and soluble. Soluble polar lipids are further divided into two sub-classes depending on whether or not they show formation of liquid crystalline structures at higher lipid concentration in bulk [56].

Table 1.2 Popular lipodic constituents used in the SEDDS formulations

Lipids	Examples
Medium chain triglycerides	Fractionated coconut oil, palm seed oil, triglycerides of caprylic/capric acid e.g., Miglyol® 812, Captex® 355
Long chain triglycerides	Vegetable oils are glyceride esters of mixed unsaturated long-chain fatty acid, commonly known as long-chain triglycerides e.g., soybean, sesame, corn, sunflower, castor, Olive, peanut, rapeseed oils.
Mixed mono, di- and triglycerides	Novel semi synthetic medium chain derivative. Esters of propylene glycol and mixture of mono- and diglycerides of caprylic/ capric acid. E.g., Imwitor® 988, Imwitor® 308, Maisine® 35-1, Peceol®, plurol Oleique® CC49, Capryol®, Myrj®
Polar oil	Some excipients which are traditionally thought of as hydrophobic surfactants, such as sorbitan fatty acid esters (span 80, 85s are very similar in physical properties to mixed glycerides are alternative polar oils.

1.8.1.2   Surfactants

A surfactant molecule is formed by two parts with different affinities for the solvents. One of them has affinity for water (polar solvents) and the other has for oil (non-polar solvents). Figure 1.5 shows structure of surfactant molecule containing hydrophilic head and hydrophobic tail group. A little quantity of surfactant molecules rests upon the water–air interface and decreases the water–surface tension value (the force per unit area needed to make available surface). That is why the surfactant name: "surface active agent". Surfactants used to stabilize microemulsion system may be: non-ionic, zwitter ionic, cationic, anionic surfactants, combinations of these, particularly ionic and non-ionic, can be very effective at increasing the extent of the microemulsion region. Selection of an appropriate emulsifier is one of the most important factors to consider for the proper design of SEDDS. The stability of dispersions formed from SEDDS to environmental stresses such as pH, ionic strength, and temperature variation is often predominantly determined by the type of emulsifier used [54]. Surfactants being amphiphilic in nature can dissolve (or solubilize) relatively high amounts of hydrophobic drug compounds. The emulsifiers from natural origin are regarded as much safer than the synthetic ones. However as the former possess only limited self-emulsification capacities; these are employed for the formulation of SEDDS [32]. It is generally acceptable that most stable emulsions are formed in the presence of surfactant combinations, in which one acts as an emulsifier and the other as a co-emulsifier, depending on their HLB values. For imparting high self-emulsifying properties to the SEDDS formulation, the emulsifier should have a relatively high HLB (i.e., high hydrophilicity) for immediate formation of o/w droplets, and/or rapid spreading of the formulation in the aqueous media. It would keep the drug at the site of absorption for a relatively prolonged period of time for effective absorption, as the precipitation of drug compound within the GI lumen can be prevented [57, 58]. Attempts have been made to rationalize surfactant behaviour in terms of the hydrophilic-lipophilic balance (HLB), as well as the critical packing parameter (CPP). Both approaches are fairly empirical but can be a useful guide to surfactant selection. The HLB takes into account the relative contribution of hydrophilic and hydrophobic fragments of the surfactant molecule. It accepted that low HLB (3–6) surfactants are favoured for the formation of w/o microemulsions whereas surfactants with high HLB (8–18) are preferred for the formation of o/w microemulsion systems. Ionic surfactants such as sodium dodecyl sulphate, which have HLBs greater than 20, often require the presence of a co-surfactant to reduce their effective HLB to a value within the range required for microemulsion formation. In contrast, the CPP relates the ability of surfactants which was depicted in Figure 1.6 to form particular aggregates to the geometry of the molecule itself. The analysis of film curvature for surfactant association leading to microemulsion formation has been explained by Isaraelachvili [54]. Table 1.4 summarizes an account of various marketed emulsifiers with high potential to be used in the SEDDS formulations. In terms of the packing ratio (P), the packing ratio provides a direct measure of HLB and is influenced by the same factors. The o/w structure are favoured if the effective polar part is more bulky then the hydrophobic part (P<1), and the interface curves spontaneously toward water (positive curvature). When the interface curves in the opposite direction (p>1, negative curvature), the w/o structures are formed. At zero curvature, when the HLB balanced (P~1), either bicontinuous or lamellar structures may form according to the rigidity of the film. According to the HLB value, surfactants are categorized as lipophilic (HLB ≤ 10) or hydrophilic (HLB > 10) surfactants. Non-ionic hydrophilic surfactants, with HLB values above 12 (Gelucire® 44/14, Gelucire® 50/13, Labrasol®, Cremophor® EL, Cremophor® RH 40, etc.) are generally required for SEDDS and SMEDDS formulation. Table 1.4 provides an account of various marketed emulsifiers with high potential to be employed in the SEDDS formulation. Non-ionic surfactants are considered as safer than the ionic ones. Usually for forming stable SEDDS, the surfactant concentration usually should range between 30% and 60% w/w as higher concentrations may be irritating to the GI mucosa. The properties of surfactants such as HLB value, cloud point, viscosity and affinity for oil phase, all have a strong influence on the emulsification process and droplet size. Inverse relationship between the droplet size and the concentration of the surfactant has also been observed [59].

Table 1.3 Classification of polar lipids by small

Class of polar lipids	Characteristics
I	Insoluble non-swelling
	Insoluble in water
	Cannot swell by taking up water
	Form stable monolayers at interfaces
	Examples: Triglycerides, diglycerides, cholesterol, long chain fatty acids

Contd...

Contd...

Class of polar lipids		Characteristics
II		Insoluble swelling
		Form stable monolayers at interfaces
		Insoluble in water
		Can incorporate water between their polar head groups, creating a swollen lipid structure (liquid crystalline state)
III	III a	Soluble
		Form unstable monolayers at interfaces
		Form micelles above CMC
		Form liquid crystalline structures at higher lipid concentrations
		Examples: lyso-phospholipids, sodium and potassium salts of long-chain fatty acids, amphiphiles, lipophilic surfactants with low HLB like Cremophor RH 40, Labrasol
	III b	Soluble
		Form micelles
		Form unstable monolayers at interfaces
		Do not form liquid crystalline structures at higher lipid concentrations
		Examples: conjugated and free bile salts, saponins, surfactants with high HLB

Table 1.4 Popular surfactants employed in the SEDDS formulations with their HLB values

Brand name	Chemical name	HLB	Manufacturer/ suppliers
Tween 20	PEG-20 sorbitan monolaurate	17	Atlas/ICI
Tween 60	PEG-20 sorbitan monostearate	15	
Tween 65	PEG-20 sorbitan tristearate	11	Atlas
Tween 80	PEG-20 sorbitan monooleate	15	Atlas
Tween 85	PEG-20 sorbitan trioleate	11	Atlas
Span 80	Sorbitan monooleate	4.3	Atlas
Cremophor-EL Cremophor-ELP	PEG-35 castor oil	12–14	BASF
Cremophor RH 40	PEG-40 hydrogenated castor oil	13	BASF
Labrasol	PEG-8 caprylic/capric glycerides	14	Gatteffose
Labrafac CM 10	PEG-8 caprylic/capric glycerides	>10	Gatteffose

Contd...

Contd...

Brand name	Chemical name	HLB	Manufacturer/ suppliers
Labrafil WL 2609 BS	PEG-8 corn oil	6–7	Gatteffose
Labrafil M 2125 CS	PEG-6 corn oil	4	Gatteffose
Labrafil M 1944 CS	PEG-6 apricot kernel oil	4	Gatteffose
Pluronic F 127	Polyoxyethylene-polyoxypropylene copolymers	18–23	BASF
Pluronic L- 64	Methyl-oxirane polymer with oxirane	12–18	BASF
Peceol	Glyceryl monooleate	3–4	Gatteffose
Vit E TGPS	(tocophersolan, D-α-tocopheryl PEG-1000 succinate)	13	Eastman
HCO-40	Polyoxyethylene hydrogenated castor oil 40	13	Nikkol
Brij-30	PEG-4 lauryl ether	9.7	Atlas
Lecithin	L-a-Phosphatidylcholine	4–9	Alfa Aesar
Emulphor El-620	Ethoxylated castor oil	12–15	Rhodia
Tagat TO	PEG-25 trioleate	11	Goldschmidt
Soluphor-P	Pyrrolidone-2	12–14	BASF
Transcutol P	Diethylene glycol monoethyl ether	4	Gatteffose
Solutol-HS 15	Macrogol 15 Hydroxystearate	14–16	BASF

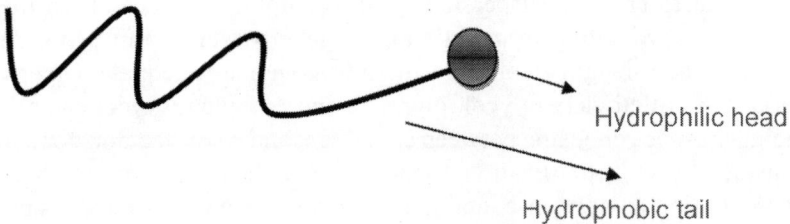

Hydrophilic head

Hydrophobic tail

Figure 1.5 Structure of surfactant molecule containing hydrophilic head and hydrophobic tail group

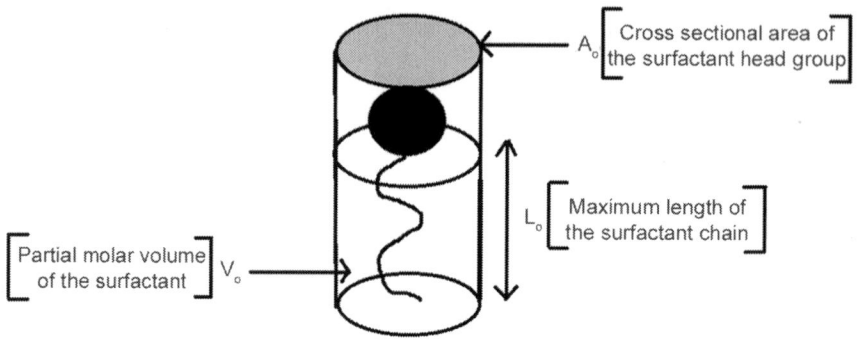

Figure 1.6 CPP of surfactant molecule

1.8.1.3   Co-surfactant and co-solvents

The production of an optimum SEDDS requires relatively high concentrations (generally more than 30% w/w) of surfactants, thus the concentration of surfactant can be reduced by incorporation of co-surfactant. Role of co-surfactant together with the surfactant is to lower the interfacial tension to a very small even transient negative value (Georgakopoulos, 1992). At this value, the interface would expand to form fine dispersed droplets, and subsequently adsorb more surfactant until their bulk condition is depleted enough to make interfacial tension positive again. This process known as 'spontaneous emulsification' forms the microemulsion. However, the use of co-surfactant in self-emulsifying systems is not mandatory for many non-ionic surfactants. The selection of surfactant and co-surfactant is crucial not only to formation of SEDDS, but also to solubilization of the drug in the SEDDS. Co-solvents in SEDDS are also used in order to increase the solubilization capacity of incorporated drugs and to enhance dispersibility of hydrophilic surfactants in the oil phase, thus promoting formulation homogeneity and stability. In general, medium-chain-length alcohols (8 to 12 C atoms) are adequate. Otherwise, derivatives of ethylene-glycol, glycerol, and propylene glycol can be also included. When choosing between co-solvent and co-surfactant, one should consider lower solubilization capacity for hydrophobic drugs observed upon diluting co-solvent-containing formulations with the aqueous phase. This is related to the large amount of co-solvent usually needed to improve the drug solubilization capacity, which in turn increases the risk of drug precipitation when the formulation is dispersed in aqueous media. While in the presence of co-surfactants, the co-administered drug is solubilized in micellar structures, systems containing co-solvents lose their solvent capacity faster due to solvent diffusion into aqueous media [60]. El-nokaly et al. summarized the role of a co-surfactant as following [61]: such as increase the fluidity of the interface, destroy liquid crystalline or gel structure which would prevent the formation of microemulsion, adjust HLB value and spontaneous curvature of the interface by changing surfactant partitioning characteristic, furthermore, use of a co-solvent increases the complexity of the SEDDS production process. SEDDS can interact with primary packaging (e.g., gelatin capsules), and therefore SEDDS without alcohols and other volatile solvents are preferred [62–64].

1.8.2   Miscellaneous excipients used in SEDDS

The addition of a polymeric precipitation inhibitor may stabilize a temporarily supersaturated state of the drug after dispersion of SEDDS in the GIT. Such an approach has led to the development of supersaturable SEDDS (S-SEDDS) which incorporate Hydroxypropyl methyl cellulose (HPMC) as a precipitation inhibitor. The addition of HPMC permits also to lower surfactant concentration in an attempt to reduce surfactant side-effects. The S-SEDDS formulation, with a HPMC load of 5%, led to a marked improvement in paclitaxel bioavailability. Furthermore, the incorporation of CsA as an inhibitor of P-gp and CYP 3A4 enzyme into the S-SEDDS further enhances the systemic exposure of paclitaxel. The charge of the oil droplet in conventional SEDDS is negative due to the existence of free fatty acids. The incorporation of a cationic lipid such as oleylamine at a concentration range of 1.0–3.0%, will result in SEDDS with a positive ζ- potential with a value of about 34–45 mV. Positively charged emulsion droplets interact with the negative charge surface component of the gastro-intestinal lumen. It was found in a bioavailability study performed on per-fused rats that administration of CsA in a positively charged SEDDS led to higher plasma levels compared to the corresponding negatively charged formulation. Additionally, comparative oral bioavailability studies in young female rats using several different liquid dosage forms of progesterone indicated that only the positively charged SEDDS could be considered a potential effective dosage form for oral administration of progesterone since it elicited the highest and most satisfactory absorption profile. Excipients such as antioxidants may be necessarily incorporated in SEEDS formulation when the oil phase is susceptible to oxidation. Antioxidants as tocopherol, ascorbyl palmitate, butylated hydroxytoluene (BHT), butylated hydroxyanisole (BHA) and propyl gallate may be used [47].

1.8.3   Recently marketed excipients and their regulatory status

In a recently published survey by Strickley, it has determined that oral lipid-based formulations have been marketed for over 2 decades and currently comprise an estimated 2–4% of the commercially available drug products surveyed in 3 markets worldwide [65]. These products accounted for approximately 2% (21 products total) of marketed drug products in the United Kingdom, 3% (27 products total) in the United States of America, and 4% (8 products total) in Japan. Strickley's survey revealed that the most frequently chosen excipients for preparing oral formulations were dietary oils composed of medium- (e.g., coconut or palm seed oil) or long-chain triglycerides (e.g., corn, olive, peanut, rapeseed, sesame, or soybean oils, including hydrogenated soybean or vegetable oils), lipid soluble solvents (e.g., polyethylene glycol 400, ethanol, propylene glycol, glycerine). Now, polyglycolyzed glycerides (PGG) with varying fatty acid and polyethylene glycol (PEG) chain lengths giving them a varied HLB value, in combination with vegetable oils have been used to solubilize poorly water-soluble drugs and improve their bioavailability [58]. According to the manufacturer, these products are derived from selected, high purity food-grade vegetable oils which are reacted with pharmaceutical grade PEG and therefore expected to be well tolerated by the body. Ethoxylated lipids are derived from castor oil that is rich in ricinoleic acid. Due to the presence of a hydroxyl group on the twelfth carbon of ricinoleic acid, glycerides containing this fatty acid can also be ethoxylated (reaction of etherification) to increase their hydrophilicity. They are widely used as surfactants to enhance bioavailability of poorly soluble drugs. Three main products representing this category are: ethoxylated castor oil (Cremophor EL) and ethoxylated hydrogenated castor oil (Cremophor RH40 and Cremophor RH60) [38].

1.9   Quality control test for lipid excipients

1.9.1   Physical analysis

Since lipid-based excipients are often processed near or above their melting points, analysis of their thermal behaviour at varying stages of formulation is of prime importance. Lipids possess complex chemical compositions that lead to broad melting ranges as opposed to a single melting point. Differential Scanning Calorimetry (DSC) permits study of the thermal behaviour of excipients: melting, crystallization, solid-to-solid transition temperatures and determination of the solid fat content of the excipient versus temperature (which can also be assessed by Nuclear Magnetic Resonance, NMR). DSC allows repeated heating/cooling cycles close to the thermal treatment; the excipients are exposed during processing. In addition, microscopic methods such as hot-stage microscopy (HSM) can be used to assess the organization of the lipid excipient during heating or cooling. Nearly all lipid excipients exist under various polymorphs. For glycerides the main crystalline structures determined by X-Ray Diffraction (XRD) is hexagonal (α), orthorhombic (β') and triclinic (β). These structures differ by their thermal properties (transition and melting temperature for example), and depend on the thermal history of the excipient. As a rule, polymorphic changes have little or no effect on the functionality of self-emulsifying systems that are readily dispersible in aqueous or physiological media. If on the other hand, the formulation matrix is slow or incapable of erosion in the dissolution media, polymorphism can significantly affect the drug release properties [66–68]. However, changes in lipid crystallinity can be controlled by adapted means: tempering at a temperature close to the melting point of the excipient [69], controlling the crystallization rate [70], adding some crystallization seeds to promote the crystallization of one chosen polymorph, or even by adding other excipients such as cellulose ethers or polysorbates to the lipid excipient [67–71].

1.9.2   Chemical analysis

The exact composition of lipid-based excipients in terms of esters, ethers and fatty acid distribution can be assayed by established HPLC and GC methods. Also, quick tests for excipient characterization are available as chemical indices likes specification value relating to the molecular weight of the fatty acid chains; iodine value as a measure of the saturation of hydrocarbon chains; hydroxyl value to determine the quantity of free hydroxyl groups from free glycerol, mono and diglycerides combined; peroxide value to quantify and monitor oxidative changes; and acid value for measuring the quantity of free (un-esterified) fatty acids.

Regular testing for peroxide value and acid value can help assess oxidative stability and potential for hydrolysis of the sensitive bonds in storage or during processing. Analysis for moisture content may also be considered especially for hygroscopic/high HLB excipients. Complete set of analytical methods are available from USP/NF, EP and may also be obtained from the manufacturer of the excipient [72].

1.10   Formulation approaches of SEDDS

The SEDDS formulation should instantaneously form a clear dispersion, which should remain stable on dilution. As the release of a drug compound from SEDDS takes place in the GI tract, the hydrophobic agent remains solubilized until the time that is relevant for its absorption [73]. Figure 1.7 illustrates the usual methodology pathways to prepare the SEDDS formulation and eventual formation of the micro/nano-emulsions following their dilution. Silva et al. found that two main factors, small particle size and polarity of resulting oil droplets, determine the efficient release of the drug compounds from SEDDS. In o/w microemulsions, however, the impact of polarity of oil droplets is not considerable because the drug compound incorporated within the oil droplets reaches the capillaries. Isotropic liquids are preferable to waxy pastes because if one or more excipient(s) crystallize(s) on cooling to form a waxy mixture, it is very difficult to determine the morphology of the materials and, most importantly, the polymorphism properties of the drug within the wax. As a general rule, it is sensible to use the simplest effective formulation, restricting the number of excipients used to a minimum. With a large variety of liquid or waxy excipients available, ranging from oils through biological lipids and hydrophobic and hydrophilic surfactants to water-soluble cosolvents, there are many different combinations that could be formulated for encapsulation in hard or soft gelatin or mixtures that disperse to give fine colloidal emulsions [74]. The following points should be considered in the formulation of a SEDDS [75] such as, the solubility of the drug in different oil, surfactants and co-solvents, the selection of oil, surfactant and cosolvents based on the solubility of the drug and the preparation of the phase diagram, the preparation of SEDDS formulation by dissolving the drug in a mix of oil, surfactant and co-solvent, the concentration of surfactant, the temperature at which self-emulsification occurs, selection of an optimized formulation and comparison of its bioavailability with a reference formulation.

1.10.1   Solubility studies

The solubility of drug in various oils, surfactants and cosurfactants is determined by using shake flask method. An excess amount of drug is added to each vial containing 1 mL of the selected vehicle, i.e. oil, surfactant or solubilizer. After sealing, the mixture is vortexed using a cyclomixer for 10 min in order to facilitate proper mixing of drug with the vehicles. Mixtures are then shaken for 72 h in an isothermal shaker maintained at $37\pm1°C$ for equilibration. Equilibrated samples are centrifuged at 5,000 rpm for 15 min, followed by filtration through membrane filter (0.22 μm). The concentrations of drug are then determined by high-performance liquid chromatography (HPLC) method [76]. Balakrishan et al. determined the solubility of Coenzyme Q10 in various oils and surfactants. After preformulation solubility studies only, oils (Labrafil M 1944 and Labrafil M 2125), surfactant (Labrasol) and cosurfactant (Lauroglycol FCC and Capryol 90) were chosen. In all the formulations, the level of Coenzyme Q10 was fixed at 6% (w/v) of the vehicle [76]. The backbone of SEDDS formulation comprises lipids, surfactants and cosolvents. The right concentration of the above three decides the self-emulsification and particle size of the oil phase in the emulsion formed [75]. Usually SEDDS are prepared in the form of liquid state and are generally enclosed by soft or hard gelatin capsules to facilitate oral administration but it produce some drawbacks like high production costs, low drug incompatibility and stability, drugs leakage and precipitation, capsule ageing. Irreversible drugs/excipients precipitation may also cause severe problem. To overcome these problems, liquid SEDDS are incorporated into a solid dosage form. Generally self-emulsifying formulations were prepared by dissolving the formulation specified amount of active medicament in the mixture of surfactant, oil and cosurfactant mixture at 60°C in isothermal water bath [70, 73].

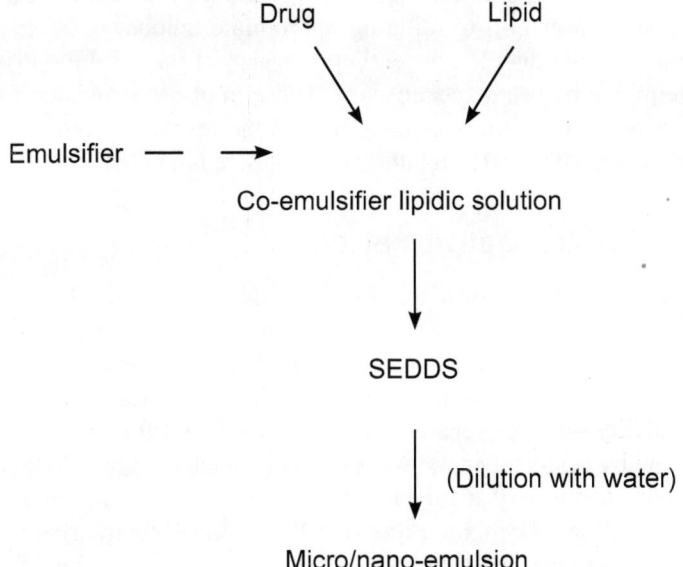

Figure 1.7 Schematic flowchart on the general strategy of formulating self-emulsifying systems and their subsequent conversion to micro/ nano emulsions.

1.10.2   Preparation of ternary phase diagram

It is useful to identify best emulsification region of oil, surfactant and co-surfactant combinations before starting the formulation [78]. Gibbs phase diagrams can be used to show the influence of changes in the volume fractions of the different phases on the phase behaviour of the system. The three components composing the system are each found at an apex of the triangle, where their corresponding volume fraction is 100%. Moving away from that corner reduces the volume fraction of that specific component and increases the volume fraction of one or both of the two other components. Each point within the triangle represents a possible composition of a mixture of the three components or pseudo-components, which may consist of one, two or three phases. These points combine to form regions with boundaries between them, which represent the "phase behaviour" of the system at constant temperature and pressure [79].

1.10.3   Dilution method

Ternary mixtures with varying compositions of surfactant, co-surfactant and oil were prepared. The percentage of surfactant, co-surfactant and oil is decided on the basis of the requirements. Compositions are evaluated for nanoemulsion formation by diluting appropriate amount of mixtures with appropriate double distilled water. Globule size of the resulting dispersions was determined by using spectroscopy. The area of nanoemulsion formation in ternary phase diagram was identified for the respective system in which nanoemulsions with desire globule size were obtaining [78].

1.10.4   Water titration method

The pseudo-ternary phase diagrams were also constructed by titration of homogenous liquid mixtures of oil, surfactant and co-surfactant with water at room temperature. Oil phase, surfactant and the co-surfactant, at Km values 1.5 and 1 (surfactant: co-surfactant ratio), oily mixtures of oil, surfactant and co-surfactant were prepared varied from 9:1 to 1:9 and weighed in the same screw-cap glass tubes and were vortexed. Each mixture was then slowly titrated with aliquots of distilled water and stirred at room temperature to attain equilibrium. The mixture was visually examined for transparency. After equilibrium was reached, the mixtures were further titrated with aliquots of distilled water until they showed the turbidity. Turbidity of the samples would indicate formation of a coarse emulsion, whereas a clear isotropic solution would indicate the formation of a micro emulsion. Percentage of oil, Smix and water at which clear mixture was formed were selected and the values were used to prepare ternary phase diagram. Clear and isotropic samples were deemed to be within the micro-emulsion region. No attempts were made to completely identify the other regions of the phase diagrams. Based on the results, appropriate percentage of oil, surfactant and co-surfactant was selected, correlated in the phase diagram and were used for preparation of SEDDS [80]. A large number of computer programs are now available for construction of ternary phase diagram which are widely used and gaining more and more acceptance in Pharma sector. Examples of such programs which have been used to design or construct the pseudo ternary phase diagrams are enlisted below: SIGMA PLOT 10.0 software (SPSS Inc., USA), SIGMA PLOT 12.0 software, TRI PLOT V14 software, TRIDRAW 4.1 software, CHEMIX software, CHEMIX 3.51 Software (Arne standnes, Norway), PRO SIM software (STRATEGE, Cedex, France), ORIGIN 7.0 and PCP Disso software ver 3.0 (M/s Pune College of Pharmacy, Pune, India).

1.11   Characterization of SEDDS

These lipidic formulations have to be evaluated and characterized using diverse in vitro, ex vivo and by in vivo procedures. A number of techniques have been employed to characterized the SEDDS and determine the feasibility of their formulation process [81].

1.11.1   Visual assessment

The primary test of self-emulsifying formulation is visual evaluation. This may provide important information about the self-emulsifying and micro emulsifying property of the mixture by measuring the turbidity to ascertain dispersion equilibrium within the reproducible time [33]. Fixed quantity of self-emulsifying system added to suitable medium (0.1 N HCl) continuous stirring at 50 rpm on magnetic hot plate at appropriate temperature. The turbidity level is monitored by turbidimeter [82].

1.11.2   Self-emulsification characterization

1.11.2.1   Dispersibility test

The efficiency of self-emulsification of oral nano/microemulsion is assessed using a standard USP dissolution apparatus II. One milliliter of each formulation was added to 500 ml of water at $37 \pm 0.5°C$. A standard stainless steel dissolution paddle rotating at 50 rpm provided gentle agitation. The in vitro performance of the formulations is visually assessed using the following grading system [42].

Grade A: Rapidly forming (within 1 min) nanoemulsion, having a clear or bluish appearance.

Grade B: Rapidly forming, slightly less clear emulsion, having a bluish white appearance.

Grade C: Fine milky emulsion that formed within 2 minutes

Grade D: Dull, grayish white emulsion having slightly oily appearance that is slow to emulsify (longer than 2 min).

Grade E: Formulation, exhibiting either poor or minimal emulsification with large oil globules present on the surface.

Grade A and Grade B formulation will remain as nanoemulsion when dispersed in GIT. While formulation falling in Grade C could be recommend for SEDDS formulation [42].

1.11.2.2   Pellets

For preliminary assessment of the self-emulsifying properties of the formulation, 0.1% Sudan Red is incorporated in the self-emulsifying pellets [84]. Pellets are then gently agitated in distilled water on a shaking water bath at 37°C at about 50 oscillations per min. Aliquots of samples are withdrawn periodically at short interval of around 30 min for microscopic examination using a light microscope with optical zoom (e.g. 50×0.70) and a high resolution eye piece (e.g., 10×20) [84–86].

1.11.3   Self-emulsification time

The self-emulsification time is determined by using USP dissolution apparatus II at 50 revolution/min, where 0.5 g of SEDDS formulations is introduced into 250 ml of 0.1N HCl or 0.5% SLS solution. The time for emulsification at room temperature is indicated as self-emulsification time for the formulation [87]. Pouton et al. quantified the efficiency of emulsification of various compositions of the Tween 85 and medium-chain triglyceride systems using a rotating paddle to promote emulsification in a crude nephelometer. This enabled an estimation of the time taken for emulsification. Once emulsification was complete, samples were taken for particle sizing by photon correlation spectroscopy, and self-emulsified systems were compared with homogenized systems. The process of self-emulsification was observed using light microscopy. It was clear that the mechanism of emulsification involved erosion of a fine cloud of small particles from the surface of large droplets, rather than a progressive reduction in droplet size [88].

1.11.4   Size and zeta potential

This is a most crucial factor in self-emulsification formulation because it determines the rate and extent of drug release as well as the stability of the emulsion. Common techniques used to determine the droplet size distributions of the resultant emulsions include photon correlation spectroscopy, laser diffraction and coulter counter [34]. A number of equipments are available for measurement of particle size, viz. Particle Size Analyzer, Mastersizer, Zetasizer, etc., which are able to measure sizes between 10 nm and 5000 nm. In many instances nanometric size range of particle is retained even after 100 times dilution with water which indicates the system's compatibility with excess water Photon correlation spectroscopy is sensitive to particles within a diameter range from 3 nm to 3 μm. Laser diffraction is effective in detecting droplets within a diameter range from 0.5 nm to 200 μm. A Coulter counter measures the change in the electrical resistance induced by the droplets of the sample flowing through a pinhole. A 30 μm pinhole is suitable for measuring the diameter of droplets within the size range of 1.5 μm to 10 μm. The advantage of this technique is that it provides an absolute droplet count within a specified size range, which is in contrast to the relative count provided by laser diffraction. Freeze-fracture electron microscopy has been used to study surface characteristics of such dispersed systems. Because of the high lability of the samples and the possibility of artifacts, electron microscopy is considered a somewhat misleading technique. Particle size analysis and low-frequency dielectric spectroscopy have been used to examine the self-emulsifying properties of Imwitor R 742 and Tween R 80 [89]. The stability of emulsion is directly related to the charge present on mobile surface, which is termed as zeta potential [34]. A Zetasizer uses light scattering techniques to measure globule size, zeta potential and molecular weight of nanoparticulate systems. The instrument measures the size and zeta-potential to optimize stability and shelf life and speed up formulation development [90]. The electrophoretic mobility of the micelles and the nanoemulsion has also been measured at 25°C with a Zetasizer. In conventional SEDDS, the charge on an oil droplet is negative because of the presence of free fatty acids [91]. However, incorporation of a cationic lipid like oleylamine will yield cationic SEDDS. The zeta potential of the dispersions is calculated by the instrument according to the Helmholtz–Smoluchowski equation (1.2).

$$U = \frac{\varepsilon \xi E_x}{\mu} \qquad \text{...(Eq. 1.2)}$$

Where, U is the electrophoretic velocity, E is the permittivity, ξ is the zeta potential, Ex is the axial electric field, μ is the viscosity [92, 93]. The value of zeta potential, as well as the size of the droplets/globules of the SEDDS formulation, can be measured using diverse techniques. A brief account of the methodologies is presented below:

1.11.4.1   Dynamic light scattering (DLS)

The DLS technique is also known as photon correlation spectroscopy. The instrument for DLS measurement is equipped with a He-Ne laser, a digital correlator and a single photon detector module. During DLS, the sample is illuminated with a laser beam and the intensity of resulting scattered light produced by the particles fluctuates at a rate that is dependent on the size of particles [94–96]. The globule size detection is carried out in a backscattering mode [97], with a high scattering angle like 173°. The reduction in this correlation function, viz. scattered light intensity and particle size with displacement time (called as "lag time"), can be used to extract information about the diffusion coefficient of a particle or droplet in solution. The measured diffusion coefficient can be used to calculate hydrodynamic radius (ρ) of the droplet using the Stokes-Einstein equation stated as Equation (1.3),

$$\rho = kT/6\Pi\eta D \qquad .... \text{(Eq. 1.3)}$$

Where, k is the Boltzmann constant, T is the absolute temperature, η is the viscosity of the continuous phase and D is the diffusion coefficient. The particle size data are first analysed by cumulant analysis to obtain an average diffusion coefficient and subsequently by a constrained inverse Laplace transform routine (i.e., CONTIN analysis [98]) to obtain information about the entire size distribution (i.e., monomodal or multimodal. Two sets of data are obtained: one measurement is usually carried out at least 4 h after preparation of the sample, while second measurement is performed after 72 h [97]. The measurement is intended to provide information on long-term stability of the microemulsion [99, 100].

1.11.4.2   Static light scattering

In a typical small angle, X-ray scattering technique (SAXS) experiment, the Fourier transformation of the electron density fluctuations in a sample is measured [101, 102]. For the nanoemulsions, the electron density is usually low in oil-cores and high in the surrounding water phase. This is exploited to determine the shape and size of the nanoemulsion droplets. More generally, the SAXS is used for investigating structures on the 1–100 nm scale. The instrument is optimized for solution scattering and a specific q-range (e.g., between 0.009 and 0.281Å) is covered with the applied setup. Measurements are generally performed on a photometer equipped with a He–Ne laser and cylindrical cells immersed in a toluene-index matched container. Intensities are recorded at different angles and normalized with respect to toluene (θ =

90°) [101, 103].

1.11.4.3   Multi-angle light scattering

Multi-angle light scattering experiments are usually performed on a spectrometer equipped with a goniometer, a uniphase He–Ne laser, an avalanche photodiode detector and a multi-angle tau correlator [97]. The temperature can be controlled using an external water bath circulator.

1.11.5   Emulsion droplet polarity

Emulsion droplet polarity is also a vital factor for characterizing emulsification efficiency [58, 104]. The HLB, chain length and degree of unsaturation of the fatty acid, molecular weight of the hydrophilic portion and the concentration of emulsifier have significant impact on the polarity of the oil droplets [105]. Polarity represents the affinity of the drug compound for oil and/or water and the type of forces formed. Rapid release of the drug into the aqueous phase is promoted by drug polarity [106].

1.11.6   Electron microscopic studies

Freeze-fracture electron microscopy has been used to study the surface characteristics of the SEDDS. However, due to the high lability of the samples and the possibility of artifacts, electron microscopy is, at times, considered as a somewhat misleading technique. Particle size analysis and low frequency dielectric spectroscopy have been utilized to examine the self-emulsifying properties of a series of Imwitor 742 (i.e., a mixture of mono- and diglycerides of capric and caprylic acids) and Tween 80 systems [89, 90].

1.11.6.1   Scanning electron microscope (SEM)

In this technique, surface of the sample is scanned with a high-energy beam of electrons in a raster scan pattern (i.e., a rectangular pattern of image capture and reconstruction in television) [107]. The electrons interact with the atoms that make up the sample producing signals that contain information about the sample's surface topography, composition and properties like electrical conductivity. This technique is quite useful for determining the parameters like powder flow and compaction which influences the production of solid dosage forms [108].

1.11.6.2   Transmission electron microscopy (TEM)

In TEM, a beam of electrons is transmitted through a thin layer of specimen. An image is formed from the interaction of the electrons transmitted through the specimen; the image is magnified and focused onto an imaging device, such as a fluorescent screen, on a layer of photographic film, or to be detected by a sensor such as a charge-coupled device (CCD) camera [109].

1.11.6.3   Cryo-transmission electron microscopy (Cryo-TEM Studies)

In Cryo-TEM studies, sometimes described as electron cryo-microscopy or cryo-electron microscopy, the sample is studied at cryogenic temperatures, generally matching those of liquid nitrogen [110]. The important merit of such studies is that it allows the observation of specimens that have not been stained or fixed in any way, showing them in their native environment, which otherwise could lead to undesirable conformational changes as in case of X-ray crystallography. The samples for the Cryo-TEM studies are usually prepared in a controlled environment vitrification system [111, 112].

1.11.7   Turbidity measurements

Turbidity is a parameter which recognizes a capable self-emulsification by establishing whether the dispersion reaches equilibrium rapidly and in a reproducible time [113]. Turbidimetric evaluation is carried out to monitor the growth of droplet after emulsification. These measurements are carried out on turbidity meters, most commonly the Hach turbidity meter and the Orbeco-Helle turbidity meter [74, 75]. Fixed quantity of SEDDS is added to fixed quantity of suitable medium (0.1 N HCl or Phosphate Buffer) under continuous stirring at 50 rpm/min on magnetic stirrer at optimum temperature and the turbidity is measured using a turbidimeter. Since the time required for complete emulsification is too short, it is not possible to monitor the rate of change of turbidity i.e. rate of emulsification [34]. Turbidity can also be observed in terms of spectroscopic characterization of optical clarity (i.e. absorbance of suitably diluted aqueous dispersion at 400 nm) [49].

1.11.8   Rheological studies

As a prelude to dissolution and permeation across GI lumen, drug molecules form the SEDDS formulations have to undergo dilution in the GI milieu to form nano/microemulsion. Accordingly, estimation of their rheological behaviour during this transition is quite vital, next only to the assessment of solubility and permeability across GI tract [114]. On dilution, the oil and emulsifier(s) present in SEDDS form an interface with GI fluids, termed as intermediate liquid crystalline phase. The rheological studies, consequently, focus on exploration of the viscoelastic properties of this intermediate liquid crystalline phase and to evaluate its effect on self-emulsification performance. Additionally, assessment of rheological properties of SEDDS is indispensable to have better understanding of their phase behaviour during extreme conditions of temperature, humidity, transportation, etc. Rheological behaviour of micro/nanoemulsions formed after dilution has been determined using digital instruments coupled with either cup and bob or a co-axial measuring device. A rotational viscometer has also been used for viscosity measurements on fresh microemulsions and those stored for long periods [115]. A typical rheometer test program for rheological characterization (i.e., flow, thixotropy, static yield, creep value) presents a data analysis for flow curves, quality control min/max limits, mathematical models, data averaging and many more analytical functions [116–118]. Generally, viscosity measurements indicate that on dilution with distilled water (e.g. 10 and 100 times), viscosity of a formulation decreases, thus construing that the drug absorption is likely to be faster from stomach. Besides, the effect of concentration of excipients on the rheological profile of SEDDS can also be examined [119].

1.11.9   Stability studies

1.11.9.1   Thermodynamic stability studies

The physical stability of a lipid-based formulation is also crucial to its performance, which can be adversely affected by precipitation of the drug in the excipient matrix. In addition, poor formulation physical stability can lead to phase separation of the excipient, affecting not only formulation performance, but visual appearance as well. In addition, incompatibilities between the formulation and the gelatin capsules shell can lead to brittleness or deformation, delayed disintegration, or incomplete release of drug [115].

Heating cooling cycle

Six cycles between refrigerator temperature 4°C and 45°C with storage at each temperature of not less than 48 h is studied. Those formulations, which are stable at these temperatures, are subjected to centrifugation test [34].

Centrifugation

Passed formulations are centrifuged thaw cycles between 21°C and 25°C with storage a teach temperature for not less than 48 h is done at 3500 rpm for 30 min. Those formulations that does not show any phase separation are taken for the freeze thaw stress test [120].

Freeze thaw cycle

The formulations are subjected to 3–4 freeze-thaw cycles, which include freezing at −4°C for 24 hours followed by thawing at 40°C for 24 hours [121]. Those formulations passed this test showed good stability with no phase separation, creaming, or cracking [122].

1.11.9.2   Robustness to dilution

Emulsions or nano-emulsions upon dilution with various dissolution media should not show any phase separations or precipitation of drug even after 12 hours of storage, that formulation is considered as robust to dilution [123–127].

1.11.10   Liquefaction time

This test is done to determine the time required by solid SEDDS formulation to melt in vivo in the absence of agitation in simulated gastric fluid. The formulation is packed in a transparent polyethylene film and tied to the bulb of thermometer by means of thread. The thermometer with attached tablets is placed in a round bottom flask containing 250 ml of simulated gastric fluid without pepsin maintained at $37 \pm 1°C$ by means of thermo-regulated heating mantle. The time taken for liquefaction is subsequently noted [128].

1.11.11   Permeation studies

For information about oral bioavailability enhancement of a formulation, one must have to perform in vitro or ex vivo studies [34].

1.11.11.1   Isolated and perfused intestinal segments

During the last decade, a wide range of isolated organ systems have been developed for biomedical and pharmaceutical research. The availability of sophisticated equipment, increased manual skills, and the routine use and standardization of models and protocols, have led to the increased reproducibility and validity of experimental results. These methods contribute to the reduction of live animal experimentation. The results are quite predictive of the in vivo situation including absorption at the organ level [129, 130]. Isolated perfused organs have the distinct advantage that the scientist works with an intact organ, where physiological cell–cell contacts and normal intracellular matrixes are preserved. The major limitation, however, is the short duration of the experiments, since changes tend to occur rapidly [130].

1.11.11.2 In situ single pass perfusion technique (SPIP)

The perfusion solution is passed through the intestinal segment (i.e., jejunum) by cannulating it at both the ends, and various permeability parameters are calculated from amount of drug unabsorbed from the intestine [131]. Besides providing experimental conditions closer to that occurring in vivo, this technique is also able to predict the exact mechanism of absorption, i.e. passive absorption, carrier-mediated absorption or active transport [132]. Figure 1.8 describes the schematic representation of a typical SPIP technique in rat. In this technique, the proximal part of the jejunum, 2–5 cm below the ligament of Trietz is cannulated with a glass cannula and connected to the reservoir [133]. The intestine segment is perfused with blank phosphate buffer (37±1°C) until perfusate is clear. The intestine is subsequently perfused with drug solution maintained at 37± 1°C at a perfusion rate of 0.2–0.3 ml/min. During the experiment, the animal is kept under a heating lamp, and the exposed abdomen is covered with a cotton pad to minimize dehydration. Steady state is usually achieved within 30 min, following which 4–5 samples are obtained at regular intervals of around 15 minutes. During the experiment, the amount of water entering into the system and water leaving the system is carefully recorded to calculate water flux [134]. Effective permeability (Peff) is calculated after correcting the outlet concentration for water flux on the basis of ratio of weight of perfusion solution collected and infused for each sampling points as mentioned in Equation (1.4).

$$P_{eff} = \frac{Q\left\{ \dfrac{C_{in}}{C_{out}} - 1 \right\}}{2\pi r l} \quad \text{...(Eq. 1.4)}$$

Where, Q is the flow rate, Cin and Cout are the respective inlet and outlet concentration, r is the radius of the intestine, and l is the length of the intestine measured after completion of perfusion. Aqueous permeability (P_{aq}) was calculated using Equation (1.5).

$$P_{aq} = A \times \left[\frac{\pi D l}{2Q} \right]^{1/3} \quad \text{...(Eq. 1.5)}$$

Where D is the Diffusion coefficient, and the factor $[\Box Dl/ 2Q]^{1/3}$, known as Graetz number (Gz), plays important role in determining the dimensionless quantity, A is the Wall permeability (P_{wall}) is calculated using Equation (1.6):

$$P_{wall} = \frac{Peff \times Paq}{Paq - Peff} \quad \text{...(Eq. 1.6)}$$

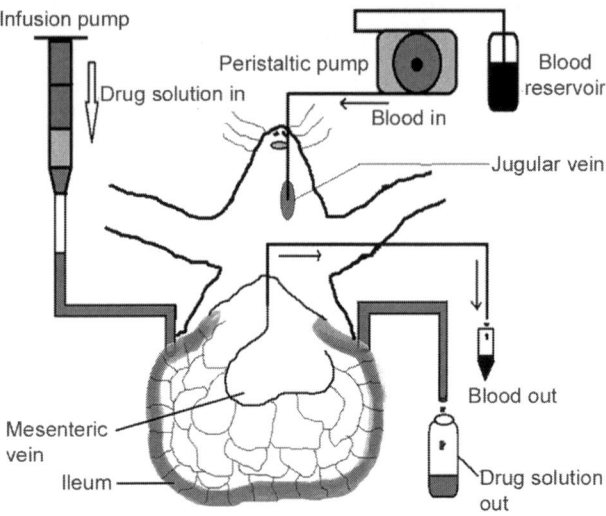

Figure 1.8 Schematic representation of single pass intestinal perfusion (SPIP) technique as employed for the SEDDS formulation in rat.

1.11.11.3   Everted sac technique

In this method, a 2–4 cm section of the intestine is tied off at one end and everted using a glass rod or a thread. This method can be used to determine kinetic parameters with high reliability and reproducibility [135]. Oxygenated tissue culture media and specific preparation techniques ensure tissue viability for up to 2 hours. The technique can be used to study drug transport across the intestine and into the epithelial cells, provided that sensitive detection methods are employed [136]. Radio-labelled compounds are most appropriate for the purpose. It is used mainly to quantify the paracellular transport of hydrophilic molecules and to estimate the effects of potent enhancers on their absorption. Molecules that cross the epithelial barrier by a transcellular route have much higher permeability which can also be accurately quantified using the everted sac system. This kind of model is also suitable for measuring absorption at different sites in the small intestine and for performing preliminary experiments on the colon [135–137]. It is also useful for estimating the first-pass metabolism of drugs in intestinal epithelial cells. Also, by using this model (everted or not), it is convenient to study the effect of Pgp on xenobiotic transport through the intestinal barrier. A potential disadvantage of this approach is the presence of the muscularis mucosa, which is not usually removed from everted sac preparations. Therefore, this model does not reflect the actual intestinal barrier, because compounds under investigation pass from the lumen into the lamina propria (where blood and lymph vessels are found) and across the muscularis mucosa. Thus, the transport of compounds with a propensity to bind to muscle cells might be underestimated [137].

1.11.11.4   Diffusion cells using tissues

In this method, diffusion across a small section of intestine representing mucosal environment is studied into a system with specific fluid pH, temperature, etc. representing secrosal environment. The buffer solution at both sides of the membrane is gassed continuously with carbogen. The same method can be used for tissues other than intestinal tissue (e.g., buccal, esophageal, gastric, rectal, nasal, lung, skin tissue, etc.). The usefulness of these cells for intestinal transport studies has long been recognized. These cells have also been used to study the intestinal metabolism of xenobiotics [135]. In this system, the drug can be exposed at either the mucosal level or the serosal level. The simplicity of these cells makes them an attractive in vitro model system for studying drug transport. This type of study may provide additional information on the pharmacological behaviour of the test compound.

1.11.12   Cell models

1.11.12.1   Caco-2 cells

Caco-2 cells are the most popular intestinal cellular model in studies on passage and transport. These cell lines are derived from human colorectal adenocarcinoma [138]. In culture, they differentiate spontaneously into polarised intestinal cells possessing an apical brush border and tight junctions between adjacent cells, and they express hydrolases and typical microvillar transporters [15]. Caco-2 cells, despite their colonic origin, express in culture the majority of the morphological and functional characteristics of small intestinal absorptive cells, including phase I and phase II enzymes, detected either by measurement of their activities toward specific substrates, or by immunological techniques [139,140].

1.11.12.2   Brush border membrane vesicles

In this approach, cell homogenates or intestinal pieces are treated by calcium chloride precipitation method using centrifugation. The final pellet contains the luminal wall-bound proteins and phospholipids, which contain most of the brush border enzymatic and carrier activity. Re-suspension of the pellet in buffer results in the formation of vesicles. These vesicles are mixed with the permeant in buffer and filtered after a fixed time, the amount of permeant taken up by the vesicles is then determined. Typically, only the apical transcellular transport is measured by this system. Despite drawbacks, like the need for a radio-labelled compound and day-to-day variation in precipitation, this method is highly useful for mechanistic studies of the drug absorption process [141].

1.11.12.3   Epithelial cell model

Uptake of a variety of substances is controlled by biological barriers like epithelial tissues. Therefore, much attention is currently paid to the use of epithelial cell cultures for studies of drug transport mechanisms. Such studies are best performed in a model that contains only absorptive cells, without the confounding contributions of mucus, the lamina propria and/or the muscularis mucosa [142].

1.11.12.4   Isolated intestinal cells

These cells, obtained from the intestine of animal or human origin, can be used as uptake systems in the assessment of oral bioavailability. However, the use of isolated intestinal epithelial cells has been relatively slow to gain popularity, because of the difficulty to culture them and their limited viability [143, 144]. Development of human cell culture systems has been limited by the loss of important in vivo anatomical and biochemical features. Attention has, therefore, turned to the use of human adenocarcinoma cell lines, such as HT-29 and Caco-2, which reproducibly display a number of properties characteristic of differentiated intestinal cells. They offer the advantage of relative simplicity and suitability for automated procedures and HTS. Nevertheless, the limitations of cell models must not be overlooked. These cell lines originate from tumors, and out of the in vivo physiological environment. Therefore, extrapolation of the data to the in vivo situation may be difficult, as is true for most of in vitro systems [145].

1.11.12.5   Non-intestinal cell systems

Madin Darby canine kidney (MDCK) cells have been isolated from dog kidneys [146, 147]. They are currently used to study the regulation of cell growth, drug metabolism, toxicity and transport at the distal renal tubule epithelial level [148, 149]. MDCK cells have been shown to differentiate into columnar epithelial cells and to form tight junctions when cultured on semi-permeable membranes [150].

1.11.13   Refractive index and percent transmittance

Refractive index and percent transmittance proved the transparency of formulation. Refractive index of the formulation is measured by refractometer by placing drop of solution on slide and then compare it with water (R.I. = 1.333). The percent transmittance of the formulation is measured at a particular wavelength using UV spectrophotometer by using distilled water as blank. If R.I. of formulation is similar to that of water and formulation having percent transmittance is greater than 99%, then the formulation are transparent in nature [151].

1.11.14   Electro-conductivity study

The SEDD system contains ionic or non-ionic surfactant, oil, and water. So, this test is used to measure the electro conductive nature of system whether SEDD system contains ionic or non-ionic surfactant, oil, and water [152]. The electro conductivity of resultant system is measured by electro conductometer [42].

1.11.15   Drug content

Drug from pre-weighed SEDDS is extracted by dissolving in suitable solvent. Drug content in the solvent extract was analyzed by suitable analytical method against the standard solvent solution of drug [33].

1.11.16   Self-emulsification time

The self-emulsification time is determined by using USP dissolution apparatus II at 50 r/min, where 0.5 g of SEDDS formulations is introduced into 250 ml of 0.1N HCL or 0.5% SLS solution. The time for emulsification at room temperature is indicated as self-emulsification time for the formulation [123].

1.11.17   Absorption studies

Male rats weighing 300 ± 20 g are used. The animals are divided into 3 groups, the first group is fasted for 12 h before drug administration; the second and third groups are continuously fed with normal diet and lipid diet for 12 h before drug administration, respectively. After anesthesia, the femoral artery is cannulated and cannula is flushed with heparin saline solution to prevent blood clotting. After rats recovered from anesthesia, SEDDS after dilution with distilled water is administered orally to rats using oral sonde. Blood samples are withdrawn at regular time intervals and frozen until analysis. The pharmacokinetic parameters like AUC, T max and C max are calculated from the plasma data [76].

1.11.18   Polydispersitivity index (PI)

PI is a measure of particle homogeneity and it varies from 0.0 to 1.0. The closer to zero the PI value the more homogenous are the particles. The PI showed that ME formulation had narrow size distribution [153].

1.11.19   Dissolution studies

The quantitative in vitro dissolution studies are carried out to assess drug release from oil phase in to aqueous phase by USP type II dissolution apparatus using 500 ml of simulated gastric fluid containing 0.5% w/v of SLS (Sodium Lauryl Sulphate) at 50 r/min and maintaining the temperature at 37 + 0.5°C. Aliquots of samples are withdrawn at regular intervals of time and volume withdrawn is replaced with fresh medium. Samples taken are then analyzed by using UV spectrophotometer or any other suitable technique [154].

1.11.20   Nuclear magnetic resonance (NMR) studies

The structure and dynamics of microemulsions can be studied by NMR techniques. Self-diffusion measurements using different tracer techniques, generally radio labelling, supply information on the mobility and microenvironment of the component. The Fourier transform pulsed-gradient spin-echo (FT-PGSE) techniques use the magnetic gradient on the samples and it allows simultaneous and rapid determination of the self-diffusion coefficients of many components. Self-diffusion coefficient (D) can be calculated using Stokes-Einstein Equation (1.7):

$$D = \frac{KT}{6\pi\eta r} \qquad .... (Eq.\ 1.7)$$

Where, K = Boltzmann constant, T = Absolute temperature, η = Viscosity of the medium, r = radius of the droplet [155, 156].

1.11.21   Temperature stability

Shelf-life as a function of time and storage temperature was evaluated by visual inspection of the SEDDS system at different time period. SEDDS was diluted with purified distilled water and to check the temperature stability of samples, they were kept at three different temperature range 2–8°C (refrigerator), room temperature and observed for any evidences of phase separation, flocculation or precipitation [158].

1.12   Delivery approaches

1.12.1   Oral drug delivery

Depending upon the final form, these systems can be broadly classified as liquid SEDDS, semi-solid SEDDS and solid SEDDS. A holistic interplay of lipidic and solidifying excipients can transform the liquid SE systems into different dosage forms like SE pellets, microspheres, controlled release tablets, minicapsules, etc. using various approaches like spray drying, porous carrier-based absorption, osmosis, eutectic phenomenon, supersaturability, etc. [3].

1.12.1.1   Solid self-emulsifying drug delivery system (S-SEDDS)

SEDDS can exist in either liquid or solid states. SEDDS are usually, however, limited to liquid dosage forms, because many excipients used in SEDDS are not solids at room temperature. In recent years, as they frequently represent more effective alternatives to conventional liquid SEDDS. From the perspective of dosage forms, S-SEDDS mean solid dosage forms with self-emulsification properties. S-SEDDS focus on the incorporation of liquid/semisolid SE ingredients into powders/ nanoparticles by different solidification techniques (e.g. adsorptions to solid carriers, spray drying, melt extrusion, Nanoparticle technology, and so on). Such powders/nanoparticles, which refer to SE nanoparticles/dry emulsions/solid dispersions, are usually further processed into other solid SE dosage forms, or, alternatively, filled into capsules (i.e. SE capsules). SE capsules also include those capsules into which liquid/semisolid SEDDS are directly filled without any solidifying excipient. To some extent, S-SEDDS are combinations of SEDDS and solid dosage forms, so many properties of S-SEDDS (e.g. excipients selection, specificity, and characterization) are the sum of the corresponding properties of both SEDDS and solid dosage forms. Conventional solid SEDDS are capsules, solid dispersions, and dry emulsions but recently, a number of other SEDDS have been prepared such as pellets, microspheres, tablets, beads, implants and suppositories [158].

1.12.1.2   Techniques for solid formulations

Spray cooling

The molten droplets are sprayed into cooling chamber, which will congeal and re-crystallize into spherical solid particles that fall to the bottom of the chamber and subsequently collected as fine powder. The fine powder may then be used for development of solid dosage forms tablets or direct filling into hard shell capsules. Many types of equipment are available to atomize the liquid mixture and to generate droplets: rotary, pressure, two-fluid or ultrasonic atomizers [159].

Spray drying

Essentially, this technique involves the preparation of a formulation by mixing lipids, surfactants, drug, solid carriers, and solubilization of the mixture before spray drying. Spray drying is defined as a process by which a liquid solution is sprayed into a hot air chamber to evaporate the volatile fraction. The solubilized liquid formulation is then atomized into a spray of droplets. The droplets are introduced into a drying chamber, where the volatile phase (e.g. the water contained in an emulsion) evaporates, forming dry particles under controlled temperature and airflow conditions. Such particles can be further prepared into tablets or capsules. Polyoxyl glycerides (lauroyl or stearoyl) have been used alone or in combination with a solid carrier (silicon dioxide) to form microparticles of etoricoxib and glibenclamide. Dry emulsion technology solves the stability problems associated with classic emulsions (phase separation, contamination by microorganism, etc.) during storage and helps also avoid using harmful or toxic organic solvents. Dry emulsions may be redispersed into water before use. Medium chain triglycerides are commonly used as oil phase for these emulsions [159, 160]. Figure 1.9 illustrates the process for preparing spray-dried SEDDS.

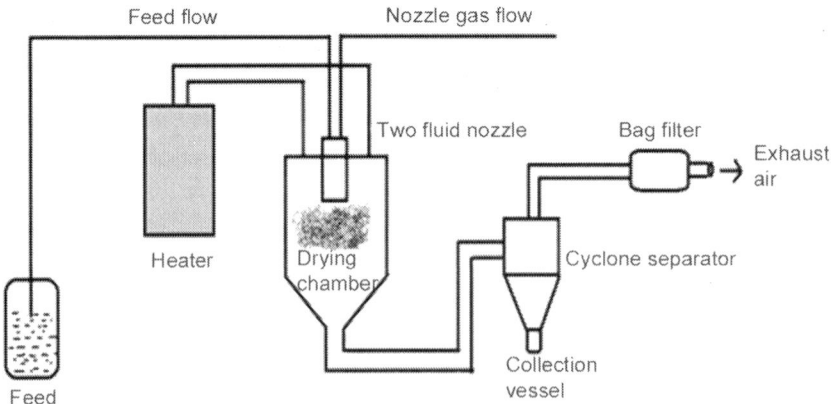

Figure 1.9 Process for preparing spray dried SEDDS.

Adsorption on solid carriers

The adsorption process is simple and involves addition of the liquid formulation onto the carrier of choice by mixing in a blender. The carriers used for this purpose include microporous calcium silicate (Florite™ RE), magnesium aluminometasilicate (Neusilin™ US2), silicon dioxide (Sylysia™ 320), or carbon nanotube. These carriers should be selected for their ability to absorb a great quantity of liquid excipients (to allow for a high drug loading and high lipid exposure) and for the flowability of the mixture after adsorption. The resulting free flowing powder may then be filled directly into capsules or alternatively mixed with suitable excipients before compression into tablets. A significant benefit of the adsorption technique is good content uniformity. SEDDS can be adsorbed at high levels (up to 70% w/w) onto suitable carriers. The adsorption technique has been successfully applied to gentamicin and erythropoietin with caprylo caproyl polyoxyl glycerides (Labrasol) formulations that maintained their bioavailability enhancing effect after adsorption on carriers [161].

Melt granulation

Melt granulation or pelletization is a one-step process allowing the transformation of a powder mixer containing the drug into granules or spheronized pellets. The melted binder forms liquid bridges with the powder particles that shape into small agglomerates (granules) which can, by further mixing under controlled conditions transform to spheronized pellets. The main parameters that control the granulation process are impeller speed, mixing time, binder particle size, and the viscosity of the binder during melt granulation. Nucleation (onset of granule formation) is largely affected by binder viscosity at high impeller speed and binder particle size at low speed. Depending on the combination of process parameters, two distinct mechanisms namely "distribution" and "immersion" may be at play in the development of granules. Fine or atomized excipients with low viscosity at high impeller speed favour a homogenous "distribution" of the binder onto the surface of the powder. Immersion of the powder on the other hand is the preferred mechanism which is assisted by combination of large binder particles possessing high viscosity and mixing under low impeller speed. The granule size distribution is controlled by the combined effect of the impeller and chopper speeds. Generally, lipids with low HLB and high melting point are suitable for sustained release applications. Semi-solid excipients with high HLB on the other hand may serve in immediate release and bioavailability enhancement. The progressive melting of the binder allows the control of the process and the selection of the granule's size. Also, the melt granulation process may be used for adsorbing semi-solid self-emulsifying systems on solid neutral carriers (mainly silica and magnesium aluminometasilicate). Hydrogen bonding with adsorbent during storage governs drug dissolution from solid dispersion granules. The main advantages of melt granulation/pelletization with lipids are process simplicity (one-step), absence of solvents, and more importantly the potential for the highest drug loading capacity 85% theoretically, and up to 66% actually reported in the literature [162, 163].

Melt extrusion or spheronisaton

Extrusion is a process of converting a raw material with plastic properties into a product of uniform shape and density by forcing it through a die under controlled temperature, product flow and pressure conditions [164]. This approach has been successfully tried for 17β-estradiol and two model drugs with surfactants such as sucrose monopalmitate, lauroylpolyoxylglycerides and polysorbate 80 (Tween® 80). Gelucire 44/14 to be used directly in the core of the formulation matrix. An innovative "system-in cylinder" molding technique was recently employed to develop a dual purpose (enhanced bioavailability and controlled release) formulation with propranolol hydrochloride. Melt extrusion is a solvent-free process that allows high drug loading as well as content uniformity for low dose high potency actives [164, 165]. The extrusion–spheronization process requires the following steps [166]: such as dry mixing of the active ingredients and excipients to achieve a homogeneous powder, wet massing with binder, extrusion into a spaghetti-like extrudate, spheronization from the extrudate to spheroids of uniform size and drying sifting to achieve the desired size distribution and coating (optional).

Supercritical fluid-based method

Lipids may be used in supercritical fluid based methods either for coating of drug particles, or for producing solid dispersions. For environmental reasons, the preferred supercritical fluid of choice is supercritical carbon dioxide. Examples include controlled release applications using glyceryl trimyristate (Dynasan™ 114) and stearoyl poly-oxylglycerides (Gelucire® 50/02) [167].

Solid lipid nanoparticles and nanostructured lipid carriers

SLN and NLC are two types of submicron size particles (50–1000 nm) composed of physiologically tolerated lipid components. SLN are produced by high-pressure homogenization of the solid matrix and drug with an aqueous solution of the glyceryl dibehenate as solid lipid matrix and poloxamers 188 or polysorbates 80 as surfactants. They typically contain a liquid lipid excipient such as medium chain triglycerides in addition to classic components of SLN. They have been mainly used for controlled-release applications in oral, intravenous or topical route [168].

1.12.1.3   Techniques for liquid or semisolid SEDDS

Capsule filling is the simplest and the most common technology to encapsulate liquid or semisolid lipid-based formulations for the oral route. A primary consideration in capsule filling is the compatibility of the excipients with the capsule shell. Prior to filling, in the case of semisolid or solid lipid-based excipients, the bulk fill reservoir should be heated to maintain the formulation molten and under stirring to avoid phase separation and sedimentation of the drug if dispersed. For filling suspensions into soft capsules, the particle size distribution must be below 250 μm and the viscosity should be controlled to ensure a homogeneous suspension and an easy filling. Numerous publications have described the use of this technology for enhancing drug solubility and absorption via the gastro-intestinal tract. For semi-solid formulations, it is a four step process: heating of the semisolid excipient to at least 20°C above its melting point, incorporation of the active substances (with stirring), capsule filling with the molten mixture and cooling to room temperature. For liquid formulations, it involves a two-step process such as filling of the formulation into the capsules followed by sealing of the body and cap of the capsule, either by banding or by microspray sealing. A primary consideration in capsule filling is the compatibility of the excipients with the capsule shell. The filling temperature is one of the key parameters for capsule filling. It should be at least 2°C above the temperature at which the apparent viscosity of the drug-excipient mixture significantly increases during cooling (temperature assessed by thermorheology studies). The maximum filling temperatures are 70°C for hard shell capsules and 40°C for soft gelatin capsules [169].

1.12.2   Non-oral delivery

The immense success of SEDDS has lately been extrapolated to routes other than oral too. Due to the promising drug delivery instances of SEDDS through oral route, these are also now favorably used for drug delivery through parenteral, implantable, vaginal, rectal and transdermal routes. The highly lipophilic drugs requiring drug absorption at local site of action are administered as self-nanoemulsifying systems through such routes. These provide drug delivery at the locally site by bypassing hepatic first-pass, efflux by P-gp and degradation by intestinal cytochrome P-450 enzymes. Ideally, several dosage forms have been developed for drug delivery including parenteral self-nanoemulsifying drug delivery systems (PSNEDDS), self-microemulsifying suppositories (SMES), self-emulsifying implantable systems (SEIS) and self-emulsifying transdermal drug delivery systems (SETDDS) [170].

1.12.2.1   Suppositories

Some investigators proved that solid-SEDDs could not only increase the GI absorption but also increase the rectal/vaginal adsorption. The dosage form of such formulations is suitable for promoting drug absorption through rectal hemorrhoidal veins for direct entry into the systemic circulation by circumnavigating the extensive hepatic first pass effect and high P-gp efflux. Upon rectal administration, the suppository bases (e.g., cocoa butter, macrogol esters and saturated polyglycolyzed glycosides) liquefy at body temperature [171]. Subsequently, the lipidic excipients undergo emulsification in the presence of rectal fluid to produce fine nano-emulsion/ microemulsion (o/w type), which can be easily absorbed through hemorrhoidal veins to augment systemic bioavailability of drugs and faster onset of action. Glycyrrhizin hardly achieves therapeutic plasma concentrations by oral route, but can achieve acceptable therapeutic levels for chronic hepatic diseases by either vaginal or rectal SE suppositories. The formulation included glycyrrhizin and a mixture of a C6–C18 fatty acid glycerol ester and a C6–C18 fatty acid macrogol ester [172].

1.12.2.2   Implants

SE implants have very much improved efficacy under application of SSEDDS, since they have short half-life. As an example, 1, 3-bis (2 chloroethyl)-1-nitrosourea (carmustine, BCNU) is a chemotherapeutic agent used to treat malignant brain tumors. In order to enhance its stability compared with that released from poly (d, 1-lactide-coglycolide) (PLGA) wafer implants, SES was formulated with tributyrin, Cremophor RH 40 (Polyoxyl40 hydrogenated castor oil) and Labrafil 1944 (polyglycolyzed glyceride). Therefore SES increased in vitro half-life of BCNU up to 130 min compared with 45 min of intact BCNU. In vitro release of BCNU from SE PLGA wafers were extended up to 7 days. Such wafers had higher in vitro antitumor activity and were less prone to hydrolysis than those wafers without of SES [173].

1.12.2.3   Injections

Also known as parenteral SEDDS, primarily containing anhydrous mixture of oil, surfactants and high amount of solubilizers (30%) (such as glycerol, sorbitol and propylene glycol) can form self-emulsifying isotropic liquid. These are suitable particularly for hydrophobic drugs which need to be administered parenterally. Such systems solubilize the drugs in formulation upon dilution in dipolar solvents (e.g., DMSO) to produce submicron particles below 0.4 μm due to spontaneous emulsification [113].

1.12.2.4   Ocular systems

Recently, the SEDDS have shown their immense utility for the ocular delivery of drugs for the treatment of pathological disorders like edema, uveitis, neovascularisation, diabetic retinopathy, etc. [174]. The SE formulations provide stellar advantage of delivering the poorly water-soluble drugs by intra-conjuctival injection to enhance their ocular bioavailability over conventional drug delivery systems. SEDDS can be a useful approach to improve bioavailability of poor water-soluble drugs by increasing drug dissolution rate or through a potential influence of surfactants on the permeability of corneal barrier. Additionally, surfactants decrease surface tension and may eliminate vision disorders. Moreover, surfactants in self-emulsifying oils should be less irritating than in the aqueous medium [175].

1.12.2.5   Transdermal systems

The SE formulations can enable the transdermal delivery of hydrolysable drugs undergoing extensive hepatic first pass effect [176]. These systems undergo phase inversion on coming in contact with the aqueous secretions of the skin to produce a supersaturated system. This phenomenon of inversion generates the driving force, (i.e., flux) for transdermal delivery of drugs through stratum corneum to enhance its systemic availability [177].

1.13   Types of SEDDS dosage forms

1.13.1   Dry emulsions

Dry emulsions are powders in which emulsion spontaneously occurs in vivo or after exposure to an aqueous solution dry emulsion technology solves the stability problems associated with classic emulsions (e.g. phase separation, and contamination by microorganisms) during storage and also helps avoid the use of harmful or toxic organic solvents. Dry emulsions may be redispersed in water before use. Medium chain triglycerides are commonly used as the oil phase for these emulsions. Dry emulsion formulations are typically prepared from oil/water (O/W) emulsions containing a solid carrier (such as lactose or maltodextrin) in the aqueous phase by rotary evaporation [177], freeze-drying [178] or spray drying [179]. Dry emulsions can be used for further preparation of tablets and capsules. To promote the bioavailability of the poorly soluble drug, amlodipine, oleyl polyoxyl glycerides (Labrafil® M 1944 CS) were used as the lipophilic phase of the dry emulsion. Most recently, nimodipine dry emulsions have been prepared using Dextran 40 as a water-soluble solid carrier. The most exciting finding in this field is the newly developed enteric-coated dry emulsion formulations, which are potentially applicable for the oral delivery of peptide and protein drugs. These formulations consist of a surfactant, a vegetable oil, and a pH-responsive polymer, and lyophilization is used [180].

1.13.2   Self-emulsifying capsules

It is a capsule containing liquid or semisolid form of SES. In the GIT, the capsules get dispersed to SES uniformly in the fluid to micron size, enhancing the bioavailability. Second type of self-emulsifying capsule is solid SES filled into capsule [181].

1.13.3   Sustained/controlled release tablets

The preparation of self-emulsifying tablets depends on combination of lipids and surfactants. Researchers evaluated some parameters before formulating self-emulsifying tablets, which are colloidal silicates X1, magnesium stearate mixing time X2, and compression force X3, on hardness and coenzyme Q10 (CoQ10) dissolution from tablets of eutectic-based SMEDDS. The optimized conditions (X1 = 1.06%, X2 = 2 min, X3 = 1670 kg) were achieved by a face-centered cubic design. The amount of solidifying excipients reduced for transformation of SEDDS into solid dosage forms, a gelled SEDDS has been developed by Patil et al. In their study, colloidal silicon dioxide (Aerosil 200) was selected as a gelling agent for the oil-based systems, which served the dual purpose of reducing the amount of required solidifying excipients and aiding in slowing down of the drug release [182].

1.13.4   Sustained/controlled release pellets

Serratoni et al. prepared SE controlled-release pellets by incorporating drugs into SES, thereby improved their rate of release, and then by coating pellets with a water-insoluble polymer that reduces the rate of drug release [183]. To formulate and prepare SEDDS, there were some basic guidelines to be considered: safety, compatibility, drug solubility, efficient self-emulsification efficiency and droplet size, etc. Pellets are multiple unit dosage forms, which may provide many advantages than conventional solid dosage forms, due to some factors like flexibility of manufacture, reducing intra subject and inter subject variability of plasma profiles and minimizing GI irritation without lowering drug bioavailability. Thus, it is very interesting to combine the advantages of pellets with those of SEDDS by SE pellets. They were prepared by extrusion/spheronization and contained two water-insoluble model drugs (methyl and propyl parabens): SES contained mono-diglycerides and Polysorbate 80 [184].

1.13.5   Nanoparticles

For the production of self-emulsifying nanoparticles, nanoparticle technology might be used successfully. Under these one of the techniques called solvent injection is used. In this technique, molten lipid mass was prepared, which contains the mixture of lipid, surfactant, and drug. And then this lipid mass transferred into a non-solvent in a drop-wise manner and mix them. And thereafter filter them and dried. By this method we get the nanoparticles (about 100 nm) with a drug loading efficiency of 74%. More recently, a novel nanoparticle drug delivery system consisting of chitosan and glyceryl monooleate (GMO) for the delivery of paclitaxel (PTX) has been developed. The SE property of GMO enhanced the solubility of PTX and provided a basis for chitosan aggregation, in the meantime causing near 100% loading and entrapment efficiencies of PTX. And one more study was to prepare a self-nanoemulsifying system (SNES) containing model lipophilic drug, felodipine (FLD), to improve its solubility. The SNES was formulated using altering amounts of Miglyol 840 (as an oil), Cremophor EL (as a surfactant), and Capmul MCM (as a co-surfactant) [185, 186].

1.13.6   Beads

SES can be formulated as a solid dosage form by using less solidifying excipient. Patil and Paradkar discovered that deposition of SES into micro porous polystyrene beads was done by solvent evaporation. Porous polystyrene beads (PPB) with complex internal void structures were typically produced by copolymerizing styrene and divinyl benzene. It is inert and stable over a wide range of pH, temperature and humidity. Geometrical features such as bead size and pore architecture of PPB was found to preside over the loading efficiency and in vitro drug release from SES loaded PPB [187].

1.13.7   Solid dispersions

Solid dispersions could increase the dissolution rate and bioavailability of poorly water-soluble drugs but some manufacturing difficulties and stability problems are arised, to overcome these problems self-emulsifying excipients like Gelucire 44/14, Gelucire 50/02, Labrasol, Transcutol and TPGS (tocopheryl polyethylene glycol 1000 succinate) have been widely used [188–

190]. Gupta et al. prepared SE solid dispersion granules using the hot-melt granulation method. Seven drugs, including four carboxylic-acid-containing drugs, a hydroxyl-containing drug, an amide-containing drug (phenacetin) and a drug with no proton-donating groups (progesterone) were chosen. Gelucire150/13 was used as the dispersion carrier; whereas Neusilin US2 was used as the surface adsorbent [191].

1.13.8   Microspheres

Zedoary turmeric oil (ZTO; a traditional Chinese medicine) shows effective pharmacological actions like tumor suppressive, antibacterial, and antithrombotic activity. You *et al.* prepared solid SE sustained release microspheres using the quasiemulsion-solvent diffusion method of the spherical crystallization technique; in this technique ZTO is used as oil phase. ZTO-released activities might be controlled by the ratio of hydroxyl propyl methylcellulose acetate succinate to Aerosil 200 in the formulation. After oral administration of such microspheres to rabbits, the plasma concentrations were achieved with increased bioavailability of 135.6% with respect to the conventional liquid SEDDS [192].

1.14   Special types of SEDDS formulations

1.14.1   Positively charged SNEDDS

Many physiological studies have proved that the apical potential of absorptive cells, as well as that of all other cells in the body, is negatively charged with respect to the mucosal solution in the lumen [93, 193–195]. The drug exposure of the positively charged SEDDS has been found to be higher vis-à-vis the conventional formulations especially for bioavailability enhancement. More recently, it has been shown that the enhanced electrostatic interactions of positively charged droplets with the mucosal surface of the everted rat intestine are mainly responsible for the preferential uptake of the drugs [93]. The binding of these cationic SNEDDS has been found to be much higher compared with the anionically charged formulation, suggesting increased adhesion of the droplets to the cell surface due to electrostatic attraction [196, 197]. Hence, studies on the successful formulation of some cationic SNEDDS have been undertaken. Olelylamine, a GRAS-approved excipient, has been preferred for SNEDDS formulation primarily owing to its resemblance to the hydrocarbon chain length of oleic acid, a known biodegradable fusogenic agent [198]. Further, oleylamine induces charge in the range of 30–35 mV which is vital for an efficient and stable emulsion/microemulsion system.

Besides oleylamine, the other agents like stearylamine, chitosan, etc., have also been employed for inducing cationic charge in various lipid-based drug delivery systems. Stearylamine-coated emulsions produce globules having zeta potential in the range of 22–26 mV, whereas chitosan-coated ones generate potential in the range of 20–23 mV. However, the use of stearylamine in drug delivery is restricted to a concentration of 0.5% w/wowing to its reported cytotoxicity. Chitosan, on the other hand, has been reported to be effective as charge inducer in lipidic drug delivery systems upto a concentration of 0.5% w/w, after which virtually it has little effect on the charge or the globule size [199].

1.14.2   Supersaturable self-emulsifying drug delivery system (S-SEDDS)

Mainly the supersaturable SEDDS (S-SEDDS) formulations have been designed and developed to reduce the surfactant effects, since high surfactant concentration here in SEDDS formulations can direct to GI side effects, and thereby this approach can provides a better toxicity/safety profile. A new class of supersaturable formulations, including the S-SEDDS approach is designed to make an extended supersaturated solution of the drug, while the formulation is released from a suitable dosage form into an aqueous medium. Supersaturation is intended to increase the thermodynamic activity of the drug further than its solubility limit, and consequently, the effect in an increased driving force for transit into and across the biological barrier, e.g. an S-SEDDS of paclitaxel (PTX) was developed using hydroxypropyl methyl cellulose (HPMC) as a precipitation inhibitor with a conventional SEDDS formulation, and a poorly soluble drug, PNU-91325, was formulated as an S-SEDDS [200]. The S-SEDDS formulations have been demonstrated to improve both the rate and extent of the oral absorption of the poorly water-soluble drugs quite effectively [94, 95, 201]. The inclusion of cellulosic polymers in the S-SEDDS formulation tends to effectively suppress the precipitation of drugs. Various viscosity grades of Hydroxypropyl methylcellulose (HPMC) are well recognized for their ability to inhibit crystallization and, thereby, generate and maintain their supersaturated state for extended time periods [202–204]. In vitro dilution of the S-SEDDS formulation results in the formation of a microemulsion, followed by slow crystallization of the drug on standing indicating that the supersaturated state of the system is prolonged by HPMC in the formulations. While in the absence of HPMC, the SEDDS formulation undergoes rapid precipitation, yielding a lower drug concentration [94]. It has been indicated that the presence of a small amount of HPMC in the formulation is critical to achieve a stabilized supersaturated state of drug upon mixing with water. Table 1.5 enlists various studies reported in the literature on the supersaturated SEDDS formulations and their respective compositions; and Table 1.6 presents the updated comprehensive account of the SEDDS formulations of various drugs along with their excipient composition.

Table 1.5 A literature update on various reports on supersaturable SEDDS

Drug	Lipids/oils used	Surfactants/ co-surfactants	Polymeric precipitation inhibitor (PPI)	References
PACLITAXEL	Capmul MCM	Cremophor EL, Pluronic F108, Tween A80 and PEG 400	PVP 12PF and K30 and HPMC E4M, E50, E5, K100 and K3	[209]
PNU 91325	Mount Olive, Glycerol monooleate and Glycerol dioleate	Cremophor EL, PEG 400, Propylene glycol, Dimethyl acetamide, Pluronic L44 and Tween 80	HPMC 2910 E5LV and E50LV	[210]
DOCETAXEL	Labrafac	Cremophor RH40 and Transcutol P	HPMC K100	[211]
AMG 517	Capmul MCM	Cremophor EL, Pluronic F108, Tween 80 and PEG 400	PVP 12PF and K30 and HPMC E 4M, E50, E5, K100 and K3	[212]
SIMVASTATIN	Capmul PG8, Ethyl oleate	Tween 80, Cremophor EL, Transcutol HP	HPMC E50LV, E5LV and K4M	[213]
INDIRUBIN	Maisine 35-1	Cremophor EL, Transcutol P	PVP K-17	[214]

Table 1.6 An updated comprehensive account of the SEDDS formulations of various drugs along with their excipients composition

Drug name	Lipid name	Surfactants	Co-surfactants	Techniques	Dosage form	References
Ibuprofen	Goat fat	Tween 60	–	Melt granulation	SE tablets	[215]
Ibuprofen	Capmul PG 8 (oil)	Cremophore EL	Capmul PG 8 (Co-solvent)	Capsule filling	SEDDS	[216]
Cinnarizine	Oleic acid	Tween 80	Capmul MCM C-8	Simple mixing method	SNEDDS	[217]
Cinnarizine	Miglyol 810, Imwitor 988, 308 Captex Soybean oil	Tween 80, Tween 85, HCO-30, Cremophore EL	Propylene glycol	Simple mixing method	SNEDDS	[218]
Cisapride	Oleic acid	Tween 80	Propylene glycol	Solubility study, water titration method	SEDDS	[219]
Glipizide	Capmul MCM	Cremophore EL	Transcutol P	Water titration method	Micro-emulsion	[220]
Domperidone	Oleic acid	Tween 60	PEG 400 (Co-surfactant)	Water titration method	SMEDDS	[221]
Efavirenz	Labrafac PG	Labrasol	Tween 80 (15-15.2)	Adsorption to solid carriers	Solid SMEDDS	[222]
Indomethacin	Castor oil	Cremophor RH 40	Capmul MCM C-8	Capsule filling	SNEDDS	[223]
Carvedilol	Miglyol 812	HCO-40	Transcutol HP	Incorporating in liquisolid tablets	SNEDDS tablets	[224]
Lumefantrine	Oleic acid	Cremophor EL	Transcutol P, Capmul MCM, PEG 400, Capmul PG 8	Capsule filling	SEDDS	[225]
Ubiquinone	Lemon oil	Cremophor EL	Capmul MCM (cosurfactant)	Capsule filling	SNEDDS	[226]
Ketotifen	Captex 200	Tween 80	Capmul MCM	Simple mixing method	Gelled SEDDS	[227]
Carbamazepine	Labrafil M 1944 CS	Cremophor RH 40	PEG 400	Simple mixing method	SMEDDS	[228]
Exemestane	Capryol 90	Cremophor ELP	Transcutol HP	Simple mixing method	SMEDDS	[229]

Contd...

Contd...

Drug name	Lipid name	Surfactants	Co-surfactants	Techniques	Dosage form	References
Candesartan cilexitil	Miglyol 812	Cremophor EL Tween 80	Labrasol	Capsule filling method	Solid SMEDDS	[230]
Valsartan	Labrasol	Tween 20	PEG 400	Simple mixing method	SNEDDS	[231]
Valsartan	Capmul MCM	Labrasol	Tween 20, PEG 600	Adsorption to solid carriers (Aerosil 200, Sylysia (350, 550, and 730) and Neusilin US2)	Solid SNEDDS	[232]
Olmesartan medoxomil	Acrysol EL 135	Tween 80	Transcutol P	Simple mixing method	SMEDDS	[233]
Tacrolimus	Rice germ oil Capmul PG 8	Cremophor EL	Transcutol P	Simple mixing method	SMEDDS	[234]
Olanzapine	Isopropyl Palmitate, Capryol 90	Cremophor RH 40, Tween 80	Transcutol HP	Simple mixing method	SNEDDS	[235]
Probucol	Capmul MCM, Captex 355	Cremophor EL	–	Simple mixing and Adsorption to solid carriers	Solid SEDDS	[236]
Nimodipine	Gelucire 44/14	Labrasol containing 80% Transcutol P	Plurol Oleique CC 497	Simple mixing	SEDDS	[237]
Simvastatin	Capryol 90	Cremophor RH 40	Transcutol HP	Simple mixing	SNEDDS	[238]
Pioglitazone HCl	Labrafac PG, Oleic acid	Tween 80	Propylene glycol (co-solvent)	Simple mixing	SEDDS	[239]
Cinnarizine	Oleic acid	Tween 80	Transcutol P	Simple mixing	SMEDDS	[240]
Curcumin	Capryol 90	Cremophor EL	Transcutol P	Capsule filling	SMEDDS	[241]
Irbesartan	Capryol 90	Cremophor EL	Carbitol	Simple mixing method	SNEDDS	[242]
Efavirenz	Softigen 707	Cremophor EL	Capmul MCM	Adsorption to solid carriers	SMEDDS	[243]
Flutamide	Sesame oil	Tween 20	PEG 400	Simple mixing followed by vortexing	SNEDDS	[244]
Lornoxicam	Capmul MCM	Tween 20	PEG 400	Spray drying	SNEDDS	[245]
Glibenclamide	Capmul MCM C8	Cremophor RH 40	Transcutol P	Adsorption technique	S-SEDDS	[246]

Contd...

Contd...

Drug name	Lipid name	Surfactants	Co-surfactants	Techniques	Dosage form	References
Docetaxel	Labrafac	Cremophor RH 40	Transcutol P	Spray drying	Super-saturable SEDDS	[247]
Glyburide	Capryol 90	Tween 20	Transcutol P	Vortexing	SMEDDS	[248]
Albendazole	Labrafac, Capmul PG8, Captex 300	Tween 80, Solutol HS	PEG 400, Propylene glycol	Simple mixing method	SEDDS	[249]
Glibenclamide	Peceol	Tween 20	Transcutol P	Vortexing	SMEDDS	[250]
Simvastatin	Captex, Lauroglycol	Cremophor EL	Capmul MCM	Simple mixing method	SEDDS	[251]
Silymarin	Ethyl oleate: MCT(1:1)	Cremophor EL	Transcutol P	Stirring and vortex mixing	SMEDDS	[252]
Aceclofenac	Peppermint oil	Solutol HS 15	Labrasol	Vortexing	SMEDDS	[253]
Atorvastatin	Oleic acid	Tween 80	PEG 400	Vortexing	SEDDS	[254]
Lutein	Peceol	Labrasol	Transcutol HP or Lutrol E-400	Simple mixing	SNEDDS	[256]
Nifedipine	Sesame oil	Span 80 and Tween 80 (3:7)	n-butanol	Vortexing	SMEDDS	[257]
Paclitaxel	Peanut oil, ethyl oleate, oleic acid	Solutol HS 15, Tween 80	PEG 400, PG, Ethanol	Simple mixing	SNEDDS	[258]
Gemfibrozil	Labrafac CM 10	Cremophor EL	PEG 400	Moderate stirring and vortex mixing	SMEDDS	[259]
Repaglinide	Oleic acid	Tween 80	Transcutol P	Vortex mixing	SEDDS	[260]
Lamotrigine	Capmul MCM C8	Labrasol	Tween 80	Adsorption to solid carriers	SMEDDS	[261]
Metronidazole	Palm Kernel oil And Palm oil	Tween 65	–	Simple mixing and capsule filling	SEDDS	[262]

1.14.3   SEDDS for traditional herbal medicine

Well accepted in the mainstream of medical care throughout the Asian continent, the traditional herbal drugs are considered as alternative medical system in much of the Western world too. The concepts of herbal medicines are quite well-defined in "Ayurveda" [205, 206], Chinese practice [207, 208] and "Unani" system of medicine. Nevertheless, the absorption of many active phytochemical constituents from these traditional herbs like Cardus Marianus (Silybin), Curcuma zedoaria (Turmeric), etc., has been reported to be not-up-to-the-mark. Several studies, accordingly, have been undertaken to formulate the self-emulsifying systems of vital constituents of such herbs to attain the desired objectives.

1.15   Regulatory issues of SEDDS

From a regulatory point of view, quality and safety issues related to preclinical and clinical studies are the main difficulties likely to be encountered in launching a lipid-based dosage form on the market. The overall drug stability and absence of immunological reactions to the oils or lipid excipients has to be demonstrated. Sufficient details explaining the use of lipid excipients and the types of dosage form, the drug release mechanism and their manufacture should be provided to convince the regulatory authorities of their acceptability. Safety assessment and the potential influence of biopharmaceutical factors on the drug or lipid excipients need to be explored. It may be difficult to predict in vivo performances of a lipid dosage form based on in vitro results obtained with conventional dissolution methods in view of the convoluted GI processing of lipid formulations. More mechanistic studies should be conducted to facilitate a better understanding of the pharmaceutical characteristics of lipid formulations and interactions between lipid excipients, drug and physiological environment. The lack of predictability for product quality and performance may be due to the nature of empirical and iterative processes traditionally employed [263]. A list of some typical excipients used in SEDDS is depicted in Table 1.7, along with their chemical name and regulatory status. With the aim of rationalizing the design of lipid formulation, and to better understand the fate of a drug after oral administration in a lipid-based formulation, a consortium, composed of academics and industrial scientists, has been created. The consortium sponsors and conducts research to develop in vitro methods to assess the performance of LBDDS during dispersion and digestion, which are critical parameters. The primary objective is to develop guidelines that rationalize and accelerate the development of drug candidates through the identification of key performance criteria, and the validation and eventual publication of universal standard tests and operating procedures. In order to establish approved guidelines, appropriate dialogue with pharmaceutical regulatory bodies (FDA, EMEA) is also foreseen [264].

Table 1.7 Regulatory status of some typical excipients used in SEDDS

Type of excipients	Trade name	Chemical name	Regulatory status
Lipids	Vegetable oil	Long-chain TAG	Oral product, GRAS, FDA IIG
	Miglyol 812	Medium-chain TAG caprylic/ capric acid	Oral product, GRAS, FDA IIG
	Tricaprylin	Medium-chain TAG	–
	Labrafac CC	Caprylic/capric TG	–
	Ethyl oleate	Ethyl ester of C18:1 FA	FDA IIG
	Captex 355	Glycerol caprylate caprate	GRAS, FDA IIG
	Isopropyl myristate	FA ester	FDA IIG
	Labrafac PG	PG dicaprylocaprate	USFA,JSFA,EP
	Peceol	Glyceryl mono-oleate	GRAS,E471,EP,USP-NF,FDA IIG
	Maisine 35-1	Glyceryl mono-linoleate	Oral product, GRAS,EP, USP-NF, E471
	Imwitor 988	Caprylic/capric glycerides	USP, Ph.Eur
Surfactants HLB < 12	Tween 85	Polyoxyethylene (20) sorbitan trioleate	UK
	Labrafil M1944CS	Oleoyl macrogolglycerides	EP, FDA IIG, USP NF
	Labrafil M2125CS	Linoleoyl macrogolglycerides	EP, FDA IIG, USP NF
	Lauroglycol 90	PG monolaurate	USFA, FCC, EFA, USP-NF
Surfactants HLB >12	Vitamin E TGPS	D-alpha-tocopheryl PEG 1000 succinate	Oral product
	Cremophor EL	Polyoxyl 35 castor oil	Oral product, USP-NF, FDA IIG
	Cremophor RH 40	Polyoxyl 40 hydrogenated castor oil	Oral product, USP-NF, FDA IIG
	Gelucire 44/14	Lauroyl macrogolglycerides	EP, USP-NF, FDA IIG
	Labrasol	Caprylocaproyl macrogol glycerides	EP, USP-NF, FDA IIG
	Polysorbate 80/ Tween 80	Polyoxyethylene (20) sorbitan monooleate	Oral product, GRAS, EP, USP-NF, FDA IIG
	Polysorbate 20/ Tween 20	Polyoxyethylene (20) sorbitan monolaurate	Oral product, GRAS, EP, USP-NF, FDA IIG
Co-solvents	Ethanol	–	Oral product, EP, USP-NF
	PEG	PEG 300 and PEG 400	Oral product, EP, USP-NF
	Transcutol P	Diethylene glycol monoethyl ester	EP, FDA IIG

PEG: Polyethylene Glycol, PG: Propylene Glycol, TAG: Triacylglyceride, MAG: 2-Monoacylglyceride, DAG: Diacylglyceride, FA: Fatty acid, GRAS: Generally Recognized As Safe, E471: European Food Additive, EP: European Pharmacopoeia, USP-NF: United States Pharmacopoeia-National Formulary, FDA IIG: FDA Inactive Ingredient Guide, Ph.Eur: Pharmacopoeia Europea, USFA: United States Food Administration, FCC: Food Chemical Codex, JSFA: Japanese Standards for Food Additives, UK: United Kingdom.

1.16   Recent patents on SEDDS

The recent patenting systems related to the self-emulsifying drug delivery systems are enlisted briefly as follows:

EP 1170 003 B1 describe a novel formulation, i.e. self-emulsifying drug delivery system for fat-soluble drugs such as tocotrienols which are incorporated into a palm-olein and a surfactant system being a combination of caprylocaproyl macrogolglycerides and polyoxyethylene 20 sorbitan monooleate, wherein the weight ratio of said first component to said second component is between 9:1 and 7: 3 [265].

EP 1333 810 B1 discloses an improved oral dosage self-emulsifying formulations of pyranone protease inhibitors comprising a lipophilic phase, a hydrophilic phase preferably with polyethylene glycol, one or more pharmaceutically acceptable surfactants and a basic amine in the amount from 0.1% to about 10% by weight of total composition [266].

EP 1340 497 A1 describe self-emulsifying drug delivery systems for poorly soluble drugs such as paclitaxel which is a taxoid and also comprises the relative proportions of vitamin E, TGPS and polyoxyl hydrogenated castor oil [267].

EP 1349 541 B1 describe self-emulsifying lipid matrix wherein the active compound is a 4-phenylpyridin derivative along with soyalecithin as emulsifier and cocoa butter as sweeteners and/or flavours [268].

EP 1648 517 B1 describe self-emulsifying and self-microemulsifying formulations for the oral administration of taxoids where its concentration is not over 10% w/w also comprising medium chain triglyceride (Miglyol 812N®), surfactant (Cremophor EL®) and co-surfactant is chosen from Peceol®, Lauroglycol 129®, Capryol 90®, Maisine 35-1® and Imwitor 988® [269].

EP 1787 638 B1 describe butyl benzene phthalein self-emulsifying drug delivery system, its preparation method and application. The present invention comprises butylphthalide and an emulsifying agent (is preferably the mixture of polyoxyethylene castor oil and polyethyleneglycol-8 glycerin caprylate/caprate in the ratio of 1:0.5 to 1:5 (by weight)), also an appropriate antioxidant such as dibutyl hydroxytoluene and a flavoring agent such as mint oil, green apple oil [270].

EP 2,062,571 A1 describe self-emulsifying pharmaceutical composition with enhanced bioavailability wherein the drug is cyclosporine, tacrolimus, ibuprofen, ketoprofen, nifedipine, amlodipine, or simvastatin, a hydrophilic carrier is selected from the group consisting of ethanol, isopropanol, polyethylene glycol (PEG), glycerin, propylene glycol, and the mixture thereof likewise a surfactant is selected from the group consisting of PEG 40 hydrogenated castor oil, polysorbate, cocamidopropyl betaine, glyceryl cocoate, PEG 6 caprylic/capric glycerides, Poloxmer, Labrafil M1944CS, Labrafil M2125CS, Labrasol, Cremophor EL, Cremophor RH, Brij, Spans, and the mixture thereof. [271]

US 7,276,113 B2 describe self-emulsifying pigments or surface-treated pigments for cosmetic products such as foundations, lip sticks, lotions, and creams having at least two surface-active agents each of which is chemically immobilized onto the surface of the pigment, wherein the first surface-active agent and second surface-active agent have HLB value of 10 or higher and 9 or lower, respectively. The first surface-active agent contains one or more hydroxyl groups or alkylene oxide moieties whereas it is devoid in second surface active agent [272].

US 7,736,666 B2 discloses a self-emulsifying drug delivery system comprising a compound of the formula consists of naproxen, a butyl spacer and a NO-releasing moiety, said three parts being linked together into one single molecule. Also consists a non-ionic surfactant (Poloxamer) and a short-chain alcohol is selected from the group consisting of ethanol, propylene glycol and glycerol [273].

US 5,965,160 describe self-emulsifiable formulation producing an oil-in-water emulsion of the drug cyclosporine-A encapsulated in a capsule comprises cationic lipid is stearylamine or oleylamine [274].

US 6,221,391 B1 describe self-emulsifying Ibuprofen solution and soft gelatin capsule for use therewith including a polyoxyethylene castor oil derivative preferably PEG 35CO and/or PEG 40HCO [275].

US 6,316,497 B1 explain self-emulsifying systems containing anticancer medicament o-(chloroacetylcarbamoyl) fumigillol, where in the pharmaceutically acceptable carrier comprises an oily constituent selected from the group consisting of alcohols, propylene glycol, propylene glycol esters, polyethylene glycol, monoglycerides, diglycerides, triglycerides, fatty acids, naturally occurring oils, and a mixture thereof and atleast one surfactant, and the stabilizing component comprises from about 1% to about 15% water relative to the weight of the stabilized self-emulsifying system [276].

US 6,436,430 B1 describe self-emulsifying compositions for drugs poorly soluble in water such as cyclosporine ibuprofen, paclitaxel and naproxen as the therapeutically active agent and surfactant is polyoxyethylene sorbitan ester, the polyethoxylated product of hydrogenated vegetable oils, polyethoxylated castor oil or polyethoxylated hydrogenated castor oil [277].

US 6,630,150 B1 describe spheronized self-emulsifying system for hydrophobic and water sensitive agents which comprises:

(I)   A first portion comprising microcrystalline cellulose; and

(II) A second portion comprising: (A) Up to 200%, based on the weight of the first portion, of an oily substance; (B) Between 2 and 100%, based on the weight of the first portion, of a surfactant; and (C) Between 2 and 1000%, based on the weight of the oily substance and the surfactants, of water; wherein the total weight of the oily substance and the surfactant is between 2% and 200% of the first portion [278].

US 7,022,337 B2 describe self-emulsifying formulations of fenofibrate and/or fenofibrate derivatives with improved oral bioavailability and/ or reduced food effect wherein the fenofibrate or fenofibrate derivative is dissolved in fibrate solubilizer containing a solvent such as the N-alkyl derivatives of 2-pyrrolidone, mono- or di- or polyethylene glycol mono ethers, C_{8-12} fatty acid mono or di-esters of propylene glycol or glycerol, or combinations thereof. Additionally the formulation contains a surfactant that may be ionic or non-ionic or a combination of both [279].

US 8,158,162 B2 describe eutectic-based self-nanoemulsified drug delivery system is formulated from polyoxyl 35 castor oil (Cremophor), medium chain mono- and diglycerides (Capmul), essential oils, and a pharmacologically effective drug such as ubiquinone (CoQ_{10}) [280].

US 8,618,168 B2 disclose self-emulsifying composition of omega-3 fatty acid, i.e. ethyl eicosapentaenoate ester, an emulsifier polyoxyethylene hydrogenated castor oil and propylene glycol or glycerin [281].

US 8,592,490 B2 describe self-microemulsifying drug delivery systems and microemulsions to enhance the solubility of pharmaceutical ingredients comprising a polyoxyethylene sorbitan fatty acid ester emulsifier, a fatty acid ester co-emulsifier and oil [282].

US 2002/0103139 A1 describe solid self-emulsifying controlled release drug delivery system composition for enhanced delivery of water-insoluble phytosterols (such as Beta-Sitosterol, sitosternol, campesterol) and other hydrophobic natural compounds for body weight and cholesterol level control [283].

US 2002/0119198 A1 describe self-emulsifying drug delivery systems for extremely water-insoluble or lipophilic drugs, i.e. 3-[(2, 4-dimethylpyrrol-5-yl) methylene] 2-indolinone and also comprises polyvinylpyrrolidone, a fatty acid and a surfactant [284].

US 2002/0131945 A1 disclose delivery of reactive agents via self-emulsification for use in shelf-stable products wherein the reactive agent is comprised of one or more reactive groups of the electrophilic, nucleophilic or protected thiol type, a water immiscible solvent, one or more surfactants. Also disclosed are methods for treating amino-acid-based substrates, and methods for bleaching, coloring and conditioning hair with these treatment compositions [285].

US 2003/0105141 A1 describe finely self-emulsifiable pharmaceutical composition comprising celecoxib, oleic acid, triethanolamine, polyethylene glycol and water wherein the composition is finely self-emulsifiable in simulated gastric fluid [286].

US 2007/0012895 A1 describe highly concentrated self-emulsifying preparations containing organopolysiloxanes and alkyl ammonium compounds and their use in aqueous systems. This invention consist of fully or partially quaternized amino-functional organopolysiloxanes, quaternary alkylammonium compounds, if appropriate an organic hydrotrope and if appropriate water [287].

US 2007/0104740 A1 describe self-microemulsifying drug delivery systems of a HIV protease inhibitor such as (3R, 3aS, 6aR)-hexahydrofuro [2,3-b]furan 3-yl (1 S,2R)-3-[[(4 aminophenyl) sulfonyl](isobutyl l) amino]-1-benzyl-2-hydroxypropyl-carbamate and also include salts, esters, polymorphic and pseudopolymorphic forms thereof and comprise as carrier a lipophilic phase, one or more surfactants, a hydrophilic solvent and a nucleation inhibitor [288].

US 2007/0104741 A1 describe delivery of tetrahydrocannabinol which is formulated as a self-emulsifying drug delivery system by dissolving it in an oily medium (e.g. triglycerides and/or mixed glycerides and/or free fatty acids containing medium and/or long chain saturated, mono-unsaturated, and/or poly unsaturated free fatty acids) together with at least one surfactant. The delivery system of the present invention can be administered as either a liquid or semi-solid matrix within a capsule shell for immediate or sustained release rates [289].

US 2004/0185068 A1 describe self-emulsifying compositions, methods of use and preparation. These self-emulsifying compositions are prepared using one or two surfactants without mechanical homogenization. Consequently, the self-emulsifying compositions are ideally suited for ophthalmic applications. Self-emulsifying compositions prepared by the disclosed method are described. The therapeutic compound selected from the group consisting of cyclosporine, prostaglandins, Brimonidine and Brimonidine salts and the surfactant component are Lumulse GRH-40 and TGPS [290].

WO 2008/142090 A1 describe a self-emulsifying formulation of Tipranavir for oral administration comprising vitamin E TGPS as a surfactant and one or more pharmaceutically acceptable solvents are propylene glycol, polypropylene glycol, polyethylene glycol (such as PEG 300,400,600, etc.), glycerol, ethanol, triacetin, dimethyl isosorbide, glycofurol, propylene carbonate, water, dimethyl acetamide or a mixture thereof [291].

WO O1/72282 A1 disclose self-emulsifying matrix type transdermal preparation comprising a polymer matrix, an oil phase, atleast one co-solvent, at least one surfactant, an aqueous phase (water) and atleast one pharmacologically active substance. The matrix preparation can be prepared by drying the resulting mixture solution after mixing and dissolving the components. The matrix preparation also includes additives such as antioxidants and preservative which are pharmaceutically acceptable [292].

WO 00/16744 describes a pharmaceutical solution for the treatment of erectile dysfunction, prepared by SEDDS formulation. It relates to a new pharmaceutical composition containing prostaglandins, or prostaglandin E1 (PGE1) using Labrafac CC, Labrafac lipophile, Labrafil M1944CS, WL 2609BS, Maisine. These dosage form administered into the urinary tract by means of an infusing device which can increase retention of the drug in the urinary tract either by gaining viscosity or by forming gels upon contact with a small amount of water [293].

EP 2207531 B1 describe self-emulsifying pharmaceutical compositions of Rhein or Diacerein or salts or esters or prodrugs thereof in one or more pharmaceutically acceptable vehicles comprising one or more emulsifier selected from the group consisting of polyoxyethylene glycerol esters of fatty acids, polyoxylated castor oil, ethylene glycol esters, propylene glycol esters, glyceryl esters of fatty acids, etc., and one or more pharmaceutically acceptable polymers, wherein the pharmaceutically acceptable polymer is selected from the group consisting of cellulosic polymers or its derivatives including hydroxypropylmethyl cellulose, hydroxyl methylcellulose, hydroxy ethyl cellulose, etc. [294]. In spite of that above information related to the SEDDS, the major literature based SEDDS are described in Table 1.8.

Table 1.8 Various patents on self-emulsifying drug delivery systems

Title	Drug	Excipients	Purpose of invention	Claims	Patent No.
Self-emulsifying systems Containing anticancer medicament	o-(chloroacetyl carbamoyl) fumigillol	Captex 200, Capmul MCM, Tween 80, Labrasol, Miglyol 812, and Lauroglycol FCC, Miglyol 840	Present invention exhibit improved bioavailability of lipophilic compounds useful against cancer conditions	A stabilized self-emulsifying system, comprising a therapeutically effective amount of o-(chloro acetylcarbamoyl) fumigillol, a pharmaceutically acceptable carrier and a stabilizing component	US 6316497 B1/ 2001 [276]

Contd...

Contd...

Title	Drug	Excipients	Purpose of invention	Claims	Patent No.
Self-emulsifying compositions for drugs poorly soluble in water	Cyclosporin Ibuprofen, paclitaxel and naproxen	Propylene glycol Monoester of Caprylic Acid, Polyoxyethylene (20), sorbitan ester, Polyoxyl 30 castor oil	directed to a method of forming a microemulsion of a pharmaceutical composition containing a lipophilic drug	Increases the solubility of the lipophilic drug in the pharmaceutical carrier, and also facilitates uniform absorption thereof in the treated mammal and enhances the bioavailability of the lipophilic drug	US 6436430 B1/ 2002 [277]
spheronized self-emulsifying system for hydrophobic and water sensitive agents		microcrystalline cellulose, polysorbate 80	To provides a process for making a solid dosage form comprising drying a mixture comprising MCC, an oily substance, surfactant and active ingredient	the active ingredient is present in an amount of between 5 and 25% based on the weight of the oily substance and surfactant	US 6630150 B1/ 2003 [278]
self-emulsifying formulations of fenofibrate and/or fenofibrate derivatives with improved oral bioavailability and/or reduced food effect	Fenofibrate	Captex 200, cremophor RH 40, Span 80, Gelucire 44/14, Gelucire 50/13	To provides an oral self-emulsifying formulation Wherein the fenofibrate is dissolved in a fibrate solubilizer	The weight ratio of the fibrate to the stabilizer is between 50:1 to about 1:10	US 7022337 B2/ 2006 [279]
Eutectic-based self-nanoemulsified drug delivery system	Ubiquinone (CoQ10	Cremophor, Capmul, essential oils, Kollidon VA 64, Maltodextrin, MCC	To overcome problems of conventional self-emulsifying vehicles	A method of preparing an orally administrated dietary supplement comprising the steps of admixing coenzyme Q10 and a sufficient amount of a volatile essential oil to reduce the melting point	US 8158162 B2/ 2012 [280]
self-emulsifying composition of omega 3 fatty acid	Ethyl eicosapentaenoate ester	an emulsifier polyoxyethylene hydrogenated castor oil and propylene glycol or glycerin	To provide a self-emulsifying composition which contains atleast one compound selected from group consisting of ω3 PUFA, its esters and an emulsifier having HLB of atleast 10	A self-emulsifying composition comprising 50-95% by weight in total of atleast one active compound	US 8618168 B2/ 2013 [281]
self-microemulsifying drug delivery systems and microemulsions	Candesartan cilexetil, celecoxib,	Imwitor 308, polysorbate 80, Miglyol 812, MCC, Crospovidone, mannitol, sorbitol, absolute ethanol,	To provide a drug delivery system which does not damage the crystal structure of the drug	A self-micro emulsifying drug delivery system consisting of a polyoxyethylene sorbitan fatty acid ester, a co-emulsifier, and an oil	US 8592490 B2/ 2013 [282]

Contd...

Contd...

Title	Drug	Excipients	Purpose of invention	Claims	Patent No.
solid self-emulsifying controlled release drug delivery system composition for enhanced delivery of water insoluble phytosterols and other hydrophobic natural compounds for body weight and cholesterol level control	β-sitosterol, sitosternol, campesterol	Octacosanol, Saw Palmetto Extract, Gugulipid Extract, Tocopherol Acetate and Coenzyme Q10	to provide a composition for body weight and cholesterol control, comprising: at least one phytosterols in an amount between 5% and 50% by Weight	A composition for body weight and cholesterol control, comprising: a self-emulsifying base providing in the body submicron particles of dissolve components including one phytosterols	US 0103139 / 2002 [283]
self-emulsifying drug delivery systems for extremely water-insoluble or lipophilic drugs	3-[(2,4-dimethylpyrrol-5-yl)methylene] 2-indolinone	Polyvinylpyrrolidone, ethanol, ascorbyl palmitate, tocopherol, oleic acid, CAPMUL® MCM, CREMOPHOR® RH40;	To provides a formulation for an extremely water-insoluble active agent which can enhanced bioavailability over conventional dosage form	the Weight ratio of said fatty acid to said polyvinyl-pyrrolidone is about 2: 1 to about 1:3, and the Weight ratio of said surfactant to said polyvinyl-pyrrolidone is about 10:1to about 1:1	US 0119198 A1/ 2002 [284]
Delivery of reactive agents via Self emulsification for use in Shelf-stable products	Conditioning Liquid Emulsifiable Concentrate	Polymer, Isopar C, Neodol 23-3E, Heptanol, Propylene Carbonate, Cyclomethicone D5	invention relates to delivery of chemically unstable reactive agents via liquid emulsifiable concentrates for use in chemically shelf stable products	composition provides a long-lasting treatment effect	US 0131945 A1/ 2002 [285]
Finely self-emulsifiable pharmaceutical composition	celecoxib	oleic acid, triethanolamine, polyethylene glycol, Tagat TO, Transcutol, Dimethylethanolamine	To provide an effective method of treatment of acute pain	The drug is present in a therapeutically effective amount of about 1% to about 75% by weight of the composition and present in the solvent liquid in dissolved or solubilized form	US 0105141 A1/2003 [286]
Highly concentrated, self-emulsifying preparations containing organopolysiloxanes and alkyl ammonium compounds and their use in aqueous systems	Organopolysiloxanes , alkyl ammonium compounds			The preparations of the present invention possess good stability to shearing stresses on high-speed textile finishing machines	US0012895 A1/ 2007 [287]
Self-microemulsifying Delivery systems of a HIV protease Inhibitor	(3R,3aS,6aR)-hexahydrofuro [2,3-b] furan-3-yl(lS, 2R)-3-[[(4-amino phenyl) sulfonyl](isobutyl) amino]-1 -benzyl-2-hydroxypropyl Carbamate	Cremophor RH 40 Capmul MCM Capryol 90 Transcutol P	To provides methods of administration and treatment of HIV infected patients or suffering from AIDS	The formulation is in a form suitable for oral administration	US 0104740 A1/2007 [288]

Contd...

Contd...

Title	Drug	Excipients	Purpose of invention	Claims	Patent No.
Delivery of tetrahydro cannabinol	Δ9-THC	Oleic Acid, Peppermint oil , sesame oil, soybean oil, capmul MCM, Cremophor RH 40, Labrasol, Labrafil M 1944 CS, ascorbyl palmitate, vitamin E, povidone K-30	To improved dissolution, stability, and bioavailability of Δ9-THC	promotes targeted chylomicron delivery and optimal bioavailability upon administration to the mammalian intestinal lumen where endogenous bile salts reside	US 0104741 A1/2007 [289]
Self-emulsifying composition, methods Of use and preparation	Cyclosporin, prostaglandins. Brimonidine, Brimonidine salts	castor oil, Lumulse GRH-40	To formulate a self-emulsifying which will be therapeutic when administered to the eye	The oil globules size of self-emulsifying formulation have an average size of less than 0.15 micron	US 0185068 A1/ 2004 [290]
self-emulsifying formulation of Tipranavir for oral administration	Tipranavir	Polyethylene Glycol 400, Propylene glycol, Capmul MCM, Vitamin E TGPS, Ascorbic acid, sucralose, buttermint 24020, butter toffee 78185-33	To form a pharmaceutically acceptable, self-emulsifying oral formulation in the form of a solution		WO142090 A1/ 2008 [291]
self-emulsifying matrix type transdermal preparation	Flurbiprofene, Estradiol	Oleic acid, propylene glycol lauric acid ester, poloxamer 124, diethylene glycol diethyl ether	To provide a self-emulsifying matrix type preparation capable of containing high concentration of active substance and capable of avoiding collapse of SE system	The self-emulsifying system form the self-emulsification at the aqueous phase ratio of 0 to 60% and its continuous self-emulsifying range is atleast 20%	WO O1/ 72282 A1/ 2001 [292]
A pharmaceutical solution for the treatment of erectile dysfunction, prepared by SEDDS formulation	Prostaglandins, or Prostaglandin E1 (PGE1)	Labrafac CC, Cremophor EL, Ethanol, cremophor ELP, Benzyl alcohol, Labrafac lipophile, Labrafil M1944CS, Labrafil WL 2609BS, Maisine	To provide a new urethral solution containing PGE1 and to overcome instability of drug to moisture	The solution is for administration into urinary tract by means of an infusing device which can increase retention of the drug by gaining viscosity	WO 00/16744/ 2000 [293]
Self-emulsifying pharmaceutical compositions of Rhein or Diacerein	Rhein or Diacerein	Labrafil, Labrasol, Tween 80, Labrafac, Triacetin, Capmul, Acconon, Cremophor, Gelucire, HPC	Diacerein cannot be completely absorbed by digestive tract which may result in undesirable side effects such as soft stools so to overcome this is formulated as self-emulsifiable composition	the composition is further filled into hard gelatin capsules, soft gelatin capsules or hydroxypropyl-methyl cellulose capsules	EP 2207 531 B1/ 2012 [294]

1.17 Conclusions

Self-emulsifying drug delivery systems (SEDDS) are a promising approach for the formulation of drug compounds with poor aqueous solubility. The oral bioavailability of hydrophobic drugs can be improved significantly by this approach. The self-emulsifying drug delivery system possesses enormous potential to be commercialized owing to their simple technology and regulatory excipient status, coupled with scale-up and validation. In addition these systems offer much more flexibility in drug loading, drug release, stability and better feasibility in producing dosage forms such as tablets, capsules, pellets and creams.

1.18   References

1. Tiwari R., Tiwari G., Srivastava B., Rai A.K., Singh P. Solid Dispersions: An Overview To Modify Bioavailability Of Poorly Water Soluble Drugs. *Int.J. PharmTech Res.,* **2009,** 1 (4):1338–1349.

2. Serajuddin A.T. Solid dispersion of poorly water-soluble drugs: early promises, subsequent problems, and recent breakthroughs. *J. Pharm. Sci.,* **1999,** 8 (8):1058–1066.

3. Craig D.Q.M. The mechanisms of drug release from solid dispersions in water-soluble polymers. *Int. J. Pharm.,* **2002,** 231:131–144.

4. Chiou W.L., and Riegelman S. Pharmaceutical applications of solid dispersion systems. *J. Pharm. Sci.,* **1971,** 60:1281–1302.

5. Matsumoto T. and Zografi G. Physical properties of solid molecular dispersions of Indomethacin with poly (vinylpyrrolidone) and poly (vinylpyrrolidone-co-vinyl acetate) in relation to Indomethacin crystallization. *Pharm. Res.,* **1999,** 16:1722–1728.

6. Bajaj H., Bisht S., Yadav M., and Singh V. Bioavailability enhancement: a review. *Int. J.Pharma and Bio Sci.,* **2011,** 2(2):202–216.

7. Araya H., Tomita M., Hayashi M. The novel formulation design of O/W microemulsion for improving the gastrointestinal absorption of poorly water soluble compounds. *Int J Pharm.,* **2005,** 305(1–2):61–74.

8. Shafiq S., Shakeel F., Talegaonkar S., Ahmad F.J., Khar R.K. and Ali M. Development and bioavailability assessment of ramipril nanoemulsion formulation. *Eur J Pharm Biopharm.,* **2007,** 66(2):227–243.

9. Palin K.J., Phillips A.J. and Ning A. The oral absorption of cefoxitin from oil and emulsion vehicles in rats. *Int J Pharm.,* **1986,** 33:99–104.

10. Aungst B.J., Nguyen N., Rogers N.J., Rowe S., Hussain M., Shum L. White. Improved oral bioavailability of an HIV protease inhibitor using Gelucire 44/14 and Labrasol vehicles. *SBT Gattefosse.,* **1994,** 87:49–54.

11. Singh B., Bandopadhyay S., Kapil R., Singh R., Katare O.P. Self-Emulsifying Drug Delivery Systems (SEDDS): Formulation Development, Characterization, and Applications. *Critical Reviews™ in Therapeutic Drug Carrier Systems,* **2009,** 26(5): 427–521.

12. O'Driscoll C.M., Griffin B.T. Biopharmaceutical challenges associated with drugs with low aqueous solubility-the potential impact of lipid-based formulations. *Adv Drug Deliv Rev.,* **2008,** 60(6):617–624.

13. Koga K., Kusawake Y., Ito Y., Sugioka N., Shibata N., Takada K. Enhancing mechanism of Labrasol on intestinal membrane permeability of the hydrophilic drug gentamicin sulfate. *Eur J Pharm Biopharm.,* **2006,** 64(1):82–91.

14. Sha X., Yan G., Wu Y., Li J., Fang X. Effect of self-microemulsifying drug delivery systems containing Labrasol on tight junctions in Caco-2 cells. *Eur J Pharm Sci.,* **2005,** 24(5): 477–486.

15. Sha X.Y. and Fang X.L. Effect of self-microemulsifying system on cell tight junctions. *Yao Xue Xue Bao.,* **2006,** 41(1):30–35.

16. Yang S., Gursoy R.N., Lambert G. and Benita S. Enhanced oral absorption of paclitaxel in a novel self-microemulsifying drug delivery system with or without concomitant use of P-glycoprotein inhibitors. *Pharm Res.,* **2004,** 21(2):261–270.

17. Humberstone A.J. and Charman W.N. Lipid-based vehicles for the oral delivery of Poorly water soluble drugs. *Adv Drug Deliv Rev.,* **1997,** 25(1):103–128.

18. Patel N., Rathva S.R., Shah V.H., Upadhyay U.M. Review on Self Emulsifying Drug Delivery System: Novel Approach for Solubility Enhancement. *Int. J.Pharma Res & Allied Sci.,* **2012,** 1(3):1–12.

19. Dressman J.B., Reppas C. In vitro-in vivo correlations for lipophilic, poorly water-soluble drugs. *Eur J Pharm Sci.,* **2000,** 11(2):S73–S80.

20. Pouton C.W. Formulation of poorly water-soluble drugs for oral administration: Physicochemical and physiological issues and the lipid formulation classification system. *Eur J Pharm Sci.,* **2006,** 29(3–4):278–287.

21. Amidon G.L., Lennernas H., Shah V.P., Crison J.R. A theoretical basis for a biopharmaceutic drug classification: the correlation of in vitro drug product dissolution and in vivo bioavailability. *Pharm Res.* **1995,** 12(3):413–420.

22. Vogt M., Kunath K., Dressman J.B. Dissolution enhancement of fenofibrate by micronization, co grinding and spray-drying: comparison with commercial preparations. *Eur J Pharm Biopharm.,* **2008,** 68(2):283–288.

23. Wan J., Yuan S., Chen J., Li T., Lin L. and Lu X. Solubility-enhanced electrokinetic movement of hexachlorobenzene in sediments: a comparison of cosolvent and cyclodextrin. *J Hazard Mater,* **2009,** 166(1):221–226.

24. Kararli T.T. and Gupta V.W. Solubilization and dissolution properties of a leukotriene-D4 antagonist in micellar solutions. *J Pharm Sci.* **1992,** 81(5): 483–485.

25. Srinarong P., Faber J.H., Visser M.R., Hinrichs W.L., Frijlink H.W. Strongly enhanced dissolution rate of fenofibrate solid dispersion tablets by incorporation of superdisintegrants. *Eur J Pharm Biopharm.,* **2009,** 73(1):154–161.

26. Papadimitriou S. and Bikiaris D. Dissolution rate enhancement of the poorly water-soluble drug Tibolone using PVP, SiO_2, and their nanocomposites as appropriate drug carriers. *Drug Dev Ind Pharm.,* **2009,** 35(9):1128–1138.

27. Wagh M., Jadhav G.V., Khairnar D.A., Chaudhari C.S. Self emulsifying drug delivery system: an emerging paradigm. Int. J. Pharma Res. Bio-Sci., **2013**, 2(6): 287–304.

28. Pouton C.W. and Porter C.J. Formulation of lipid-based delivery systems for oral administration: materials, methods and strategies. *Adv Drug Deliv Rev.*, **2008**, 60(6): 625–637.

29. McEvoy G.K. AHFS Drug Information. *American Society of Health-System Pharmacists.* **2009**, 1 ed: 4000.

30. The Merck Index. *Merck & Co., Inc. USA.* **2006**, 14 ed.: 2520.

31. Sweetman S.C. Martindale: The Complete Drug Reference. *Pharma Press.*, **2009**, 36 ed: 3694.

32. Pouton C.W. Formulation of poorly water-soluble drugs for oral administration: Physicochemical and physiological issues and the lipid formulation classification system. *Eur J Pharm Sci.* **2006**, 29(3–4): 278–287.

33. Solanki T., Thakar D.M., Bharadia P.D., Pandya V.M., Modi D.A. Self-Emulsifying Drug Delivery System: An Alternative Approach for Poorly Water Soluble Drugs. *IJPI's J Pharmaceutics and Cosmetology.*, **2011**, 1(5):99–109.

34. Sapra K., Sapra A., Singh S.K., Kakkar S. Self Emulsifying Drug Delivery System: A Tool in Solubility Enhancement of Poorly Soluble Drugs. *Indo Global J Pharma Sci.*, **2012**, 2(3):313–332.

35. Yeole S.E., Pimple S.S., Gale G.S., Gonarkar A.G., Nigde A.T., Randhave A.K., Chaudhari P.D. A Review-Self-Micro Emulsifying Drug Delivery Systems (SMEDDSs). *Indo American J Pharma Res.*, **2013**, 3(4): 3031–3043.

36. Porter C.J.H., Trevaskis N.L., Charman W.N. Lipids and lipid-based formulations: optimizing the oral delivery of lipophilic drugs. *Nat. Rev. Drug Discov.*, **2007**, 6: 231–248.

37. Chen M.L. Lipid excipients and delivery systems for pharmaceutical development: A regulatory perspective. *Adv Drug Deliv Rev.*, **2008**, 60(6): 768–77.

38. Constantinides P.P. Lipid microemulsions for improving drug dissolution and oral absorption: physical and biopharmaceutical aspects. *Pharm Res.*, **1995**, 12(11): 1561–1572.

39. Dabros T., Yeung A., Masliyah J., Czarnecki J. Emulsification through Area Contraction. *J Colloid Interface Sci.*, **1999**, 210(1): 222–4.

40. Wakerly M.G., Pouton C.W., Meakin B.J. and Morton F.S. Self-emulsification of vegetable oil-non-ionic surfactant mixtures. *ACS Symp Series*, **1986**, 311: 242–55.

41. Sarpal K., Pawar Y.B., Bansal A.K. Self-emulsifying drug delivery systems: A strategy to improve oral bioavailability. Curr Res Information on Pharma Sci., **2010**, 11(3): 42–49.

42. Pujara N.D. Self-emulsifying drug delivery system: A novel approach. *Int J Curr Pharm Res.*, **2012**, 4(2):18–23.

43. Khedekar K., Mittal S. Self-emulsifying drug delivery system: A review. *Int J Pharma Sci Res.*, **2013**, 4(12): 4494–4507.

44. Shukla J.B. Self micro emulsifying drug delivery system. *J Pharmacy Pharmaceutical Sci.*, **2010,** 1(2): 13–33.

45. Pandey V., Kohli S. Self micro emulsifying drug delivery systems (smedds): as a promising approach in enhancing bioavailability of poorly aqueous soluble drugs. *J Medical Pharma Allied Sci.*, **2013,** 2 (2): 7–28.

46. Singh B. Self-emulsifying drug delivery systems (SEDDS): formulation development, characterization, and applications. *Crit. Rev. Ther. Drug Carrier Syst.*, **2009,** 2 (6): 427–521.

47. Chen M.L. Lipid excipients and delivery systems for pharmaceutical development: regulatory perspective. *Adv. Drug Deliv Rev.*, **2008,** 60 (2): 768–777.

48. Swenson E.S. Intestinal permeability enhancement: efficacy, acute local toxicity and reversibility. Pharm Res., **1994,** 11(4): 1132–1142.

49. Kohli K., Chopra S., Dhar D., Arora S., Khar R.K. Self emulsifying drug delivery system: An approach to enhance the oral bioavailability. *Drug Discov Today.*, 2010, 15: 958–965.

50. Hai Rong Shen, Ming Kang Zhong. Preparation and evaluation of self microemulsifying drug delivery systems (SMEDDS) containing atorvastatin. *J Pharma Pharmacol.*, **2006,** 58:1183–1191.

51. Halle P.D., Sakhare R.S., Dadge K.K., Nabde M.K., Raut D.B. A Review: Self Emulsifying Drug Delivery System. *Ind J Universal Pharma Life Sci.*, **2013,** 3(2): 155–175.

52. Mohsin K., Shahba A.A., Alanazi F.K. Lipid based Self-emulsifying Formulations for poorly water soluble drugs-An Excellent Opportunity. *Ind J Pharm Edu Res.*, **2012,** 46 (2): 88–96.

53. Gursoy R.N., Benita S. Self-emulsifying drug delivery systems (SEDDS) for improved oral delivery of lipophilic drugs. *Biomedicine & Pharmacotherapy.*, **2004,** 58:173–182.

54. Cerpnjak K., Zvonkar A., Gasperlin M., Vrecer F. Lipid-based systems as a promising approach for enhancing the bioavailability of poorly water-soluble drugs. *Acta Pharm.*, **2013,** 63:427–445.

55. Eid A.M., Baie S.H., Arafat O.M. The Effect of Surfactant Blends on the Production of Self-Emulsifying System. *Int J Pharma Frontier Res.*, **2012,** 2 (2):21–31.

56. D.M. Small, Surface and bulk interactions of lipids and water with a classification of biologically active lipids based on these interactions. *Fed. Proc.*, **1970,** 29:1320–1326.

57. Serajuddin A.T., Sheen P.C., Mufson D., Bernstein D.F., Augustine M.A. Effect of vehicle amphiphilicity on the dissolution and bioavailability of a poorly water-soluble drug from solid dispersions. *J Pharm Sci.*, **1988,** 77(5): 414–7.

58. Shah N.H., Carvajal M.T., Patel C.I., Infeld M.H., Malick A.W. Self-emulsifying drug delivery systems (SEDDS) with polyglycolyzed glycerides for improving in vitro dissolution and oral absorption of lipophilic drugs. *Int J Pharm.*, **1994,** 106(1): 15–23.

59. Patil R.V., Patil K.K., Mahajan V.R., Dhake A.S. Self Emulsifying Therapeutic System - A Review. *Int J Pharma Bio Archives,* **2012,** 3(3):481–486.

60. Reddy S., Katyayani T., Navatha A., Ramya G. Review on self micro emulsifying drug delivery systems. *Int J.Res. Pharm. Sci.,* **2011,** 2 (3): 382–392.

61. Patel M.J., Patel S.S., Patel N.M., Patel M.M. A self-microemulsifying drug delivery system. *Int J Pharma Sci Rev Res.,* **2010,** 4(3): 29–35.

62. Zvonar A., Gasperlin M., Kristl J. Self (micro) emulsifying systems – alternative approach for improving bioavailability of lipophilic drugs, *Farm. Vestn,* **2008,** 59: 263–268.

63. Kale A.A., Patravale V.B. Design and evaluation of self-emulsifying drug delivery systems (SEDDS) of nimodipine. *AAPS Pharm SciTech.,* **2008,** 9(1): 191–6.

64. Borhade V.B., Nair H.A., Hegde D.D. Development and characterization of self-microemulsifying drug delivery system of tacrolimus for intravenous administration. *Drug Dev Ind Pharm.,* **2009,** 35(5): 619–30.

65. Rahman A., Hussain A., Hussain S., Mirza A., Iqbal Z. Role of excipients in successful development of self-emulsifying/microemulsifying drug delivery system(SEDDS/ SMEDDS). *Drug development and Industrial Pharmacy, Informa healthcare,* **2013,** 39 (1): 1–19.

66. Khan N., Craig D.Q. The influence of drug incorporation on the structure and release properties of solid dispersions in lipid matrices. *J Control Release,* **2003,** 93:355–368.

67. Freitas C., Müller R.H. Correlation between long-term stability of solid lipid nanoparticles (SLN) and crystallinity of the lipid phase. *Eur J Pharm Biopharm.,* **1999,** 47:125–132.

68. Brubach J.B., Ollivon M., Jannin V., Mahler B., Bougaux C., Lesieur C., Roy P. Structural and thermal characterization of mono and diacyl polyoxyethylene glycol by infra red spectroscopy and X-ray diffraction coupled to differential calorimetry. *J Phys Chem.,* **2004,** 108, 17721–17729.

69. Choy Y.W., Khan N., Yuen K.H. Significance of lipid matrix aging on in vitro release and in vivo bioavailability. *Int J Pharm.,* **2005,** 299:55–64.

70. Brubach J.B., Jannin V., Mahler B., Bourgaux C., Lessieur P., Roy P. Structural and thermal characterization of glyceryl behenate by X-ray diffraction coupled to differential calorimetry and infrared spectroscopy. *Int J Pharm.,* **2007,** 336:248–256.

71. Jenning V., Schäfer-Korting M., Gohla S. Vitamin A-loaded solid lipid nanoparticles for topical use: drug release properties. *J Control Release,* **2000,** 66:115–126.

72. Jannin V., Musakhanian J., Marchaud D. Approaches for the development of solid and semi-solid lipid-based formulations. *Adv Drug Deliv Rev.,* **2008,** 60:734–746.

73. Spernath A., Aserin A. Microemulsions as carriers for drugs and nutraceuticals. *Adv Colloid Interface Sci.,* **2006,** 128–130: 47–64.

74. Patel V.P., Desai T., Kapupura P., Atara S., Keraliya R. Self emulsifying drug delivery system: a conventional and Alternative approach to improve oral bioavailability of Lipophilic drugs. *Int. J. Drug Dev. & Res.,* **2010,** 2(4): 859–870.

75. Kohli K., Chopra S., Dhar D., Arora S., Khar R.K. Self-emulsifying drug delivery systems: an approach to enhance oral bioavailability. *Drug Discov. Today,* **2010,** 15(2): 958–965.

76. Mehta K., Borade G., Rasve G., Bendre A. Self-emulsifying drug delivery system: Formulation and Evaluation. *Int. J. Pharm. Bio. Sci.,* **2011,** 2 (4): 398–412.

77. Balakrishnana P., Leeb B.J., Oha D.H., Kima J.O., Leea Y.I., Kimc D.D. Enhanced oral bioavailability of Coenzyme Q10 by self-emulsifying drug delivery systems. *Int. J. Pharm.,* **2009,** 374: 66–72.

78. Khedekar K., Mittal S. Self-emulsifying drug delivery system. *Int. J. Pharma. Sci. & Res.,* **2013,** 4 (12): 4494–4507.

79. Deokate U.A., Shinde N., Bhingare U. Novel approaches for development and characterization of SMEDDS: Review. *Int. J. Curr Pharm Res.,* **2013,** 5 (4): 5–12.

80. Sharma V. SMEDDS: A novel approach for lipophilic drugs. *Int. J.Pharma Sci. & Res.,* **2012,** 3(8): 2441–2450.

81. Singh A.K., Chaurasiya A., Singh M., Upadhyay S.C., Mukherjee R., Khar R.K. Exemestane loaded self-microemulsifying drug delivery system (SMEDDS): development and optimization. *AAPS Pharm SciTech.,* **2008,** 9(2): 628–34.

82. Gowri R., Balaji P. Harshapriya G., Siji M., Shaik B.S. Self emulsifying drug delivery system-a promising tool to enhance dissolution of poorly soluble drugs. *Int. J. Chem & Pharm Sci.,* **2012,** 3(4): 19–23.

83. Tuleu C., Newton M., Rose J., Euler D., Saklatvala R., Clarke A., Booth S. Comparative bioavailability study in dogs of a self-emulsifying formulation of progesterone presented in a pellet and liquid form compared with an aqueous suspension of progesterone. *J Pharm Sci.* **2004;** 93(6): 1495–502.

84. Abdalla A., Mader K. Preparation and characterization of a self-emulsifying pellet formulation. *Eur J Pharm Biopharm.* **2007,** 66(2): 220–6.

85. Abdalla A., Klein S., Mader K. A new self-emulsifying drug delivery system (SEDDS) for poorly soluble drugs: characterization, dissolution, in vitro digestion and incorporation into solid pellets. *Eur J Pharm Sci.,* **2008,** 35(5): 457–64.

86. Iosio T., Voinovich D., Grassi M., Pinto J.F., Perissutti B., Zacchigna M., Quintavalle U., Serdoz F. Bi-layered self-emulsifying pellets prepared by co-extrusion and spheronization: influence of formulation variables and preliminary study on the in vivo absorption. *Eur J Pharm Biopharm.,* **2008,** 69(2): 686–97.

87. Zaghloul A., Khattab I., Nada A. Preparation, characterization and optimization of Probucol self-emulsified drug delivery system to enhance solubility and dissolution. *Pharmazie,* **2008,** 63: 654–660.

88. Pouton, C.W. Formulation of self emulsifying drug delivery systems. *Adv. Drug Deliv. Rev.,* **1997,** 25(3): 47–58.

89. Craig D.Q.M., Lievens H.S.R., Pitt K.G., Storey D.E. An investigation into the physicochemical properties of self emulsifying systems using low frequency dielectric spectroscopy, surface tension measurements and particle size analysis. *Int J Pharm.,* **1993,** 96:147–155.

90. Craig D.Q.M., Barker S.A., Banning D., Booth S.W. An investigation into the mechanisms of self-emulsification using particle size analysis and low frequency dielectric spectroscopy. *Int J Pharm.*, **1995**, 114(1): 103–10.

91. Gershanik T., Benita S. Positively charged self-emulsifying oil formulation for improving oral bioavailability of progesterone. *Pharm Dev Technol.*, **1996**, 1:147–157.

92. Gershanik T., Benita S. Self-dispersing lipid formulations for improving oral absorption of lipophilic drugs. *Eur J Pharm Biopharm.*, **2000**, 50(1): 179–88.

93. Gershanik T., Benzeno S., Benita S. Interaction of a self-emulsifying lipid drug delivery system with the everted rat intestinal mucosa as a function of droplet size and surface charge. *Pharm Res.*, **1998**, 15(6): 863–869.

94. Gao P., Morozowich W. Development of supersaturatable self-emulsifying drug delivery system formulations for improving the oral absorption of poorly soluble drugs. *Expert Opin Drug Deliv,* **2006**, 3(1): 97–110.

95. Morozowich W., Gao P., Charton M. Speeding the development of poorly soluble/poorly permeable drugs by SEDDS/S-SEDDS formulations and prodrugs, Part 1. *Am Pharm Rev.*, **2006**, 9: 110–114.

96. Ditner C., Bravo R., Imanidis G., Kuentz M. A systematic dilution study of self-microemulsifying drug delivery systems in artificial intestinal fluid using dynamic laser light backscattering. *Drug Dev Ind Pharm.*, **2009**, 35(2): 199–208.

97. Goddeeris C., Cuppo F., Reynaers H., Bouwman W.G., Van den Mooter G. Light scattering measurements on microemulsions: Estimation of droplet sizes. *Int J Pharm.*, **2006**, 312(1–2): 187–195.

98. Ju R.T.C., Frank C.W., Gast A.P. CONTIN analysis of colloidal aggregates. *Langmuir.* **1992**, 8(9): 2165–2171.

99. Laia C.A.T., López-Cornejo P., Costa S.M.B., d'Oliveira J., Martinho J.M.G. Dynamic Light Scattering Study of AOT Microemulsions with Nonaqueous Polar Additives in Oil Continuous Phase. *Langmuir,* **1998**, 14(13): 3531–7.

100. Kuentz M., Cavegn M. Critical concentrations in the dilution of oral self-microemulsifying drug delivery systems. *Drug Dev Ind Pharm.* **2009** (In press).

101. Fatouros D., Deen G., Arleth L., Bergenstahl B., Nielsen F., Pedersen J., Mullertz A. Structural Development of Self Nano Emulsifying Drug Delivery Systems (SNEDDS) During In Vitro Lipid Digestion Monitored by Small-angle X-ray Scattering. *Pharm Res.* **2007**, 24(10): 1844–53.

102. Fatouros D.G., Deen G.R., Arleth L., Bergenstahl B., Nielsen F.S., Pedersen J.S., Mullertz A. Structural development of self nano emulsifying drug delivery systems (SNEDDS) during in vitro lipid digestion monitored by small-angle X-ray scattering. *Pharm Res.* **2007**, 24(10): 1844–53.

103. Fatouros D.G., Karpf D.M., Nielsen F.S., Mullertz A. Clinical studies with oral lipid based formulations of poorly soluble compounds. *Ther Clin Risk Manag,* **2007**, 3(4): 591–604.

104. Kommuru T.R., Gurley B., Khan M.A., Reddy I.K. Self-emulsifying drug delivery systems (SEDDS) of coenzyme Q10: formulation development and bioavailability assessment. *Int J Pharm.*, **2001**, 212(2): 233–46.

105. Gursoy R.N. and Benita S. Self-emulsifying drug delivery systems (SEDDS) for improved oral delivery of lipophilic drugs. *Biomed Pharmacother.*, **2004**, 58(3): 173–82.

106. Chambin O., Karbowiak T., Djebili L., Jannin V., Champion D., Pourcelot Y., Cayot P. Influence of drug polarity upon the solid-state structure and release properties of self-emulsifying drug delivery systems in relation with water affinity. *Colloids Surf B Biointerfaces,* **2009**, 71(1): 73–8.

107. Goldstein G.I., Newbury D.E., Echlin P., Joy D.C., Fiori C., Lifshin E., Scanning electron microscopy and x-ray microanalysis. **1981,** *New York: Plenum Press.*

108. Chambin O., Jannin V., Champion D., Chevalier C., Rochat-Gonthier M.H., Pourcelot Y. Influence of cryogenic grinding on properties of a self-emulsifying formulation. *Int J Pharm.* **2004**, 278(1): 79–89.

109. Reimer L., Kohl H. Transmission Electron Microscopy: Physics of Image Formation. **2008:** Springer.

110. Frank J., Three-Dimensional Electron Microscopy of Macromolecular Assemblies. **2006,** New York: Oxford University Press.

111. Fatouros D.G., Bergenstahl B., Mullertz A. Morphological observations on a lipid-based drug delivery system during in vitro digestion. *Eur J Pharm Sci.* **2007,** 31(2): 85–94.

112. Shah R.B., Zidan A.S., Funck T., Tawakkul M.A., Nguyenpho A., Khan M.A. Quality by design: Characterization of self-nano-emulsified drug delivery systems (SNEDDs) using ultrasonic resonator technology. *Int J Pharm.* **2007,** 341(1–2): 189–94.

113. Sunitha R., Sireesha D.S., Aparna M.V.L. Novel self-emulsifying drug delivery system- an approach to enhance bioavailability of poorly Water soluble drugs. *Int. J. Res. Pharm & Chem.,* **2011,** 1(4): 828–838.

114. Groves M.J., de Galindez D.A. Rheological characterisation of self-emulsifying oil/surfactant systems. *Acta Pharm Suec.,* **1976,** 13(4): 353–60.

115. Cirri M., Mura P., Mora P.C. Liquid spray formulations of xibornol by using self-microemulsifying drug delivery systems. *Int J Pharm.* **2007,** 340 (1–2): 84–91.

116. Biradar S.V., Dhumal R.S., Paradkar A. Rheological investigation of self-emulsification process. *J Pharm Pharm Sci.,* **2009,** 12(1): 17–31.

117. Newton J.M., Bazzigialuppi M., Podczeck F., Booth S., Clarke A. The rheological properties of self-emulsifying systems, water and microcrystalline cellulose. *Eur J Pharm Sci.* **2005,** 26(2): 176–83.

118. Lee S., Lee J., Choi Y.W. Design and evaluation of prostaglandin E1 (PGE1) intraurethral liquid formulation employing self-microemulsifying drug delivery system (SMEDDS) for erectile dysfunction treatment. *Biol Pharm Bull.* **2008,** 31(4): 668–72.

119. Patel D., Sawant K.K. Oral bioavailability enhancement of acyclovir by self-microemulsifying drug delivery systems (SMEDDS). *Drug Dev Ind Pharm.,* **2007,** 33(12): 1318–26.

120. Patel N.N., Rathva S.R., Shah V.H., Upadhyay U.M. Review on Self Emulsifying Drug Delivery System: Novel Approach for Solubility Enhancement. *Int. J. Pharma Res. Allied Sci.* **2012,** 1(3): 1–12.

121. Shukla P., Prajapati S.K., Sharma U.K., Shivhare S., Akhtar A. A review on self micro-emulsifying drug delivery system: An approach to enhance the oral bioavailability of poorly water soluble drugs. *Int. Res. J. Pharm.,* **2012,** 3(9): 1–6.

122. Shafiq S. Development and bioavailability assessment of ramipril nanoemulsion formulation. *Eur. J. Pharm. Biopharm.,* **2007,** 66: 227–243.

123. Gupta A.K., Mishra D.K., Mahajan S.C. Preparation and *in vitro* evaluation of SEDDS of anti hypertensive drug valsartan. *Int. J. Pharm. life Sci.,* **2011,** 2:633–639.

124. Patel D., Sawant K.K. Oral bioavailability enhancement of acyclovir by self micro emulsifying drug delivery systems (SMEDDS). *Drug Dev. Ind. Pharm.,* **2007,** 33: 1318–1326.

125. Date A.A., Nagarsenker M.S. Design and evaluation of self nanoemulsifying drug delivery systems (SNEDDS) for cefpodoxime proxetil. *Int. J. Pharm.,* **2007,** 329: 166–172.

126. Gershanik T., Benita S. Positively charged self-emulsifying oil formulation for improving oral bioavailability of progesterone. *Pharm. Dev. Tech.,* **1996,** 1:147–157.

127. Borhade V., Nair H., Hegde D. Design and evaluation of self micro emulsifying drug delivery system (SMEDDS) of tacrolimus. *AAPS Pharm. Sci. Tech.,* **2008,** 9: 13–21.

128. Attama A.A., Nzekwe I.T., Nnamani P.O. The use of solid self emulsifying system in the delivery of diclofenac. *Int. J. Pharm.,* **2003,** 262: 23–28.

129. Barthe L., Woodley J., Houin G. Gastrointestinal absorption of drugs: methods and studies. *Fundam Clin Pharmacol.* **1999,** 13(2): 154–68.

130. Levet-Trafit B., Gruyer M.S., Marjanovic M., Chou R.C. Estimation of oral drug absorption in man based on intestine permeability in rats. *Life Sci.* **1996,** 58(24): PL359–63.

131. Yao J., Lu Y., Zhou J.P. Preparation of nobiletin in self-microemulsifying systems and its intestinal permeability in rats. *J Pharm Pharm Sci.* **2008,** 11(3): 22–9.

132. Sharma P., Varma M.V., Chawla H.P., Panchagnula R. In situ and in vivo efficacy of peroral absorption enhancers in rats and correlation to in vitro mechanistic studies. *Farmaco.* **2005,** 60(11–12): 874–83.

133. Ho Y.F., Lai M.Y., Yu H.Y., Huang D.K., Hsueh W.C., Tsai T.H., Lin C.C. Application of rat in situ single-pass intestinal perfusion in the evaluation of presystemic extraction of indinavir under different perfusion rates. *J Formos Med Assoc.* **2008,** 107(1): 37–45.

134. Lane M.E., Levis K.A., Corrigan O.I. Effect of intestinal fluid flux on ibuprofen absorption in the rat intestine. *Int J Pharm.* **2006,** 309(1–2): 60–6.

135. Barthe L., Woodley J.F., Kenworthy S., Houin G. An improved everted gut sac as a simple and accurate technique to measure paracellular transport across the small intestine. *Eur J Drug Metab Pharmacokinet.* **1998,** 23(2): 313–23.

136. Leppert P.S., Fix J.A. Use of everted intestinal rings for in vitro examination of oral absorption potential. *J Pharm Sci.* **1994,** 83(7): 976–81.

137. Chowhan Z.T., Amaro A.A. Everted rat intestinal sacs as an in vitro model for assessing absorptivity of new drugs. *J Pharm Sci.* **1977,** 66(9): 1249–53.

138. Gershanik T., Haltner E., Lehr C.M., Benita S. Charge-dependent interaction of self-emulsifying oil formulations with Caco-2 cells monolayers: binding, effects on barrier function and cytotoxicity. *Int J Pharm.* **2000,** 211(1–2): 29–36.

139. Gershanik T., Haltner E., Lehr C.M., Benita S. Charge-dependent interaction of self-emulsifying oil formulations with Caco-2 cells monolayers: binding, effects on barrier function and cytotoxicity. *Int J Pharm.* **2000,** 211(1–2): 29–36.

140. Buyukozturk F., Benneyan J.C., Carrier R.L. Impact of emulsion-based drug delivery systems on intestinal permeability and drug release kinetics. *J Control Release.* **2009** (In press).

141. Osiecka I., Porter P.A., Borchardt R.T., Fix J.A., Gardner C.R. In Vitro Drug Absorption Models. I. Brush Border Membrane Vesicles, Isolated Mucosal Cells and Everted Intestinal Rings: Characterization and Salicylate Accumulation. *Pharm. Res.* **1985,** 02 (10): 284–93.

142. Daum N., Neumeyer A., Wahl B., Bur M., Lehr C.M. In vitro systems for studying epithelial transport of macromolecules. *Methods Mol Biol.* **2009,** 480: 151–64.

143. Roediger W.E., Truelove S.C. Method of preparing isolated colonic epithelial cells (colonocytes) for metabolic studies. *Gut.* **1979,** 20(6): 484–8.

144. Chopra D.P., Yeh K., Brockman R.W. Isolation and characterization of epithelial cell types from the normal rat colon. *Cancer Res.* **1981,** 41(1): 168–75.

145. Simon-Assmann P., Bouziges F., Daviaud D., Haffen K., Kedinger M. Synthesis of glycosaminoglycans by undifferentiated and differentiated HT29 human colonic cancer cells. *Cancer Res.* **1987,** 47(16): 4478–84.

146. Madin S.H., Darby N.B., Jr. Established kidney cell lines of normal adult bovine and ovine origin. *Proc Soc Exp Biol Med.* **1958,** 98(3): 574–6.

147. Madin S.H., Andriese P.C., Darby N.B. The in vitro cultivation of tissues of domestic and laboratory animals. *Am J Vet Res.* **1957,** 18(69): 932–41.

148. Liu Y., Zeng S. Advances in the MDCK-MDR1 cell model and its applications to screen drug permeability. Yao Xue Xue Bao. **2008,** 43(6): 559–64.

149. Lavelle J.P., Negrete H.O., Poland P.A., Kinlough C.L., Meyers S.D., Hughey R.P., Zeidel M.L. Low permeabilities of MDCK cell monolayers: a model barrier epithelium. *Am J Physiol.* **1997,** 273(1 Pt 2): F67–75.

150. Gonzalez J.E., Digeronimo R.J., Arthur D.E., King J.M. Remodeling of the tight junction during recovery from exposure to hydrogen peroxide in kidney epithelial cells. *Free Radic Biol Med.* **2009.**

151. Patel P.A., Chaulang G.M., Akolkotkar A., Mutha S.S., Hardikar S.R., Bhosale A.V. Self-emulsifying drug delivery system: a review. *Res. J. Pharm. and Tech,* **2008,** 1(4):313–323.

152. Gershanik T., Benita S. Positively Charged Self-Emulsifying Oil Formulation for Improving Oral Bioavailability of Progesterone. *Pharma Dev Technol.,* **1996,** 1(3):147–154.

153. Mi-Jin P., Shan R., Beom-Jin L. In vitro and in vivo comparative study of itraconazole bioavailability when formulated in highly soluble self-emulsifying system and in solid dispersion. *Biopharm Drug Dispos.* **2007,** 28: 199–207.

154. Singh B., Khurana L., Kapil R. Development of optimized self nano emulsifying drug delivery systems of carvedilol with enhanced bioavailability potential. *Drug Del.,* **2011,** 18:599–612.

155. Parker Jr W.O., Genova C., Carignano G. Study of micellar solutions and microemulsions of an alkyl oligoglucoside via NMR spectroscopy, *Colloids Surfaces A. Physicochem Eng Aspects,* **1993,** 72: 275–284.

156. Corswant C.V., Engstrom S., Soderman O. Microemulsions based on soybean phosphatidylcholine and triglycerides Phase behaviour and microstructure. *Langmuir,* **1997,** 13:5061–5070.

157. Patel M.J. "A Self-Microemulsifying Drug Delivery System (SMEDDS)". *Int. J. Pharma Sci. Rev. & Res,* **2010,** 4(3), 29–35.

158. Shete Y.A. Development of solid self-emulsifying drug delivery system and dosage forms. *Int. J. Pharma Res & Dev.* **2013,** 5 (05): 95–103.

159. Abdalla A., Klein S., Mader K. A new self-emulsifying drug delivery system (SEDDS) for poorly soluble drugs: characterization, dissolution, in vitro digestion and incorporation into solid pellets. *Eur J Pharm Sci.* **2008,** 35(5): 457–64.

160. Yi T., Wan J., Xu H., Yang X. A new solid self-microemulsifying formulation prepared by spray-drying to improve the oral bioavailability of poorly water soluble drugs. *Eur J Pharm Biopharm.* **2008,** 70(2): 439–44.

161. Abdalla A., Mader K. Preparation and characterization of a self-emulsifying pellet formulation. *Eur J Pharm Biopharm.* **2007,** 66(2): 220–6.

162. Gupta M.K., Goldman D., Bogner R.H., Tseng Y.C. Enhanced drug dissolution and bulk properties of solid dispersions granulated with a surface adsorbent. *Pharm Dev Technol.* **2001,** 6(4): 563–72.

163. Gupta M.K., Tseng Y.C., Goldman D., Bogner R.H. Hydrogen bonding with adsorbent during storage governs drug dissolution from solid-dispersion granules. *Pharm Res.* **2002,** 19(11): 1663–72.

164. Breitenbach J. Melt extrusion: from process to drug delivery technology. *Eur. J. Pharm. Biopharm.* **2002,** 54: 107–117.

165. Iosio T., Voinovich D., Grassi M., Pinto J.F., Perissutti B., Zacchigna M., Quintavalle U., Serdoz F. Bi-layered self-emulsifying pellets prepared by co-extrusion and spheronization: influence of formulation variables and preliminary study on the in vivo absorption. *Eur J Pharm Biopharm.* **2008,** 69(2): 686–97.

166. Newton M., Petersson J., Podczeck F., Clarke A., Booth S. The influence of formulation variables on the properties of pellets containing a self-emulsifying mixture. *J. Pharm. Sci.* **2001,** 90: 987–995.

167. Tayal A., Jamil F., Sharma R., Sharma S. Self-emulsifying drug delivery system: a review. *Int. Res. J. Pharm.* **2012,** 3(5): 32–36.

168. Souto E.B., Muller R.H. "SLN and NLC for topical delivery of Ketoconazole". *J Microencapsulation.* **2005,** 22:501–10.

169. Jannin V., Musakhanian J., Marchaud D. Approaches for the development of solid and semi-solid lipid based formulations. *Adv. Drug Deliver. Rev.* **2008,** 60: 734–746.

170. Singh B., Bandopadhyay S., Beg S., Katare O.P. Handling poorly bioavailable drugs using nanoemulsifying drug delivery systems. *The Pharma Rev.* **2011,** 91–98.

171. Kim J.Y., Ku Y.S. Enhanced absorption of indomethacin after oral or rectal administration of a self-emulsifying system containing indomethacin to rats. *Int. J. Pharm.* **2000,** 194: 81–89.

172. Takada K., Murakami M. Glycyrrhizin preparations for transmucosal absorption. US Pat 6890547.**2005.**

173. Loomis G.L. Bioresorbable compositions for implantable prostheses. US5854382 A.**1998.**

174. Yu Z., Huth S.W. Stable ophthalmic oil-in-water emulsions with sodium hyaluronate for alleviating dry eye. US/2007/0036829.**2007.**

175. Dor P.J., Mudumba S., Nivaggioli T., Weber D.A. Formulations for ocular treatment. US/2006/0185068.**2004.**

176. Nanotech for transdermal sildenafil (Viagra). Science daily. http:// www.sciencedaily. com/releases/2011/07/110727083659.htm (Accessed on February 20, **2014**).

177. Christensen K.L., Pedersen G.P., Kristensen H.G. Technical optimization of redispersible dry emulsions. *Int. J. Pharm.* **2001,** 212: 195–202.

178. Bamba J., Cave G., Bensouda Y., Tchoreloff P., Pulsieux F., Couarrraze G. Cryoprotection of emulsions in freeze-drying: freezing process analysis. *Drug Dev. Ind. Pharm.* **1995,** 21: 1749–1760.

179. Hansen T., Holm P., Schultz K. Process characteristics and compaction of spray-dried emulsions containing a drug dissolved in lipid. *Int. J. Pharm.* **2004,** 287: 55–66.

180. Jang D.J., Jeong E.J., Lee H.M., Kim B.C., Lim S.J., Kim C.K. Improvement of bioavailability and photostability of amlodipine using redispersible dry emulsion. *Eur. J. Pharm. Sci.* **2006,** 28: 405–411.

181. Shaji J., Jadhav D. Newer approaches to self-emulsifying drug delivery system. *Int. J. Pharm & Pharma Sci.* **2010,** 2 (1): 37–42.

182. Pouton C.W. SEDDS: Assessment of the efficiency of emulsification. *Int J Pharm.***1985,** 27:335–348.

183. Zang P., Liu Y., Feng N., Xu J., Preparation and evaluation of self microemulsifying drug delivery system of oridonin. *Int. J. Pharm.* **2008,** 355: 269–276.

184. Gandhi R. Extrusion and spheronization in the development of oral controlled-release dosage forms. *PSTT.* **1999,** 2: 160–170.

185. Attama A.A., Nkemnele M.O. In vitro evaluation of drug release from self micro-emulsifying drug delivery systems using a biodegradable homo lipid from *Capra hircus*. *Int J Pharm*. **2005**, 304: 4–10.

186. Trickler W.J.A. Novel nanoparticle formulation for sustained paclitaxel Delivery. *AAPS Pharm Sci Tech*. **2008**, 9(2): 486–493.

187. Patil P., Paradkar A. Porous polystyrene beads as carriers for Self emulsifying system containing Loratadine. *AAPS Pharm Sci. Tech.*, **2006**, 7: E 28

188. Serajuddin A.T. Solid dispersion of Poorly Water soluble Drugs: early promises, subsequent problems & recent breakthroughs. *J. Pharm. Sci.*, **1999**, 88: 1058–1066.

189. Vasanthavada M., Serajuddin A.T. Lipid based self emulsifying solid dispersions. In Oral lipid based formulations: Enhancing bioavailability of Poorly Water Soluble drugs (Hauss DJ. ed.) *pp informa Healthcare*, **2007**, 149–184.

190. Serajuddin A.T., Sheen P.C., Mufson D. Effect of vehicle amphiphilicity on the dissolution and bioavailability of a poorly water soluble drug from solid dispersions. *J. Pharm. Sci.*, **1998**, 77: 414–417.

191. Gupta M.K., Tseng Y.C., Goldman D. Hydrogen Bonding with Adsorbent during storage governs drug dissolution from solid dispersion granules. *Pharm. Res.*, **2002**, 19: 1663–1672.

192. You J., Cui F.D., Li Q.P. A Novel formulation design about water insoluble oily drug: Preparation of ZTO microspheres with self emulsifying ability & evaluation in Rabbits. *Int. J. Pharm.*, **2005**, 288: 315–323.

193. Barry R.J., Eggenton J. Membrane potentials of epithelial cells in rat small intestine. *J Physiol*. **1972**, 227(1): 201–16.

194. Corbo D.C., Liu J.C., Chien Y.W. Characterization of the barrier properties of mucosal membranes. *J Pharm Sci*. **1990**, 79(3): 202–6.

195. Rojanasakul Y., Wang L.Y., Bhat M., Glover D.D., Malanga C.J., Ma J.K. The transport barrier of epithelia: a comparative study on membrane permeability and charge selectivity in the rabbit. *Pharm Res*. **1992**, 9(8): 1029–34.

196. Humberstone A.J., Charman W.N. Lipid-based vehicles for the oral delivery of poorly water soluble drugs. *Adv Drug Deliv Rev*. **1997**, 25(1): 103–28.

197. Gershanik T., Haltner E., Lehr C.M. and Benita S. Charge-dependent interaction of self-emulsifying oil formulations with Caco-2 cells monolayers: binding, effects on barrier function and cytotoxicity. *Int J Pharm*. **2000**, 211(1–2): 29–36.

198. Bruckdorfer K.R., Cramp F.C., Goodall A.H., Verrinder M., Lucy J.A. Fusion of mouse fibroblasts with oleylamine. *J Cell Sci*. **1974**, 15(1): 185–99.

199. Jain V., Prasad V., Jadhav P., Mishra P.R. Preparation and performance evaluation of saquinavir laden cationic submicron emulsions. *Drug Deliv*. **2009**, 16(1): 37–44.

200. Jannin V. Approaches for the development of solid and Semi-solid lipid-based formulations. *Adv Drug Delivery Rev*. **2008**, 60: 734–746.

201. Gao P., Charton M. and Morozowich W. Speeding the development of poorly soluble/ poorly permeable drugs by SEDDS/S-SEDDS formulations and prodrugs, Part 2. *Am Pharm Rev*. **2006**, 9: 16–23.

202. Pellett M.A., Davis A.F., Hadgraft J. Effect of supersaturation on membrane transport: 2. Piroxicam. *Int J Pharm.* **1994,** 111(1): 1–6.

203. Pellett M.A., Roberts M.S., Hadgraft J. Supersaturated solutions evaluated with an in vitro stratum corneum tape stripping technique. *Int J Pharm.* **1997,** 151(1): 91–8.

204. Raghavan S.L., Trividic A., Davis A.F., Hadgraft J. Effect of cellulose polymers on supersaturation and in vitro membrane transport of hydrocortisone acetate. *Int J Pharm.* **2000,** 193(2): 231–7.

205. Sharma A.K. "Panchkarma Therapy in Ayurvedic Medicine", Scientific Basis for Ayurvedic Therapies, ed. L.C. Mishra. **2003:** CRC Press.

206. Mukherjee P.K., Wahile A. Integrated approaches towards drug development from Ayurveda and other Indian system of medicines. *J Ethnopharmacol.* **2006,** 103(1): 25–35.

207. Holland A. Voices of Qi: An Introductory Guide to Traditional Chinese Medicine. **2000:** North Atlantic Books.

208. Porkert M. The Theoretical Foundations of Chinese Medicine **1974:** MIT Press.

209. Gao P., Rush B.D., Pfund W.P., Huang T., Bauer J.M., Morozowich W., Kuo M.S., Hageman M.J. Development of a supersaturable SEDDS (S-SEDDS) formulation of paclitaxel with improved oral bioavailability. *J Pharm Sci.* **2003,** 92(12): 2386–98.

210. Gao P., Guyton M.E., Huang T., Bauer J.M., Stefanski K.J., Lu Q. Enhanced oral bioavailability of a poorly water soluble drug PNU-91325 by supersaturatable formulations. *Drug Dev Ind Pharm.* **2004,** 30(2): 221–9.

211. Chen Y., Chen C., Zheng J., Chen Z., Shi Q., Liu H. Development of a Solid Supersaturatable Self-Emulsifying Drug Delivery System of Docetaxel with Improved Dissolution and Bioavailability. *Biol. Pharm. Bull.* **2011,** 34(2): 278–286.

212. Gao P., Akrami A., Alvarez F., Hu J., Li L., Ma C., Surapaneni S. Characterization and optimization of AMG 517 supersaturatable self-emulsifying drug delivery system (S-SEDDS) for improved oral absorption. *J Pharm Sci.* **2009,** 98(2): 516–28.

213. Dehal S., Bandyopadhyay S., Singh B. Improving Bioavailability of simvastatin using various types of Self emulsifying formulations, Proc. 30th Annual Conference of Indian Association of Biomedical Scientists, Defence Institute of High Altitude Research, Chandigarh, India, 18–20 November, 2009. Abstract accepted. **2009.**

214. Chen Z.Q., Liu Y., Zhao J.H., Wang L., Feng N.P. Improved oral bioavailability of poorly water-soluble indirubin by a supersaturable self-microemulsifying drug delivery system. *Int. J. Nanomed.* **2012,** 7: 1115–1125.

215. Sri U.B., Kiranmai M., Ibrahim M., Muzib I.Y. Self-emulsifying tablets of Ibuprofen: Design, optimization and evaluation. *Res. J. Pharm Bio & Chem. Sci.* **2012,** 3(4): 1400–1407.

216. Akhter S., Hossain Md.I. Dissolution enhancement of Capmul PG8 and Cremophor EL based Ibuprofen Self Emulsifying Drug Delivery System (SEDDS) using Response surface methodology. *Int. Curr. Pharm J,* **2012,** 1(6): 138–150.

217. Suresh P.K., Sharma S. Formulation and In-vitro characterization of Self-nanoemulsifying drug delivery system of cinnarizine. *Pharmacie Globale,* **2011,** 9(08): 1–6.

218. Shabha A.A.W., Mohsin K., Alanazi F.K. Novel Self-emulsifying drug delivery systems(SNEDDS) for oral delivery of cinnarizine: Design, optimization and in-vitro assessment. *AAPS Pharm SciTech,* **2012,** 13(3): 967–977.

219. Porwal P., Bhargava S., Bhaduria R.S., Shukla S.S., Daharwal S.J. Formulation and In-vitro characterization of self-emulsifying drug delivery system of Cisapride. *Adv. Res. Pharma & Bio,* **2012,** 2(IV): 324–328.

220. Sarkar B.K., Hardenia S.S. Microemulsion drug delivery system: for oral bioavailbility enhancement of glipizide. *J. Adv. Pharm Edu & Res.* **2011,** 1(4): 195–200.

221. Sharma S., Suresh P.K. Formulation, In-vitro characterization and stability studies of self-microemulsifying drug delivery systems of Domperidone. *Int. J. Innovative Pharm Res.* **2010,** 1(4): 66–73.

222. Kumar V.K., Devi M.A., Bhikshapathi D.V.R.N. Development of solid self emulsifying drug delivery systems containing Efavirenz: In-vitro and in-vivo evaluation. *Int J Pharm Bio,* **2013,** 4(1): 869 – 882.

223. Mowafy H.A. Response surface methodology for the development of self-nanoemulsified delivery system (SNEDDS) of Indomethacin. *J. Applied Sci. Res.* **2009,** 5(10): 1772–1779.

224. Mahmoud E.A., Bendas E.R., Mohamed M.I. Preparation and evaluation of self-nanoemulsifying tablets of carvedilol. *AAPS Pharm Sci Tech,* **2009,** 10(1): 183–192.

225. Patel K., Vidur S., Vavia P. Design and evaluation of Lumefantrine-oleic acid self-nanoemulsifying ionic complex for enhanced dissolution. *DARU J. Pharm. Sci.* **2013,** 21–27.

226. Nazzal S., Khan M.A. Response surface methodology for the optimization of ubiquinone self-nanoemulsified drug delivery system. *AAPS Pharm Sci Tech,* **2002,** 3(1): 1–9.

227. Patil P., Joshi P., Paradkar A. Effect of Formulation Variables on Preparation and Evaluation of Gelled Self-emulsifying Drug Delivery System (SEDDS) of Ketoprofen. *AAPS Pharm Sci Tech,* **2004,** 5 (3): 1–8.

228. Kokare C.R., Kumbhar S.A., Patil A. Formulation and evaluation of self-emulsifying drug delivery system of carbamazepine. *Ind. J. Pharm Edu.Res,* **2013,** 47(2): 172–177.

229. Singh A.K., Chaurasiya A., Singh M., Upadhyay S.C., Mukherjee R., Khar R.K. Exemestane loaded self-microemulsifying drug delivery system (SMEDDS): Development and optimization. *AAPS Pharm Sci Tech,* **2008,** 9(2): 628–634.

230. Nekkanti V., Karatgi P., Prabhu R., Pillai R. Solid self-microemulsifying formulation for candesartan cilexetil. *AAPS Pharm Sci Tech,* **2010,** 11(1): 9–17.

231. Rajnikanth P.S., Keat N.W., Garg S. Self-nanoemulsifying Drug Delivery Systems of Valsartan: Preparation and *In-Vitro* Characterization. *Int. J of Drug Delivery,* **2012,** 4(2): 153–163.

232. Beg S., Swain S., Singh H.P., Patra C.N., Rao M.E.B. Development, Optimization, and Characterization of Solid Self-Nanoemulsifying Drug Delivery Systems of Valsartan Using Porous Carriers. *AAPS Pharm Sci Tech,* **2012,** 13(4): 1416–1427.

233. Prajapati S.T., Joshi H.A., Patel C.N. Preparation and Characterization of Self-Microemulsifying Drug Delivery System of Olmesartan Medoxomil for Bioavailability Improvement. *J Pharm.***2012**, 13: 1–9.

234. Pawar S.K., Vavia P.R. Rice germ oil as multifunctional excipient in preparation of self-microemulsifying drug delivery system (SMEDDS) of tacrolimus. *AAPS Pharm Sci Tech,* **2012**, 13 (1): 254–261.

235. Einien M.A., Mahrouk G., Elkasabgy N. Design and in-vitro evaluation of Olanzapine-loaded self nanoemulsifying Drug delivery system. *Int. J. Institutional Pharma life Sci,* **2012**, 2(3): 12–32.

236. Gumaste S.G., Dalrymple D.M., Serajuddin A.T.M. Development of Solid SEDDS, V: Compaction and Drug Release Properties of Tablets Prepared by Adsorbing Lipid-Based Formulations onto Neusilin® US2. *Pharm Res,* **2013**, 30:3186–3199.

237. Kale A.A., Patravale V.B. Design and evaluation of self-emulsifying drug delivery systems (SEDDS) of Nimodipine. *AAPS Pharm Sci Tech,* **2008**, 9(1): 191–196.

238. Mahmoud H., Suwayeh S., Elkadi S. Design and optimization of self-nanoemulsifying drug delivery systems of simvastatin aiming dissolution enhancement. *Afr J Pharm Pharmcol,* **2013**, 7(22): 1482–1500.

239. Bhikshapathi D., Madhukar P., Kumar B.D., Kumar G.A. Formulation and characterization of Pioglitazone HCl self emulsifying drug delivery system. *Der Pharmacia Lettre,* **2013**, 5 (2):292–305.

240. Suresh P.K., Sharma S. Formulation and in-vitro characterization of self-nanoemulsifying drug delivery system of cinnarizine. *Int. J. Comprehensive Pharm,* **2011**, 9(08): 1–6.

241. Zhang L., Zhu W., Yang C., Guo H., Yu A., Ji J., Gao Y., Sun M., Zhai G. A novel folate-modified self-microemulsifying drug delivery system of curcumin for colon targeting. *Int. J Nanomed,* **2012**, 7: 151–162.

242. Patel J., Patel A., Mihir R., Sheth N. Formulation and development of a self nanoemulsifying drug delivery system of Irbesartan. *J Adv. Pharm. Tech. Res,* **2011**, 2(1): 9–16.

243. Kumar V.K., Devi M.A., Bhikshapathi D.V.R.N. Development Of Solid Self Emulsifying Drug Delivery Systems Containing Efavirenz: In Vitro And In Vivo Evaluation. *Int. J. Pharma & Bio Sci,* **2013**, 4(1):869–882.

244. Jeevana J.B., Sreelakshmi K. Design and Evaluation of Self-Nanoemulsifying Drug Delivery System of Flutamide. *J. Young Pharm.,* **2011**, 3(1): 4–8.

245. Chopade V.V., Chaudhari P.D. Development and evaluation of self-emulsifying drug delivery system for lornoxicam. *Int. J. Res. & Development Pharm & Life Sci,* **2013**, 2(4): 531–537.

246. Saifee M., Zarekar S., Rao V.U., Zaheer Z., Soni R., Burande Z. Formulation and In-vitro evaluation of solid self-emulsifying drug delivery system (SEDDS) of Glibenclamide. *Am J. Advanced drug delivery,* **2013**, 1(3): 323–340.

247. Chen Y., Chen C., Jianling Z., Zhiyu C., Shi Q., Liu H. Development of a solid supersaturable self-emulsifying drug delivery system of Docetaxel with improved dissolution and bioavailability. *Biol.Pharm.Bull,* **2011**, 34(2):278–286.

248. Sharma S., Khinchi M.P., Sharma N., Agrawal D., Gupta M.K. Formulation, In-vitro evaluation and stability studies of self micro emulsifying drug delivery system of glyburide. *Int. J. Pharm & Drug Res*, **2013**, 1:57–66.

249. Meena A.K., Sharma K., Kandaswamy M., Rajagopal S., Mullangi R. Formulation development of an Albendazole self-emulsifying drug delivery system (SEDDS) with enhanced systemic exposure. *Acta Pharm*, **2012**, 62: 563–580.

250. Patil P., Patil V., Paradkar A. Formulation of a self-emulsifying system for oral delivery of simvastatin: In vitro and in-vivo evaluation. *Acta Pharm*, **2007**, 57: 111–122.

251. Nawale R.B., Mehta B.N. Glibenclamide loaded self-emulsifying drug delivery system (SMEDDS): Development and optimization. *Int. J. Pharm & Pharma Sci*, **2013**, 5(2): 325–330.

252. Liu L., Pang X., Zhang W., Wang S. Formulation design and in vitro evaluation of silymarin-loaded self-microemulsifying drug delivery systems. *Asian J. Pharma Sci*, **2007**, 2(4): 150–160.

253. Pol A.S., Patel P.A., Hegde D. Peppermint oil based drug delivery system of Aceclofenac with improved anti-inflammatory activity and reduced ulcerogenecity. *Int. J. Pharm. Biosci. Technol*, **2013**, 1(2): 89–101.

254. Khan F., Islam S., Roni M.A., Jalil R. Systematic development of self-emulsifying drug delivery systems of atorvastatin with improved bioavailability potential. *Sci. Pharma*, **2012**, 80: 1027–1043.

256. Yoo J., Baskaran R., Yoo B.K. Self-nanoemulsifying drug delivery system of lutein: physico-chemical properties and effect of bioavailability of warfarin. *Biomol Ther*, **2013**, 21(2): 173–179.

257. Kumar M.S., Shailaja P., Murthy K.V.R. Improvement of oral bioavailability of nifedipine through self microemulsifying drug delivery systems. *J. Global Trends Pharm Sci*, **2011**, 2(3): 364–388.

258. Sun M., Han J., Guo X., Li Z., Yang J., Zhang Y., Zhang D. Design, preparation and in-vitro evaluation of paclitaxel-loaded self-nanoemulsifying drug delivery system. *Asian J Pharm Sci*, **2011**, 6(1): 18–25.

259. Patel R.N., Tbaviskar D., Rajput A.P. Development and in-vitro characterization of SMEDDS (Self-microemulsifying drug delivery system) for Gemfibrozil. *Int. J. Pharm Pharm. Sci*, **2013**, 5(3): 793–800.

260. Gangineni R., Dontha P., Konda S.K. Formulation, characterization and evaluation of self-emulsifying drug delivery system of Repaglinide. *JPR: BioMedRx: An Int. J.*, **2013**, 1(7): 665–672.

261. Goli R., Katariya C., Chaudhari S. Formulation and evaluation of solid self-micro emulsifying drug delivery system of Lamotrigine. *Am. J. Pharm Tech Res*, **2012**, 2(5): 662–679.

262. Obitte N.C., Ezeiruaku H., Onyishi V.I. Preliminary studies on two vegetable oil based self emulsifying drug delivery system (SEDDS) for the delivery of Metronidazole, a poorly water soluble drug. *J. Applied Sci*, **2008**, 8(10): 1950–1955.

263. Maincent P. The regulatory environment: the challenges for lipid -based formulations. *Bulletin technique Gattefosse*, **2007**, 100: 47–49.

264. Chen M.L. Lipid excipients and delivery systems for pharmaceutical development: a regulatory perspective. *Adv Drug Deliv Rev,* **2008,** 60(6): 768–77.

265. David H.S.S., Hay Y.K., Ping Y.S. Self-emulsifying drug delivery system for fat-soluble drugs. EP1170003 B1.**2006.**

266. Chen S., Gunn J.A. Oral dosage self-emulsifying formulations of pyranone protease inhibitors. EP 1333810 B1.**2007.**

267. Gregory L., Alain R., Jean-sebastian G., Shicheng Y., Neslihan G., Simon B. Self emulsifying drug delivery systems for poorly soluble drugs. EP 1340497A1.**2003.**

268. Martin K., Dieter R. Self-emulsifying lipid matrix. EP 1349541 B1.**2006.**

269. Peracchia M.T., Cote S., Gaudel G. Self-emulsifying and self-microemulsifying formulations for the oral Administration of taxoids. EP 1648517 B1.**2008.**

270. Liu Z., Yang L., Yang H., Gao Y., Shen D., Guo W., Feng X. Butylbenzene Phthalein Self-Emulsifying Drug Delivery System, Its Preparation Method And Application. EP 1787638 B1.**2010.**

271. Hao W.H., Wang J.J., Hsu C.S. Self-emulsifying pharmaceutical composition with enhanced bioavailability. EP 2062571 A1.**2009.**

272. Mark G.L.P., William Z., Shigeru K., Yoshlaki K. Self-emulsifying Pigments. US 7276113 B2.**2007.**

273. Holmberg C., Siekmann B. Self-emulsifying drug delivery system. US 7736666 B2. **2010.**

274. Benita S., Kleinstern J., Gershanik T. Self-emulsifiable formulation producing an oil-in-water emulsion. US 5965160.**1999.**

275. Rouffer M.T. Self-emulsifying ibuprofen solution and soft gelatin capsule for use therewith. US 6221391 B1.**2001.**

276. Liu R.R., Wang Z. Self-emulsifying systems containing anticancer medicament. US 6316497.**2001.**

277. Mulye N. Self-emulsifying compositions for drugs poorly soluble in water. US 6436430 B1.**2002.**

278. Booth S.W., Charke A., Newton M. Spheronized self-emulsifying System for hydrophobic and water Sensitive agents. US 6630150 B1.**2003.**

279. Liang L., Shojaei A., Ibrahim S.A., Burnside B.A. Self-emulsifying formulations of Fenofibrate and/or fenofibrate Derivatives with improved oral Bioavailability and/or reduced Food effect. US 7022377 B2.**2006.**

280. Khan M.A., Nazzal S. Eutectic-based self-nanoemulsified drug delivery system. US 8158162 B2.**2012.**

281. Fujji H., Yamagata M. Self-emulsifying composition of omega 3 fatty acid. US 8618168 B2.**2013.**

282. Legen I., Kere J., Jurkovie P. Self-microemulsifying drug delivery systems. US 8592490 B2.**2013.**

283. Weisspapir M., Schwarz J. Solid self-emulsifying controlled release drug delivery system composition for enhanced delivery of water insoluble phytosterols and other hydrophobic natural compounds for body weight and cholesterol level control. US 2002 /0103139A1.**2002**.

284. Gao P., Morozowich W., Shenoy N. Self-emulsifying drug delivery systems for extremely water-insoluble or lipophilic drugs. US 2002/0119198 A1.**2002**.

285. Glenn R.W., Dunbar J.C., Tadros T. Delivery of reactive agents via Self emulsification for use in Shelf-stable products. US 2002/ 0131945 A1.**2002**.

286. Gao P., Karim A., Hassan F., Forbes J.C. Finely self-emulsifiable pharmaceutical composition. US 2003/0105141 A1.**2003**.

287. Sandner B., Stanica C., Jiang L. Highly concentrated, self-emulsifying preparations containing organopolysiloxanes and alkyl ammonium compounds and their use in aqueous systems. US 2007/0012895 A1.**2007**.

288. Voorspoels J.F.M. self-microemulsifying drug Delivery systems of a HIV protease Inhibitor. US 2007/0104740 A1.**2007**.

289. Murty R.B., Murty S.B. Delivery of Tetrahydrocannabinol. US 2007/0104741 A1. **2007**.

290. Yu Z.H., Huth S. Self-emulsifying compositions, methods of use and preparation. US 2004/0185068 A1.**2004**.

291. Sara A., Paula B.M., Francisco J., Fernanda M. self-emulsifying formulation of Tipranavir for oral administration. WO 2008/142090 A1.**2008**.

292. Hong C., Shin H.J., Ki M.H., Lee S.K. Self-emulsifying matrix type transdermal preparation. WO O1/ 72282 A1.**2001**.

293. Lee S.S., Lee S.K. A pharmaceutical solution for the treatment of erectile dysfunction, prepared by SEDDS formulation. WO 00/16744.**2000**.

294. Nakhat P., Mandogade P., Jain G.K., Talwar M. Self-emulsifying pharmaceutical compositions of Rhein or Diacerein. EP 2207 531 B1.**2012**.

2

Lipid-based nanostructured drug delivery systems for oral bioavailability enhancement of poorly water-soluble drugs

Sarwar Beg[1*], Gajanand Sharma[2], Mahfoozur Rahman[3], Suryakanta Swain[4]
[1]*Department of Pharmaceutics, Roland Institute of Pharmaceutical Sciences, Berhampur, Odisha, India*
[2]*Formulation Research and Tech Transfer, IPCA Laboratories Limited, Kadhivali (W), Mumbai, India*
[3]*Department of Pharmaceutical Sciences, Faculty of Health Sciences, Sam Higginbottom Institute of Agriculture, Technology & Sciences (SHIATS), Allahabad, India*
[4]*Department of Pharmaceutics, Southern Institute of Medical Sciences, College of Pharmacy, SIMS group of Institutions, Mangaldas Nagar, Vijyawada Road, Guntur-522 001, Andhra Pradesh, India.*

2.1   Introduction

The domain of nanomedicine has brought forth several promising drug delivery technologies using diverse nano-scale carriers through diverse routes of administration. Oral drug delivery systems (DDS) find the widest acceptance among patients and manufacturers for treating diverse pathological ailments. Development of such products poses major challenges to pharmaceutical development scientist, as more than one-half of new molecular entities exhibit poor and inconsistent bioavailability (Hauss 2007). Ostensible causes for these bioavailability issues encompass poor aqueous solubility, extensive hepatic first-pass effect, acid-lability in gastric fluid, restricted intestinal permeability, gut wall metabolism by cytochrome P450 (CYP450) family of isozymes and high P-glycoprotein (P-gp) efflux (Beg et al. 2011; Fasinu et al. 2011).

Popular formulations approaches like, micronization, co-crystals, inclusion complexes and solid dispersions primarily improve dissolution performance of the drugs only (Aungst 1993). The potential of lipid-based nanostructured DDS for oral intake has been explored for the purpose owing to their stellar merits like reduced gut wall metabolism by CYP450 group of enzymes, circumnavigation of extensive hepatic first-pass effect, reduced P-gp efflux and lowering intra/inter-subject inconsistencies in gastrointestinal absorption, besides the remarkable increase in dissolution performance, solubility and permeability (Porter et al. 2007; Mu et al. 2013).

2.2   Lipid-based nanostructured systems for oral drug delivery

Lipids are considered as one of the safe and ultimately biocompatible materials for drug delivery. Conventional lipidic systems, consisting of lipidic emulsions and microparticles, possess low stability related to particle size distribution, high degradation under extreme temperature conditions and toxic effects imposed by high concentration of surfactants (Porter et al. 2008). Nanostructured lipid-based DDS primarily include a variety of vesicular and non-vesicular systems, as categorized in Figure 2.1.

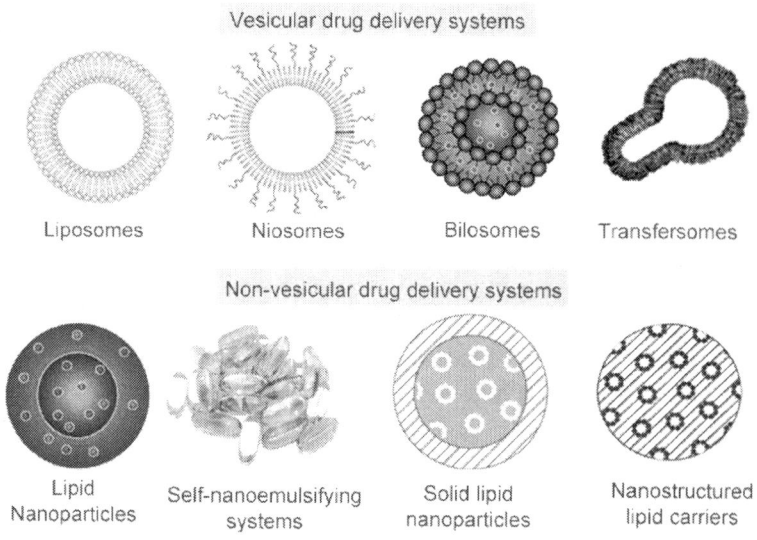

Figure 2.1 Pictorial depiction of types of vesicular and non-vesicular drug delivery systems

These lipid-based nanoformulations encompass a plethora of diverse functional and nonfunctional excipients like lipids, surfactants, cosurfactants, and at times, cosolvents, where drug(s) solubilizes in the vicinity of lipidic excipients, leading to substantial increase in the drug absorption, with reduced variability and elimination of food effects (Dahan and Hoffman 2008). Moreover, the lipid-based systems can easily be prepared at larger industrial and/or commercial scale with excellent dug loading, and high stability characteristics. Due to their numerous merits, there are several lipid-based nanostructured systems available in the global market for oral intake today (Table 2.1).

Table 2.1 Key instances of marketed oral lipid-based drug delivery technologies

Trade name	Drug	Indication	Manufacturer
Agenerase®	Amprenavir	HIV antiviral	Glaxo SmithKline
Rocaltrol®	Calcitriol	Calcium regulator	Roche
Cipro®	Ciprofloxacin	Antibiotic	Bayer
Neoral®	Cyclosporin A/I	Immuno-suppressant	Novartis
Gengraf®	Cyclosporin A/III	Immuno-suppressant	Abott
Accutane®	Isotretinoin	Anti-comedogenic	Roche
Kaletra®	Lopinavir, Ritonavir	HIV antiviral	Abott
Norvir®	Ritonavir	HIV antiviral	Abott
Lamprene®	Clofazamine	Treatment of Leprosy	Alliance
Sustiva®	Efavirenz	HIV antiviral	Bristol-Meyers
Fenogal®	Finofibrate	Anti hyperlipproteinomic	Genus
Restandol®	Testosterone undecanoate	Hormone replacement therapy	Organon laboratories
Convulex®	Valproic acid	Anti epileptic	Pharmacia
Juvela®	Tocopherol nicotinate	Hypertension, hyperlipidemic	Eisai Co.

Lipid formulations can be rationally classified into four major categories such as Type I, II, III and IV formulations. These formulations primarily contain varying proportions of lipid (liquid/solid), surfactant (water soluble/insoluble) and cosolvents for solubilization of highly lipophilic drug molecules (Kuentz 2012). Figure 2.2 represents a general phase diagram depicting the existence of possible lipidic phases generated during the interplay of excipients like lipids, surfactant and aqueous phase. Formation of stable nano-lipidic systems can, therefore, be achieved by judicious selection, variation in the lipid-to-surfactant ratio and optimization of the process parameters involved (Kohli et al. 2010).

Important factors like nature of the lipid, physiological and molecular properties of the drug dynamically affect the particulate nature of the lipidic systems along with permeability and drug absorption potential.

Formation of stable lipidic systems can be achieved by judicious selection of ratio of lipids-surfactants and optimizing the process parameters involving in homogenization technique. Rationally, lipids are selected based on their intrinsic capacity to solubilize the drug molecules in it, whereas surfactants are usually chosen according to their HLB values. Israelachvili's surfactant packing parameter (SPP) is used for identifying the optimum quantity of surfactant required in particular, as per the Equation (2.1):

$$SPP \; = \qquad\qquad\qquad\qquad ...(2.1)$$

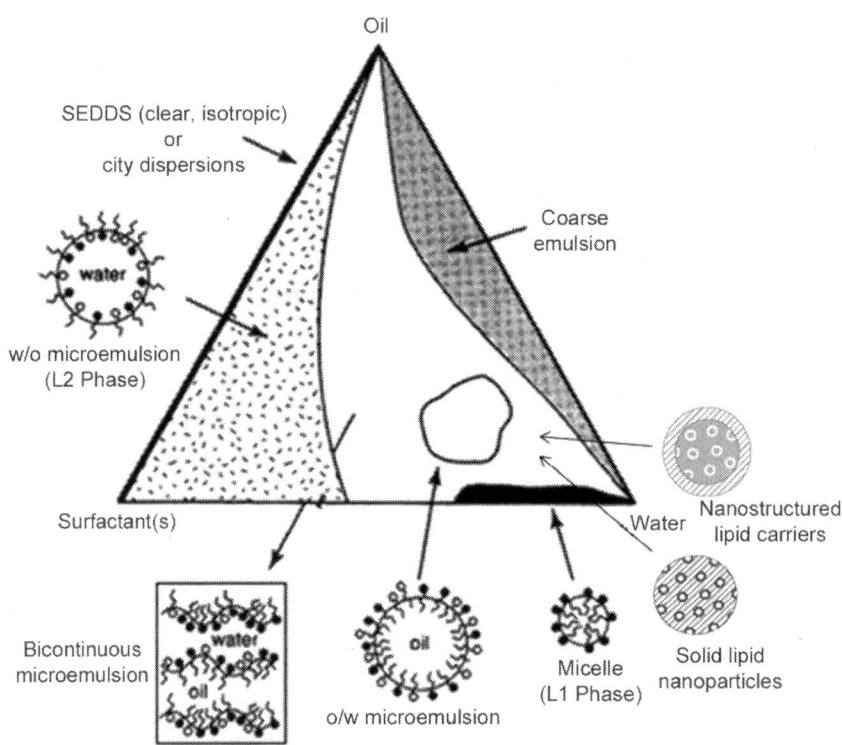

Figure 2.2 Ternary phase diagram depicting regions for formation of diverse lipid-based drug delivery systems

Where, V is the molar volume of the hydrophobic moiety of the surfactant molecules, A the cross-sectional area of the hydrophilic portion when situated at the interface, and L the critical length of the hydrophobic chain. For lipid aggregates dispersed in an aqueous phase, a SPP value between 1/3 and 1 is considered optimal. Table 2.2 illustrates the physiochemical properties and applicability of diverse lipid-based nanocarriers.

Table 2.2 Characteristics of various lipid-based nanocarriers employed for oral drug delivery

Characteristics	Liposomes	LNPs	NLCs	SLNs	LDCs	SEDDS
Ability to deliver lipophilic drugs	Yes	Yes	Yes	Yes	Yes	Yes
Ability to deliver hydrophilic drugs	Yes	Yes	Yes	Yes	Yes	Yes
Physical stability	++	+++	+++	+++	++	++++
Biological stability	++	+++	+++	+++	++	+++
Biocompatibility	+++	++	+++	+++	++	+++
Ease of sterilization	++	++	++	++	++	–
Drug targetting	++++	+++	+++	+++	+++	–
Ease of commercialization	++	+	++	++	+	++++
Ability to deliver Biotechnological therapeutics	+++	++	++	++	++	+
Regulatory acceptance of excipients	++	++	+++	+++	+	++++

Lipid nanoparticles (LNPs); Solid Lipid Nanoparticles (SLNs); Nanostructured Lipid Carriers (NLCs); Lipid Drug Conjugates (LDCs); +++: High; ++: Moderate; +: Low

2.3 Types of lipid-based nanostructured drug delivery systems

2.3.1 Vesicular systems

Vesicular systems are a novel means of drug delivery that can enhance bioavailability of encapsulated drug and provide therapeutic activity in a controlled manner for a prolonged period of time (Kamboj *et al.* 2013). Such systems invariably contain phospholipids as the integral part of the carrier along with cholesterol and surfactants for vesicle stabilization. Liposomes are the most promisingly vesicular carriers composed of lipidic bilayer, surfactants and are prepared by disrupting biological membranes using sonication (Zhou *et al.* 2014). Niosomes are vesicles composed of non-ionic surfactants, which are biodegradable, relatively non-toxic, more stable and inexpensive alternative to liposomes (Jadon *et al.* 2009). Similarly, bilosomes as novel bile salt stabilized vesicles acting as envelopes to protect their contents from the harsh environment of the gut enabling oral administration, have proved effective in the delivery of vaccines, biologicals and traditional small molecules (Shukla *et al.* 2009). Transferosomes are the flexible membrane vesicles containing high amount of edge activator and provides enhanced permeation characteristic for delivery of loaded cargo through oral route (Huang *et al.* 2011). Besides, implication of the aforesaid nanocarriers for diverse drug delivery, surface engineering techniques have also been employed for attaining effective targeting of the encapsulated drug molecules. Table 2.3 illustrates the literature instances on diverse nanostructure vesicular systems explored for oral bioavailability enhancement of drugs.

Table 2.3 Select literature instances on varied vesicular drug delivery systems employed for oral drug delivery

Drug	System	Key excipients	Key inference	Reference(s)
Liposomes				
Breviscapine	Pluronic P85-coated liposomes	Distearoyl phosphatidylcholine	5.6-fold enhancement in oral bioavailability	(Zhou *et al.* 2014)
Insulin	Bile salt conjugated liposomes	Sodium glycocholate, Sodium taurocholate, Sodium deoxycholate	8.5% and 11% improvement in the oral bioavailability, and faster onset of blood glucose level	(Niu *et al.* 2012)
Cyclosporine A	Bile salt conjugated liposomes	Sodium deoxycholate, Soya phosphatidylcholine, Cholesterol	120.3% improvement in the oral bioavailability and elimination half-life	(Guan *et al.* 2011)
Cefotaxime	Folic acid-coupled liposomes	Soya phosphotidylcholine, Cholesterol	1.4-2-times improvement in AUC and 1.2-1.8-times higher Cmax	(Ling *et al.* 2009)
Fenofibrate	Bile salt conjugated liposomes	Soya phosphotidylcholine, Sodium deoxycholate	5.13 and 3.28-fold improvement in the oral bioavailability	(Chen *et al.* 2009)
Calcitonin	Carbopol/ Chitosan coated liposomes	Dipalmitoyl phosphatidylcholine, Stearylamine	2.4 to 2.8-times higher oral bioavailability	(Takeuchi *et al.* 2003)

Contd...

Contd...

Drug	System	Key excipients	Key inference	Reference(s)
Niosomes				
Isradipine	Proniosomes	Span 80, Cholesterol	2.3-fold enhancement in the oral bioavailability vis-à-vis pure drug	(Veerareddy and Bobbala 2012)
Celecoxib	Proniosomes	Span 80, Cholesterol	172.06% improvement in the oral bioavailability compared to the pure drug	(Nasr 2010)
Recombinant hepatitis B surface antigen	NanoBilosomes	Dipalmitoyl phosphatidyl Ethanolamine, Span 40, Cholesterol	Enhanced antigenic response by developed of antibody titer	(Shukla *et al.* 2009)
Cholera toxin B subunit		Dipalmitoyl phosphatidyl ethanolamine-MCC, Span 65, Cholesterol	Significant enhancement in Serum IgG titers after single dose	(Singh *et al.* 2004)
Diphtheria toxoid		Cholesterol, Dicetyl phosphate	Significant improvement in IgA and IgG antibody response	(Shukla *et al.* 2011)
Transfersomes				
(±) Catechin	Elastic liposomes	Soy phosphatidylcholine, Cholesterol, Tween 80	Significantly improved bioavailability from elastic liposome vis-à-vis the conventional liposome	(Huang *et al.* 2011)

Diverse manufacturing techniques are used for preparation of vesicular carriers including ethanol injection, ether injection, thin-film hydration, extrusion, super critical reverse phase evaporation, etc. Each method of preparation has its own advantage for encapsulation of drug molecules in it along with attainment of size them.

2.3.2   Non-vesicular systems

2.3.2.1   Self-emulsifying drug delivery systems (SEDDS)

These are relatively newer lipid-based technological innovations with immense promise in oral bioavailability enhancement of drugs. The SEDDS encompass oils, surfactants, co-surfactants and/or co-solvents, which upon emulsification in presence of aqueous phase produces ultrafine micro/nanoemulsion (Figure 2.3). These are optically stable isotropic mixtures, easy to manufacture and for subsequent scale-up. Depending upon the globule size and the low free energy of the system, the SEDDS can be categorized as self-microemulsifying drug delivery systems (SMEDDS; ranging between 150 nm and 200 nm) or self-nanoemulsifying drug delivery systems (SNEDDS; less than 100 nm) (Singh et al. 2009; Kohli *et al.* 2010).

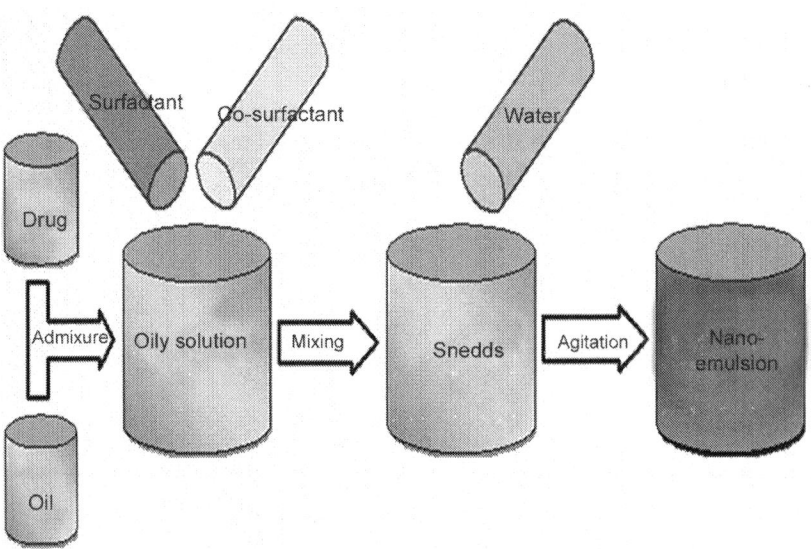

Figure 2.3 Pictorial depiction of formulation of SEDDS

Besides the pivotal advantages of SNEDDS like ease of production, enhanced solvent capacity, increased stability, and easier scalability in industrial milieu, another remarkable feature of SNEDDS is the resilience and amenability to adopt changes in the supramolecular structure and composition. Hence, technological innovations can encompass special type of lipid(s), emulgents(s), or process(es) leading to various newer and advanced modification in the SNEDDS to handle diverse type of process and issue with the drugs. Figure 2.4 portrays the instances of commonly encountered SEDDS for oral delivery of drugs.

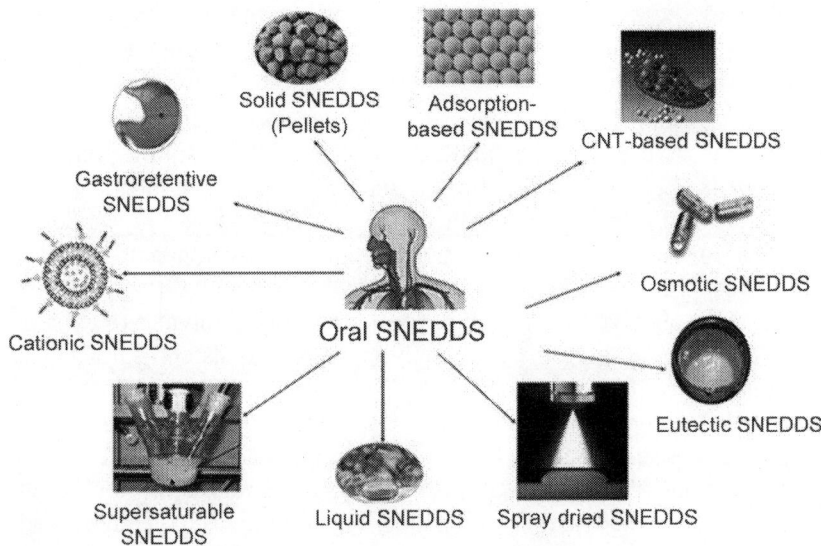

Figure 2.4 Pictorial depiction various types of SEDDS for oral drug delivery

On the basis of their final form, these systems can be broadly classified as liquid SEDDS, semi-solid SEDDS and solid SEDDS. The latter kind of SE systems invariably contain a holistic interplay of lipidic and solidifying excipients to transform the liquid SE systems into different dosage forms like SE pellets, microspheres, controlled release tablets, minicapsules, etc., using diverse approaches like spray drying, porous carrier-based absorption, osmosis, eutectic phenomenon, supersaturability, etc.

Besides, incorporation of cationic charge inducers or polar lipids for the preparation of positively charged SEDDS help in augmenting the oral bioavailability of drugs owing to increase in intestinal permeability (Gershanik and Benita 1996; Bhalani and Patel 2010). Addition of polymeric blends of hydrophilic polymers, diblock copolymers and gelling agents in the SE formulations have shown to provide improved drug stability and remarkable augmentation in the oral bioavailability, ostensibly due to the inhibition of precipitation of the drugs (Holmberg and Siekmann 2010). As lipid digestion is considered to be a major rate-limiting step in the drug absorption, innovations have lately led to the development of enzyme-based liquid SE formulations which facilitate the digestion of the lipidic triglycerides liberating the free fatty acids and drug molecules for absorption into the systemic circulation (Shlieout *et al.* 2005). Apart from the liquid SEDDS, increased number of literature findings on the semisolid and/or the solid SEDDS have already popularized these as alternative formulation approaches, with significant advantages of simpler formulation techniques, easier scale-up, and improved stability and transportability.

The semisolid SEDDS are formed *in situ* using similar lipidic constituents as used for the liquid SEDDS, but with melting point above the room temperature. Exceptionally, such formulations contain lipids and surfactants without any co-surfactants. Lauryl macrogel-glycerides like Gelucire 44/14, Gelucire 50/13, polyoxyethylene hydrogenated castor oil derivatives like Nikkol® HCO50, cetyl alcohol derivative (e.g., Emulcire® 61WL) and polyoxyethylene polyoxypropylene block polymer (e.g., Lutrol® F127, Lutrol® F188) are the commonly employed lipids and surfactants for the preparation of semisolid SEDDS. These formulations exhibit viscosity higher than the corresponding liquid SEDDS, and thus provide improved drug stability and better portability during handling. Invariably, however, under in vivo conditions, these formulations tend to suffer from poor emulsification efficiency due to the presence of high melting point lipids, potentially leading to inconsistent drug absorption profiles. Literature reports demonstrate a few instances on the semisolid SEDDS like, carvedilol (Singh *et al.* 2012) and atorvastatin (Breitkreitz *et al.* 2013) for enhancing their respective oral bioavailability.

The solid SEDDS (S-SEDDS) are relatively recent technological breakthroughs to surmount the ostensible problems associated with liquid SEDDS like lower formulation stability, plausible interaction of excipients with capsule shell, irritating effect of surfactant on GI mucosa, etc. (Schwarz 2004; Tang *et al.* 2008). These can be prepared by one of more among the several possible approaches like adsorption on the inert carriers, spray drying, extrusion-spheronization, and melt granulation producing dry powders, granules, microspheres and nanoparticles (Weisspapir and Schwarz 2000). Such systems have additional meritorious visages such as low production costs, better portability, high stability, safety and improved patient compliance. In addition, the S-SEDDS formulations allow possible reduction in the volume of administration and enhance precise dosing. The various solidification techniques available for the purpose tend to differ from each other, in terms of drug loading capacity and recovery of the solid formulation after transformation from liquid to solid state. Thus, selection of a characteristic solidification technique would eventually depend on the nature of liquid formulation, properties of the drug, batch size, thereof (Tan *et al.*

2013). Owing to the innumerable merits, the S-SEDDS have attracted much attention compared to that of liquid SEDDS. Table 2.4 provides a classified account on literature reports on diverse SEDDS formulations employed for oral bioavailability enhancement of drug.

Table 2.4 Select literature instances on self-emulsifying drug delivery systems employed for oral delivery

Drug	Excipient(s)	Bioavailability enhancement	Reference formulation	Reference
Liquid self-emulsifying drug delivery systems				
Ondansetron	Capmul MCM, Labrasol, Tween 20	3.01 fold	Conventional tablet	(Beg *et al.* 2013)
Etoposide	Caproyl PGMC, Cremophor RH 40, Transcutol P	3.2 fold 7.9 fold	Etosid® Drug suspension	(Akhtar *et al.* 2013)
Valsartan	Capmul MCM, Labrasol, Tween 20	2.5 fold	Pure drug	(Beg *et al.* 2012)
Puerarin	Castor oil, Cremophor EL, 1,2-propanediol	2.6 fold	Conventional tablets	(Quan *et al.* 2007)
Carvedilol	Capmul PG8, Cremophor EL, Transcutol P	3-4 fold	Pure drug and commercial tablet	(Singh *et al.* 2011)
Tacrolimus	TPGS, Cremophor EL40	8 fold	Aqueous solution	(Wang *et al.* 2011)
Simvastatin	Cremophore EL, Transcutol P	1.5 fold	Zocor® tablets	(Dehal *et al.* 2009)
Torcetrapib	Medium chain triglycerides, Triacetin, Polysorbate 80, Capmul MCM	2 fold	Capsules	(Perlman *et al.* 2008)
Nimodipine	Gelucire 44/14, Labrasol, Transcutol P, Plurololeique	4.6 fold 1.91 fold 1.53 fold	Suspension oily solution micellar solution	(Kale and Patravale 2008)
Celecoxib	PEG-8 caprylic/ capric glycerides, Tween 20, PG	1.32 fold	Conventional capsule	(Subramanian *et al.* 2004)
Halofantrine	Structured medium and long chain triglycerides	7.49 fold	Conventional tablets	(Holm *et al.* 2003)
Solid self-emulsifying drug delivery systems				
Tacrolimus	Labrafac, Labrasol, Lauroglycol	2.2 fold	Commercial product	(Seo *et al.* 2014)
Fenofibrate	Migloyl 812, Cremophor RH40, Tween 80, Aerosil 200	1.2 fold	Pure drug	(Han and Han 2011)

Contd...

Contd...

Drug	Excipient(s)	Bioavailability enhancement	Reference formulation	Reference
Curcumin	Lauroglycol FCC, Labrasol and Transcutol HP, Aerosil 200	4.6-7.6 fold	Pure drug	(Yan *et al.* 2011)
Dexibuprofen	Labrasol, Capryol 90, Labrafil M 1944 CS, Aerosil 200	1.9 fold	Powder suspension	(Balakrishnan *et al.* 2009)
Landolphia owariensis latex	Palm Kernel oil, Tween 80 and Span 85	1.5 fold	Pure drug	(Obitte *et al.* 2010)
Cationic self-emulsifying drug delivery systems				
Simvastatin	Cremophor RH 40, Capmul MCM, Captex 300, Epikuron 200	10 fold	Conventional SNEDDS	(Thomas *et al.* 2013)
Silybin	Labrafac CC, Labrasol, Cremophor RH40, HPMC E50 LV	3 fold	Conventional SEDDS	(Wei *et al.* 2012)
Halofantrine	Capmul MCM, Captex 300, soybean oil, Cremophor, ethanol, PVP	4–8 fold	Capsules	(Khoo *et al.* 1998)
Carbamazepine	Migloyl, Cremophor, Tween 80, PEG 400, PVP	5 fold	Commercial tablet	(Nan *et al.* 2012)
Supersaturable self-emulsifying drug delivery systems				
Lovastatin	Gelucire 50/13, Cremophor EL, oleylamine	3.2 fold	Suspension	(Singh *et al.* 2010)
Carvedilol	Lauroglycol, Gelucire 44/14, Tween 20, oleylamine	4.5 fold	Conventional tablet	(Singh *et al.* 2009)
Vinpocetine	Labrafac, oleic acid, Cremophor EL, Transcutol P, oleylamine	1.85–1.91 fold	Powder suspension	(Chen *et al.* 2008)
Fluorescent dye DilC18	Ethyl oleate, Polysorbate 80, oleylamine	7.03 fold	Simple SEDDS	(Gershanik *et al.* 2000)
Cyclosporine A	Ethyl oleate, Tween 80, dithiotreitol, ethanol, oleylamine	5 to 6 fold	Conventional SEDDS	(Gershanik *et al.* 1998)
Progesterone	Ethyl oleate, PEG 300, Tween 80, oleylamine	2.77 fold	Suspension	(Gershanik and Benita 1996)

2.3.2.2   Nanoparticulate systems

Lipid-based nanoparticulate systems include a diverse variety of DDS, including lipidic nanoparticles, solid lipid nanoparticles (SLNs), nanostructured lipidic carriers (NLCs), lipid-drug nanoconjugates and nanomixed micelles, widely explored for oral bioavailability enhancement of drugs. The pictographical depiction of lipidic nanoparticulate-based delivery systems along with their molecular dynamics is shown in Figure 2.5.

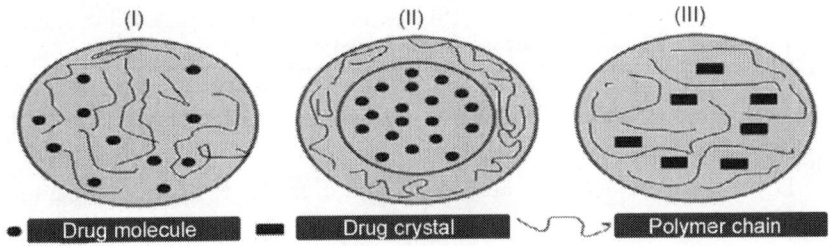

Figure 2.5 (I) Lipid nanoparticle, (II) solid lipid nanoparticle, (III) nanostructured lipid carriers

Lipidic nanoparticles (LNPs)

Since decades, LNPs and lipidic nanopellets have been successfully employed for oral drug delivery as first- and second-generation nano-structured lipidic systems containing lipid matrices, respectively. LNPs or lipid pellets are the first generation micro-structured lipidic systems containing lipid matrices introduced by Boehringer Ingelheim, Germany, almost three decades back. Subsequent to these, the second-generation "lipid nanopellets" were developed for oral administration in the mid-eighties (Sun *et al.* 2014). These systems could not be further developed owing to various reasons, primarily by non-existential patent protection in several countries. Albeit such systems these are obsolete from the research arena, yet these have provided a tangible inkling for development of alternative derivatized systems for brain drug delivery (Cerpnjak *et al.* 2013).

Solid lipid nanoparticles (SLNs)

SLNs are the third generation nanostructured colloidal carriers ranging in size between 1 nm and 1000 nm which are made up of biocompatible and biodegradable materials (i.e., lipids), frequently used as an alternative over traditional polymeric nanoparticles for incorporating both hydrophilic as well as lipophilic drugs. Their colloidal dimension provides high drug pay load, controlled drug release, drug targeting, drug protection and stability (Mehnert and Mader 2001). Upon peroral delivery, SLNs are readily taken up by intestinal lymphatics owing to their lipidic nature and nanosized structure to enhance the oral bioavailability. The increasing attention in SLNs is due to their combined biodegradable and bio-acceptable nature coupled with efficient drug delivery characteristics and hassle-free method of preparation vis-à-vis the polymeric nanoparticles. The general surface morphological structure of SLNs contains a solid hydrophobic core and a monolayer coating of phospholipid. The solid core may contain the drug in the solid high melting fat with hydrophobic end of the phospholipid chains embedded in the fat matrix (Das and Chaudhury 2010). These SLNs are primarily constituted of a blend of lipidic constituents dispersed in the emulgents and water. Additionally, presence of charge modifiers and stealth agents can prolong the circulation time and targeting ability of the SLNs through oral route by avoiding uptake of the drug molecules by reticuloendothelial system.

Diverse techniques have been described in the literature for preparation of SLNs including hot microemulsion, hot/cold homogenization and high-pressure homogenization are of considerable importance. Owing to the numerous meritorious visages of the SLNs, these have been explored for oral delivery of drugs, biomolecules and phytoconstituents (Mehnert and Mader 2001). Table 2.5 illustrates the drug delivery instances employing SLNs with oral bioavailability enhancement applications.

Table 2.5 Select literature instances on solid lipid nanoparticles (SLNs) employed for oral bioavailability enhancement

Solid lipid nanoparticles (SLNs)				
Efavirenz	Conventional SLNs	GMS, Precirol ATO-5, Stearic acid, Tween 80	5.32-fold and 10.98-fold increase in Cmax and AUC comparison to efavirenz suspension	(Gaur et al. 2014)
Candesartan	Conventional SLNs	Trimyristin, Poloxamer 188, egg lecithin E80	2.75-fold increase in oral bioavailability compared to free drug	(Dudhipala and Veerabrahma 2014)
Raloxifene	Conventional SLNs	Stearic acid, Tween 80, Polaxamer 188	308% and 270% enhancement in the AUC compared to free drug suspension	(Tran et al. 2014)
γ-tocotrienol	Conventional SLNs	Compritol 888 ATO, Lutrol F127, Taurocholate	4.1-fold improvement in the oral bioavailability with respect to pure drug	(Abuasal et al. 2012)
Cryptotanshinone	Conventional SLNs	Soy lecithin, Tween 80, Compritol 888 ATO, Sodium dehydrogencholate	1.86 and 2.05 times higher AUC vis-à-vis	(Hu et al. 2010)
Pentoxifylline	Conventional SLNs	Cetyl alcohol, Lecithin, Tween 20	Improved the oral bioavailability	(Varshosaz et al. 2010)
Praziquantel	Conventional SLNs	Glyceryl monosterate, Tween 80,	4.1-higher oral bioavailability compared to free drug suspension	(Yang et al. 2009)

Nanostructured lipidic carriers (NLCs)

NLCs are considered as another third generation of nanoparticles, with size ranging between 10 nm and 1000 nm, consisted of mixture of solid and liquid lipids. Incorporation of liquid lipid can improve the loading capacity of drugs in the NLCs (Iqbal *et al.* 2012). Such NLCs tend to overcome the limitations of SLNs such as risk of gelation, uncontrolled release, decreased chemical stability and drug leakage during storage due to polymorphism of lipids. Commonly employed methods to produce NLCs are high-pressure homogenization or hot melt-emulsification.

Based upon the variegated formulation strategies to optimize their nanostructures, the NLCs are categorized as type I, II, III or IV. In imperfect type I, NLC's are prepared by mixing spatially different lipids which provide imperfections in the crystal order of lipidic nanoparticles (Figure 2.6a) (Radtke et al. 2005). Mixing small amounts of chemically quite different liquid lipids (oils) with solid lipids in order to achieve the highest incompatibility leads to the highest drug payload. While amorphous type II NLC's can be achieved by mixing solid lipids with special lipids (Figure 2.6b). However, in multiple type III NLC's oil in fat in water (O/F/W) is present where drugs can be accommodated in the solid, but at increased solubility in the oily parts of the lipid matrix (Figure 2.6c). Nevertheless, in type IV NLC's, water-soluble drugs are conjugated by salt formation or covalent linkage with a lipid, thus forming a water-insoluble lipidic conjugate. Drug release from lipid particles takes place by diffusion and by lipid particle degradation in the body.

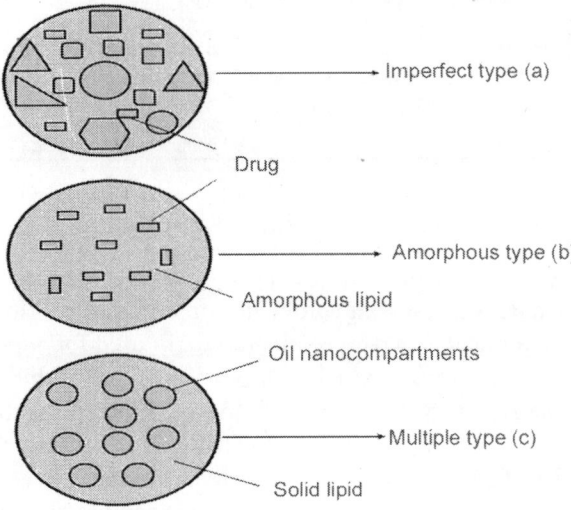

Figure 2.6 Types of NLC: (a) imperfect type, (b) amorphous type and (c) multiple type

Several instances pertaining to applications of NLCs for oral drug delivery has been enlisted in Table 2.6.

Table 2.6 Select literature instances on nanostructured lipid carriers (NLCs) employed for oral bioavailability enhancement

Nanostructured lipid carriers (NLCs)				
Isoliquiritigenin	Lyophillized NLCs	Miglyol 812, Tween 80, Polaxamer 188	545% improvement in relative oral bioavailability compared to the pure drug	(Zhang *et al.* 2014)
Fenofibrate	Lyophillized NLCs	Compritol 888 ATO, Labrafil M 1944CS, Soya lecithin, CTween 80	4-fold augmentation in the AUC compared to the oral solution	(Tran *et al.* 2014)
Silymarin	Monolithic NLCs	Precirol ATO-5, Oleic acid, Lecithin, Tween 80	2.54- and 3.10-fold enhancement in oral bioavailability compared to the marketed formulation	(Shangguan *et al.* 2013)
Testosterone undecanoate	Homogenized NLCs	Dynasan 118, Stearic acid, Tween 80,	2-times higher AUC compared to Andriol Testocaps®	(Muchow *et al.* 2013)
Quercetin	Chitosan-coated NLCs	Compritol 888 ATO, Labrasol, Polysorbate 80	Increased quercetin stability in plasma and gastrointestinal environment	(Sun *et al.* 2014)
Tripterine	Phospholipid coated NLCs	Monostearin, Soybean oil, Soya lecithin, PEG-40, DSPE-PEG	2.1 to 2.7-fold improvement in effective permeability vis-à-vis oral solution	(Zhou *et al.* 2012)
	Cell-penetrating peptide coated liposome	Precirol ATO-5, Labrafil M 1944CS, Vit-E TPGS	4.8-fold improvement in AUC vis-à-vis the drug suspension	(Chen *et al.* 2012)
Etoposide	Monolithic NLCs	PEG 40, Stearic acid	1.8 to 3.5-fold improvement in relative oral bioavailability compared to pure drug	(Zhang *et al.* 2011)

2.3.3   Other lipidic nanostructured systems

The utility of lipidic systems have paved the way for delivery of diverse pharmaceutical agents possessing problems of erratic oral bioavailability. Lipidic nanomixed micelles have been increasingly used for augmenting the oral bioavailability of drugs marked with poor aqueous solubility owing to their nano globule size, higher biocompatibility and biomimetic characteristic (Singh *et al.* 2013). Besides, lipidic nanovectors and nanocapsules have also been recently identified as relatively novel systems for oral delivery, especially the biomolecules like proteins, peptides and nucleic acids (Muller *et al.* 2002). These nanocarriers are made up of lipids dispersed in the nonionic surfactant to form a hollow shell-like structure with size ranging between 2 nm and 100 nm. These exhibit promising features for delivery of drugs through systemic route and excellent physicochemical stability ostensibly owing to the presence of nonionic surfactants imparting them with the "stealth" characteristics.

2.4   Characterization of lipid-based nanostructured systems

Various test methodologies have been employed for characterization and finding the robust lipidic formulations. Literature recounts the use of various *in vitro*, *ex vivo* and in vivo approaches for performance evaluation of the lipid-based nanoformulations. Besides, instrumental techniques like dynamic light scattering (DLS), scanning/transmission electron microscopy (SEM/ TEM), rheology, differential scanning calorimetry (DSC), fourier transformed infrared (FTIR) spectroscopy, are also widely used for characterization (Rahman *et al.* 2011).

2.4.1   Globule size and zeta potential measurement

A Zetasizer has been used to measure globule size, zeta potential and molecular weight of nanoparticulate systems. It works upon light scattering principle. The electrophoretic mobility of the micelles and the nanoemulsion has also been measured at 25°C with a Zetasizer. The zeta potential of the dispersions is calculated by the instrument according to the Helmholtz–Smoluchowski equation. The electrophoretic velocity is given by

$$U = \frac{\varepsilon \xi E_X}{\mu}$$
...(2.2)

where, E is the permittivity, ζ is the zeta potential, Ex is the axial electric field and μ is the viscosity. The value of zeta potential, as well as the size of the droplets/globules of the SNEDDS formulation, can be measured using diverse techniques.

2.4.2   Electron microscopy studies

Electron microscopy technique is used for identifying the surface morphology characteristics of the dispersions. However, due to the high lability of the samples and the possibility of artifacts, electron microscopy is, at times, considered as a somewhat misleading technique. The most commonly employed electron microscope techniques are scanning electron microscope (SEM), transmission electron microscopy (TEM), cryo-transmission electron microscopy (Cryo-TEM), etc.

2.4.3   *In vitro* drug release studies

The *in vitro* drug release studies are used to characterize the release behavior of the dosage forms employing dissolution studies. The dosage forms are sealed in the dialysis tubing of known molecular weight (i.e., 1,000–10,000 kDa) in presence of biorelevant-simulated fluids. At predetermined time intervals, the samples are withdrawn followed by estimation employing suitable analytical technique to assess the drug release profile. Further, it is imperative to assess the drug release kinetics of the developed formulations.

2.4.4   *Ex vivo* permeation studies

Ex vivo permeation studies are used by non-everted/everted gut sac technique in unisex Wistar/Sprague-Dawley rats previously abstained from solid food at least 12 h prior to the study (Ruan *et al.* 2006; Alam *et al.* 2011). After sacrificing the rats by cervical dislocation, the abdomen is incised midway and medial jejunum segment is cut for investigational purposes. The segment was rinsed with Kreb's Ringer Buffer (KRB) solution and immediately placed in the freshly prepared KRB solution at $37 \pm 1°C$ for continuous aeration. The intestinal segments are stripped off by removing adipose tissue. The test formulation is ligated at one end to form a hollow sac with opening at the other end, where the test formulation is injected. A weight of one gram was tied to the closed end of the segment and fitted into the in-house fabricated assembly containing 50 mL of KRB, continuously bubbled with atmospheric air at 10–15 bubbles per min. The intestinal sac present in the holding assembly has provision for outer water jacket for maintaining the temperature of the gut sac at $37 \pm 0.5°C$. Aliquot samples are withdrawn at different time-points and analyzed suitably with the help of analytical techniques to determine the percent drug permeated in 45 minutes. This provides an insight about the permeation potential of the drug candidate when administered in the prepared lipid-based nanostructured drug formulations.

2.4.5   *In vitro* lipolysis rate

The *in vitro* lipolysis studies are used for more accurate characterization of drug release from lipid-based formulations. It gives about idea on the *in vitro* lipid digestion in presence of biorelevant dissolution media along with pancreatic lipolytic enzymes (lipase/colipase) and bile salts. It provides a more accurate description of the process of monitoring the ability of a formulation to maintain the drug in a solubilized state during dispersion in the stomach and subsequent processing of the formulation in the presence of pancreatic and biliary fluids (Dahan and Hoffman 2006; Brogard *et al.* 2007). In this method, lipid-based formulations are introduced into a vessel containing digestion buffer along with bile salts, phospholipids and lipase/colipase enzymes representing the simulated gastric fluid. The digestion of lipid is initiated by these enzymes causes liberation of triglycerides and free fatty acids, which causes a transient drop in pH. The amount of fatty acids liberated can be quantified by titration with an equimolar concentration of sodium hydroxide. This gives idea about the exact amount of drug release after digestion of lipids. Thus, *in vitro* lipolysis provides more accurate determination of *in vitro* drug release over conventional dissolution testing.

2.4.6   *In situ* Single Pass Perfusion Technique (SPIP)

In this technique, perfusion solution is passed through the intestinal segment (i.e., jejunum) by cannulating it at both the ends, and various permeability parameters (i.e., effective permeability, apparent permeability, aqueous permeability and wall permeability) are calculated from amount of drug unabsorbed from the intestine (Yao *et al.* 2008). Besides providing experimental conditions closer to that occurring *in vivo*, this technique is also able to predict the exact mechanism of absorption, i.e. passive absorption, carrier-mediated absorption or active transport (Sharma *et al.* 2005). Lately, the SPIP studies have also used to evaluate the P-gp efflux and role of transporters (MRP, BCRP2) in reducing the oral bioavailability of drugs (Berggren *et al.* 2004).

2.4.7   *In vitro* cell line studies

2.4.7.1   *Intestinal colorectal adenocarcinoma (Caco-2) cells*

Caco-2 cells are the most popular intestinal cellular model in studies on passage and transport. These cell lines are derived from human colorectal adenocarcinoma (Gershanik *et al.* 2000). In culture, they differentiate spontaneously into polarized intestinal cells possessing an apical brush border and tight junctions between adjacent cells, and they express hydrolases and typical microvillar transporters (Sha and Fang 2006). Caco-2 cells, despite their colonic origin, express in culture the majority of the morphological and functional characteristics of small intestinal absorptive cells, including phase I and phase II enzymes, detected either by measurement of their activities toward specific substrates, or by immunological techniques (Gershanik *et al.* 2000; Sha *et al.* 2005; Buyukozturk *et al.* 2009 (In press)).

4.7.2   Non-intestinal cell systems

Madin Darby Canine Kidney (MDCK) cells have been isolated from dog kidneys (Madin et al. 1957; Madin and Darby 1958). They are currently used to study the regulation of cell growth, drug metabolism, toxicity and transport at the distal renal tubule epithelial level (Lavelle et al. 1997; Liu and Zeng 2008). MDCK cells have been shown to differentiate into columnar epithelial cells and to form tight junctions when cultured on semi-permeable membranes (Gonzalez et al. 2009).

2.5   Oral bioavailability enhancement by lipid-based drug delivery systems

Often, absorption and subsequently oral bioavailability of a drug molecule is limited by its solubility or permeability. Accordingly as per the Biopharmaceutical Classification Scheme (BCS), a drug, on the basis of these solubility and permeability characteristics, can be classified in Class I to IV categories (Amidon *et al.* 1995). The Class I drugs, being highly soluble and permeable, do not normally pose any problem of rate and extent of bioavailability, with more than 90% of drug absorption (Dressman and Reppas 2000). Bioavailability of poorly soluble Class II drugs, on the contrary, is dependent on their aqueous solubility/dissolution rate (Pouton 2006). As these drugs tend to exhibit dissolution-limited bioavailability, the in vivo physiological performance correlates well with their *in vitro* dissolution, resulting eventually in good *in vitro/in vivo* correlations (IVIVC) (1997; Singh *et al.* 2010). Absorption of Class III drugs is a distinct function of the permeability across GI barriers. Class IV drug compounds have neither sufficient solubility nor permeability for oral absorption to be complete (Amidon et al. 1995).

Upon oral administration, the lipid-based nanostructured drug formulations are transported across the GI lumen to the systemic circulation and subsequently undergo digestive, absorptive and circulatory phases. Figure 2.7 pictorially depicts different pathways through which the drug molecules absorbed from the lipidic systems tend to get absorbed into the circulatory system. During the digestive phase, the lipidic formulation disperses into a coarse emulsion. This emulsion undergoes enzymatic hydrolysis leading to the formation of lipid digestion products along with the undigested lipids. Digestion is completed by the action of the enzymes (i.e., secretin, pancreatic lipase and co-lipase) that act on the surface of the emulsified triglyceride droplets to produce the corresponding two molecules of monoglyceride and one molecule of fatty acid (Carey et al. 1983; Porter and Charman 2001).

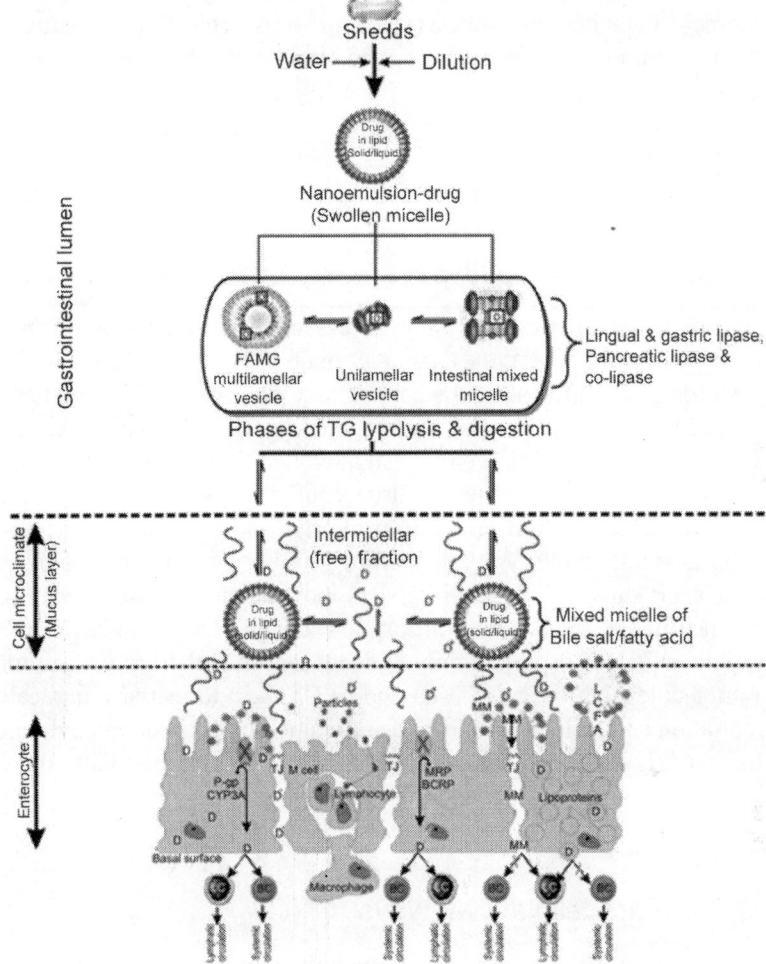

FAMG: Fatty acid monoglyceride, LCFA: Long chain fatty acid, D: Drug, D&D⁺: Ionized drug, TJ: Tight junction, MM: Macromolecules,
LC.: Lymphatic cells, BC: Blood cells, MRP: Multiple resistance proteins, Breast cancer resistance protein, P-gp: P-glycoprotein

Figure 2.7 Flow layout biofate of lipid-based drug delivery systems through oral route

In the absorptive phase, the colloidal species produced as a result of lipid digestion (i.e., chylomicrons) are taken up by passive diffusion, facilitated diffusion or active transport through the enterocyte membrane. These chylomicrons are relatively large colloidal systems capable of selective intestinal lymphatic transport of the lipophilic compounds (Charman and Porter 1996). Chylomicrons travel through the lacteals to join lymphatic vessels from other parts of the body, and enter the systemic circulation via the thoracic duct into the subclavian vein, thus protecting the drug from hepatic first-pass metabolism. During the circulatory phase, the blood-borne chylomicrons rapidly disassemble releasing the encapsulated drug. The residual constituent lipids are utilized throughout the body. After completion of the lipid digestion, the drug molecules solubilized in the lipidic formulations are absorbed through various pathways such as:

2.5.1   Lymphatic pathways

The intestinal lymphatics also play an essential role in the absorption of products from lipid digestion, e.g. long chain fatty acids. This system has the potential of circumventing the hepatic first-pass metabolism and hence to target specific disease states known to spread via the lymphatics. At cellular level, three pathways have been investigated and their choice depends upon the physicochemical properties of drug candidate and the design of drug delivery systems (Hussain *et al.* 2001; Porter and Charman 2001; Swartz 2001; O'Driscoll 2002; Trevaskis *et al.* 2008). Possible mechanisms of drug transport through intestinal barriers include, an increase in membrane fluidity facilitating transcellular absorption, opening of the tight junction to allow paracellular transport, mainly relevant for ionized drugs or hydrophilic macromolecules, inhibition of P-gp and/or CYP450 to increase intracellular concentration and residence time, and stimulation of lipoprotein/chylomicron production (Figure 2.8). The latter two mechanisms are potentially the most promising for intestinal lymphatic drug targeting using lipid-based vehicles (O'Driscoll 2002).

2.5.2   Paracellular pathways

Due to the porous structure of the lymphatic capillaries and the large size of the macromolecules, targeting into the lymphatics may be possible. Surfactant(s), one of the vital components of SNEDDS, may open up the paracellular route, resulting eventually in the increased permeability of hydrophilic molecule or macromolecule conjugate(s). This route has an additional advantage for the delivery of protein and peptide drugs owing to the existence of lower enzymatic activity. Nevertheless, there have been certain limitations too. The transport capacity of this route is limited owing to relatively low surface area and safety issues, especially due to chronic use of absorption enhancers.

2.5.3   Gut associated lymphoid tissues (GALT)

Secondly, access to the lymphatics may be gained via GALT. This route has been mainly explored in the delivery of particulates with significant potential for vaccine delivery as well (Hawley et al. 1995).

2.5.4   Transcellular pathways

Intestinal lipid transport system via the transcellular route is the third major mechanism of delivery of lipophilic compounds when formulated as lipid-based vehicles (Porter and Charman 2001). The precise mechanism of drug transport via transcellular route is not fully understood. At the cellular level, stimulation of chylomicrons production is a key factor in enhancing lymphatic transport of lipophilic compounds (Thi et al. 2009).

2.6   Conclusions

Of late, assortment of lipid-based nanostructured drug delivery systems has stimulated enormous global interest for the oral usage by the pharma industry. The SNEDDS, SLNs and NLCs, in this context, possess immense promise to be commercialized for meeting the patients' needs owing to their simple technology and regulatory excipient status, coupled with straight forward scale-up and validation. Besides, these systems offer much more flexibility in drug loading, modulation of drug release and improved performance in producing final dosage forms such as tablets, capsules, creams and injectables. Notwithstanding their stellar benefits in nanomedicine, the potential toxicity, economics and federal compliance of such nano-technological products are the major concerns for the pharmaceutical scientists to address.

2.7   References

Guidance for industry. Extended release oral dosage forms: Development, evaluation and application of in vitro/in vivo correlations. U.S. Department of Health and Human Services, Center for Drug Evaluation and Research (CDER). Washington, DC, 1997.

Abuasal B.S., Lucas C., Peyton B., Alayoubi A., Nazzal S., Sylvester P.W. and Kaddoumi A. Enhancement of intestinal permeability utilizing solid lipid nanoparticles increases gamma-tocotrienol oral bioavailability. *Lipids*. 2012, 47: 461–9.

Akhtar N., Talegaonkar S., Khar R.K. and Jaggi M. Self-nanoemulsifying lipid carrier system for enhancement of oral bioavailability of etoposide by P-glycoprotein modulation: In vitro cell line and in vivo pharmacokinetic investigation. *J Biomed Nanotehnol*. 2013, 9: 1–14.

Alam M.A., Al-Jenoobi F.I. and Al-Mohizea A.M. Everted gut sac model as a tool in pharmaceutical research: limitations and applications. *J Pharm Pharmacol.* 2011, 64: 326–36.

Amidon G.L., Lennernas H., Shah V.P. and Crison J.R. A theoretical basis for a biopharmaceutic drug classification: the correlation of in vitro drug product dissolution and in vivo bioavailability. *Pharm Res.* 1995, 12: 413–20.

Aungst B.J. Novel formulation strategies for improving oral bioavailability of drugs with poor membrane permeation or presystemic metabolism. *J Pharm Sci.* 1993, 82: 979–87.

Balakrishnan P., Lee B.J., Oh D.H., Kim J.O., Hong M.J., Jee J.P., Kim J.A., Yoo B.K., Woo J.S., Yong C.S. and Choi H.G. Enhanced oral bioavailability of dexibuprofen by a novel solid self-emulsifying drug delivery system (SEDDS). *Eur J Pharm Biopharm.* 2009, 72: 539–45.

Beg S., Jena S.S., Patra Ch N., Rizwan M., Swain S., Sruti J., Rao M.E. and Singh B. Development of solid self-nanoemulsifying granules (SSNEGs) of ondansetron hydrochloride with enhanced bioavailability potential. *Colloids Surf B Biointerfaces.* 2013, 101: 414–23.

Beg S., Swain S., Rizwan M., Irfanuddin M. and Malini D.S. Bioavailability enhancement strategies: basics, formulation approaches and regulatory considerations. *Curr Drug Deliv* 2011, 8: 691–702.

Beg S., Swain S., Singh H.P., Patra Ch N. and Rao M.E. Development, optimization, and characterization of solid self-nanoemulsifying drug delivery systems of valsartan using porous carriers. *AAPS Pharm Sci Tech.* 2012, 13: 1416–27.

Berggren S., Hoogstraate J., Fagerholm U. and Lennernas H. Characterization of jejunal absorption and apical efflux of ropivacaine, lidocaine and bupivacaine in the rat using in situ and in vitro absorption models. *Eur J Pharm Sci.* 2004, 21: 553–60.

Bhalani V. and Patel S. (2010). Pharmaceutical composition for lipophilic drugs. US5514673.

Breitkreitz M.C., Sabin G.P., Polla G. and Poppi R.J. Characterization of semi-solid Self-Emulsifying Drug Delivery Systems (SEDDS) of atorvastatin calcium by Raman image spectroscopy and chemometrics. *J Pharm Biomed Anal.* 2013, 73: 3–12.

Brogard M., Troedsson E., Thuresson K. and Ljusberg-Wahren H. A new standardized lipolysis approach for characterization of emulsions and dispersions. *J Colloid Interface Sci.* 2007, 308: 500–7.

Buyukozturk F., Benneyan J.C. and Carrier R.L. Impact of emulsion-based drug delivery systems on intestinal permeability and drug release kinetics. *J Control Release.* 2009 (In press).

Carey M.C., Small D.M. and Bliss C.M. Lipid digestion and absorption. *Annu* Rev Physiol. 1983, 45: 651–77.

Cerpnjak K., Zvonar A., Gasperlin M. and Vrecer F. Lipid-based systems as a promising approach for enhancing the bioavailability of poorly water-soluble drugs. *Acta Pharm.* 2013, 63: 427–45.

Charman W.N. and Porter C.J.H. Lipophilic prodrugs designed for intestinal lymphatic transport. *Adv Drug Deliv Rev.* 1996, 19: 149–169.

Chen Y., Li G., Wu X., Chen Z., Hang J., Qin B., Chen S. and Wang R. Self-microemulsifying drug delivery system (SMEDDS) of vinpocetine: formulation development and in vivo assessment. *Biol Pharm Bull.* 2008, 31: 118–25.

Chen Y., Lu Y., Chen J., Lai J., Sun J., Hu F. and Wu W. Enhanced bioavailability of the poorly water-soluble drug fenofibrate by using liposomes containing a bile salt. *Int J Pharm.* 2009, 376: 153–60.

Chen Y., Yuan L., Zhou L., Zhang Z.H., Cao W. and Wu Q. Effect of cell-penetrating peptide-coated nanostructured lipid carriers on the oral absorption of tripterine. *Int J Nanomed.* 2012, 7: 4581–91.

Dahan A. and Hoffman A. Use of a dynamic in vitro lipolysis model to rationalize oral formulation development for poor water-soluble drugs: correlation with in vivo data and the relationship to intra-enterocyte processes in rats. *Pharm* Res. 2006, 23: 2165–74.

Dahan A. and Hoffman A. Rationalizing the selection of oral lipid based drug delivery systems by an in vitro dynamic lipolysis model for improved oral bioavailability of poorly water soluble drugs. *J Control Release.* 2008, 129: 1–10.

Das S. and Chaudhury A. Recent advances in lipid nanoparticle formulations with solid matrix for oral drug delivery. *AAPS PharmSciTech.* 2010, 12: 62–76.

Dehal S., Bandyopadhyay S. and Singh B. Improving Bioavailability of simvastatin using various types of Self emulsifying formulations, Proc. 30th Annual Conference of Indian Association of Biomedical Scientists, Defence Institute of High Altitude Research, Chandigarh, India, 18–20 November, 2009.

Dressman J.B. and Reppas C. In vitro-in vivo correlations for lipophilic, poorly water-soluble drugs. *Eur J Pharm Sci.* 2000, 11: S73–80.

Dudhipala N. and Veerabrahma K. Candesartan cilexetil loaded solid lipid nanoparticles for oral delivery: characterization, pharmacokinetic and pharmacodynamic evaluation. *Drug Deliv.* 2014, [Epub ahead of print].

Fasinu P., Pillay V., Ndesendo V.M., du Toit L.C. and Choonara Y.E. Diverse approaches for the enhancement of oral drug bioavailability. *Biopharm Drug Dispos.* 2011, 32: 185–209. doi: 10.1002/bdd.750. Epub 2011 Apr 7.

Gaur P.K., Mishra S., Bajpai M. and Mishra A. Enhanced Oral Bioavailability of Efavirenz by Solid Lipid Nanoparticles: In Vitro Drug Release and Pharmacokinetics Studies. *BioMed Res Int.* 2014, 2014: 1–9.

Gershanik T. and Benita S. Positively charged self-emulsifying oil formulation for improving oral bioavailability of progesterone. *Pharm Dev Technol.* 1996, 1: 147–57.

Gershanik T., Benzeno S. and Benita S. Interaction of a self-emulsifying lipid drug delivery system with the everted rat intestinal mucosa as a function of droplet size and surface charge. *Pharm Res.* 1998, 15: 863–9.

Gershanik T., Haltner E., Lehr C.M. and Benita S. Charge-dependent interaction of self-emulsifying oil formulations with Caco-2 cells monolayers: binding, effects on barrier function and cytotoxicity. *Int J Pharm.* 2000, 211: 29–36.

Gonzalez J.E., Digeronimo R.J., Arthur D.E. and King J.M. Remodeling of the tight junction during recovery from exposure to hydrogen peroxide in kidney epithelial cells. *Free Radic Biol Med.* 2009, 47: 1561–9.

Guan P., Lu Y., Qi J., Niu M., Lian R., Hu F. and Wu W. Enhanced oral bioavailability of cyclosporine A by liposomes containing a bile salt. *Int J Nanomed.* 2011, 6: 965–974.

Han C.C. and Han L. Improving oral bioavailability of agaricoglycerides by solid lipid-based self-emulsifying drug delivery system. *J Appl Pharm Sci.* 2011, 1: 195–199.

Hauss D.J. Oral lipid-based formulations. Adv Drug Deliv Rev. 2007, 59: 667–76.

Hawley A.E., Davis S.S. and Illum L. Targeting of colloids to lymph nodes: influence of lymphatic physiology and colloidal characteristics. *Adv Drug Deliv Rev.* 1995, 17: 129–148.

Holm R., Porter C.J., Edwards G.A., Mullertz A., Kristensen H.G. and Charman W.N. Examination of oral absorption and lymphatic transport of halofantrine in a triple-cannulated canine model after administration in self-microemulsifying drug delivery systems (SMEDDS) containing structured triglycerides. *Eur J Pharm Sci.* 2003, 20: 91–7.

Holmberg C. and Siekmann B. (2010). Self-emulsifying drug delivery system. US 7736666.

Hu L., Xing Q., Meng J. and Shang C. Preparation and enhanced oral bioavailability of cryptotanshinone-loaded solid lipid nanoparticles. *AAPS PharmSciTech.* 2010, 11: 582–7.

Huang Y.B., Tsai M.J., Wu P.C., Tsai Y.H., Wu Y.H. and Fang J.Y. Elastic liposomes as carriers for oral delivery and the brain distribution of (+)-catechin. *J Drug Target.* 2011, 19: 709–18.

Hussain N., Jaitley V. and Florence A.T. Recent advances in the understanding of uptake of microparticulates across the gastrointestinal lymphatics. *Adv Drug Deliv Rev.* 2001, 50: 107–142.

Iqbal M.A., Md S., Sahni J.K., Baboota S., Dang S. and Ali J. Nanostructured lipid carriers system: recent advances in drug delivery. *J Drug Target.* 2012, 20: 813–30.

Jadon P.S., Gajbhiye V., Jadon R.S., Gajbhiye K.R. and Ganesh N. Enhanced oral bioavailability of griseofulvin via niosomes. *AAPS PharmSciTech.* 2009, 10: 1186–92.

Kale A.A. and Patravale V.B. Design and evaluation of self-emulsifying drug delivery systems (SEDDS) of nimodipine. *AAPS PharmSciTech.* 2008, 9: 191–6.

Kamboj S., Saini V., Magon N., Bala S. and Jhawat V. Vesicular drug delivery systems: A novel approach for drug targeting. *Int J Drug Deliv.* 2013, 5. 121–130.

Khoo S.M., Humberstone A.J., Porter C.J.H., Edwards G.A. and Charman W.N. Formulation design and bioavailability assessment of lipidic self-emulsifying formulations of halofantrine. *Int J Pharm.* 1998, 167: 155–164.

Kohli K., Chopra S., Dhar D., Arora S. and Khar R.K. Self-emulsifying drug delivery systems: an approach to enhance oral bioavailability. *Drug Discov Today.* 2010, 15: 958–65.

Kuentz M. Lipid-based formulations for oral delivery of lipophilic drugs. *Drug Discov Today Technol.* 2012, 9: e71–e174.

Lavelle J.P., Negrete H.O., Poland P.A., Kinlough C.L., Meyers S.D., Hughey R.P. and Zeidel M.L. Low permeabilities of MDCK cell monolayers: a model barrier epithelium. *Am J Physiol.* 1997, 273: F67–75.

Ling S.S.N., Yuen K.H., Magosso E. and Barker S.A. Oral bioavailability enhancement of a hydrophilic drug delivered via folic acid-coupled liposomes in rats. *J Pharm Pharmacol.* 2009, 61: 445–449.

Liu Y. and Zeng S. Advances in the MDCK-MDR1 cell model and its applications to screen drug permeability. *Yao Xue Xue Bao.* 2008, 43: 559–64.

Madin S.H., Andriese P.C. and Darby N.B. The in vitro cultivation of tissues of domestic and laboratory animals. *Am J Vet Res.* 1957, 18: 932–41.

Madin S.H. and Darby N.B., Jr. Established kidney cell lines of normal adult bovine and ovine origin. *Proc Soc Exp Biol Med.* 1958, 98: 574–6.

Mehnert W. and Mader K. Solid lipid nanoparticles: production, characterization and applications. *Adv Drug Deliv Rev.* 2001, 47: 165–96.

Mu H, Holm R and Mullertz A. Lipid-based formulations for oral administration of poorly water-soluble drugs. *Int J Pharm.* 2013; 453: 215-24.

Muchow M., Maincent P., Müller R.H. and Keck G.M. Testosterone undecanoate – increase of oral bioavailability by nanostructured lipid carriers (NLC). *J Pharm Technol Drug Res.* 2013, 2: 1–10.

Muller R.H., Radtke M. and Wissing S.A. Nanostructured lipid matrices for improved microencapsulation of drugs. *Int J Pharm.* 2002, 242: 121–8.

Nan Z., Lijun G., Tao V. and Dongqin Q. Evaluation of Carbamazepine (CBZ) Supersaturatable Self-Microemulsifying (S-SMEDDS) Formulation In-vitro and In-vivo. *Iranian J Pharm Res.* 2012, 11: 257–264.

Nasr M. In vitro and in vivo evaluation of proniosomes containing celecoxib for oral administration. *AAPS PharmSciTech.* 2010, 11: 85–9.

Niu M., Lu Y., Hovgaard L., Guan P., Tan Y., Lian R., Qi J. and Wu W. Hypoglycemic activity and oral bioavailability of insulin-loaded liposomes containing bile salts in rats: the effect of cholate type, particle size and administered dose. *Eur J Pharm Biopharm.* 2012, 81: 265–72.

O'Driscoll C.M. Lipid-based formulations for intestinal lymphatic delivery. *Eur J Pharm Sci.* 2002, 15: 405–415.

Obitte N.C., Chukwu A., Onyishi V.I. and Obitte B.C.N. The physicochemical evaluation and applicability of landolphia owariensis latex as a release modulating agent in its admixture with carbosil® in ibuprofen loaded self-emulsifying oil formulations. *Int J Appl Res Nat Prod.* 2010, 2: 27–43.

Perlman M.E., Murdande S.B., Gumkowski M.J., Shah T.S., Rodricks C.M., Thornton-Manning J., Freel D. and Erhart L.C. Development of a self-emulsifying formulation that reduces the food effect for torcetrapib. *Int J Pharm.* 2008, 351: 15–22.

Porter C.J. and Charman W.N. In vitro assessment of oral lipid based formulations. *Adv Drug Deliv Rev.* 2001, 50: S127–47.

Porter C.J. and Charman W.N. Intestinal lymphatic drug transport: An update. Adv Drug Deliv Rev. 2001, 50: 61–80.

Porter C.J., Pouton C.W., Cuine J.F. and Charman W.N. Enhancing intestinal drug solubilisation using lipid-based delivery systems. *Adv Drug Deliv Rev.* 2008, 60: 673–91.

Porter C.J., Trevaskis N.L. and Charman W.N. Lipids and lipid-based formulations: optimizing the oral delivery of lipophilic drugs. *Nat Rev Drug Discov.* 2007, 6: 231–48.

Pouton C.W. Formulation of poorly water-soluble drugs for oral administration: Physicochemical and physiological issues and the lipid formulation classification system. *Eur J Pharm Sci.* 2006, 29: 278–287.

Quan DQ, Xu GX and Wu XG. Studies on preparation and absolute bioavailability of a self-emulsifying system containing puerarin. *Chem Pharm Bull.* 2007; 55: 800-3.

Radtke M., Souto E.B. and Muller R.H. Nanostructured lipid carriers: A novel generation of solid lipid drug carrier. *Pharm Tech.* 2005, 17: 45–50.

Rahman M.A., Harwansh R., Mirza M.A., Hussain S. and Hussain A. Oral lipid based drug delivery system (LBDDS): formulation, characterization and application: a review. *Curr Drug Deliv.* 2011, 8: 330–45.

Ruan L.P., Chen S., Yu B.Y., Zhu D.N., Cordell G.A. and Qiu S.X. Prediction of human absorption of natural compounds by the non-everted rat intestinal sac model. *Eur J Med Chem.* 2006, 41: 605–10.

Schwarz J. (2004). Solid self-emulsifying dosage form for improved delivery of poorly soluble hydrophobic compounds and the process for preparation thereof. US 20030072798

Seo Y.G., Kim D.W., Cho K.W., Yousaf A.M., Kim D.S., Kim H., Kim J.O., Yong C.S. and Choi H.G. Preparation and pharmaceutical evaluation of new tacrolimus-loaded solid self-emulsifying drug delivery system. *Arch Pharm Res.* 2014, 38: 223–8.

Sha X., Yan G., Wu Y., Li J. and Fang X. Effect of self-microemulsifying drug delivery systems containing Labrasol on tight junctions in Caco-2 cells. *Eur J Pharm Sci.* 2005, 24: 477–486.

Sha X.Y. and Fang X.L. Effect of self-microemulsifying system on cell tight junctions. *Yao Xue Xue Bao.* 2006, 41: 30–5.

Shangguan M., Lu Y., Qi J., Han J., Tian Z., Xie Y., Hu F., Yuan H. and Wu W. Binary lipids-based nanostructured lipid carriers for improved oral bioavailability of silymarin. *J Biomater Appl.* 2013, 28: 887–96.

Sharma P., Varma M.V., Chawla H.P. and Panchagnula R. In situ and in vivo efficacy of peroral absorption enhancers in rats and correlation to in vitro mechanistic studies. *Farmaco.* 2005, 60: 874–83.

Shlieout G., Boedecker B., Schaefer S., Thumbeck B. and Gregory P. (2005). Pharmaceutical compositions of lipase-containing products, in particular of pancreation.

Shukla A., Katare O.P., Singh B. and Vyas S.P. M-cell targeted delivery of recombinant hepatitis B surface antigen using cholera toxin B subunit conjugated bilosomes. *Int J Pharm.* 2009, 385: 47–52.

Shukla A., Singh B. and Katare O.P. Significant systemic and mucosal immune response induced on oral delivery of diphtheria toxoid using nano-bilosomes. *Br J Pharmacol.* 2011, 164: 820–7.

Singh B., Bandopadhyay S., Kapil R., Singh R. and Katare O. Self-emulsifying drug delivery systems (SEDDS): formulation development, characterization, and applications. *Crit Rev Ther Drug Carrier Syst.* 2009, 26: 427–521.

Singh B., Dehal S. and Kapil R. (2010). In vitro/in vivo correlations (IVIVC): Role in biowaivers and product development of drug delivery systems. Nanocolloidal Carriers: Site-Specific and Controlled Drug Delivery. S. P. Vyas. New Delhi (In press), CBS Publishers.

Singh B., Khurana L., Bandyopadhyay S., Kapil R. and Katare O.O. Development of optimized self-nano-emulsifying drug delivery systems (SNEDDS) of carvedilol with enhanced bioavailability potential. *Drug Deliv.* 2011, 18: 599–612.

Singh B., Singh R., Bandyopadhyay S., Kapil R. and Garg B. Optimized nanoemulsifying systems with enhanced bioavailability of carvedilol. *Colloids Surf B Biointerfaces.* 2012, 101: 465–474.

Singh P., Prabakaran D., Jain S., Mishra V., Jaganathan K.S. and Vyas S.P. Cholera toxin B subunit conjugated bile salt stabilized vesicles (bilosomes) for oral immunization. *Int J Pharm.* 2004, 278: 379–90.

Singh S.K., Verma P.R. and Razdan B. Development and characterization of a lovastatin-loaded self-microemulsifying drug delivery system. *Pharm Dev Technol.* 2010, 15: 469–83.

Singh S.K., Verma P.R.P. and Razdan B. Development and characterization of a carvedilol-loaded self-microemulsifying delivery system. *Clin Res Reg Affairs.* 2009, 26: 50–64.

Singh V., Khullar P., Dave P.N. and Kaur N. Micelles, mixed micelles, and applications of polyoxypropylene (PPO)-polyoxyethylene (PEO)-polyoxypropylene (PPO) triblock polymers. *Int J Ind Pharm.* 2013, 4: 12–18.

Subramanian N., Ray S., Ghosal S.K., Bhadra R. and Moulik S.P. Formulation design of self-microemulsifying drug delivery systems for improved oral bioavailability of celecoxib. *Biol Pharm Bull.* 2004, 27: 1993–9.

Sun M., Wang S., Nie S. and Zhang J. Enhanced oral bioavailability of quercetin by nanostructured lipid carriers. *The FASEB J.* 2014, 28: 24–31.

Swartz M.A. The physiology of the lymphatic system. *Adv Drug Deliv Rev.* 2001, 50: 3–20.

Takeuchi H., Matsui Y., Yamamoto H. and Kawashima Y. Mucoadhesive properties of carbopol or chitosan-coated liposomes and their effectiveness in the oral administration of calcitonin to rats. *J Control Release.* 2003, 86: 235–42.

Tan A., Rao S. and Prestidge C.A. Transforminglipid-based oral drug delivery systems into solid dosage forms: An overview of solid carriers, physicochemical properties, and biopharmaceutical performance. *Pharm Res.* 2013, 30: 2993–3017.

Tang B., Cheng G., Gu J.C. and Xu C.H. Development of solid self-emulsifying drug delivery systems: preparation techniques and dosage forms. *Drug Discov Today.* 2008, 13: 606–12.

Thi T.D., Van Speybroeck M., Barillaro V., Martens J., Annaert P., Augustijns P., Van Humbeeck J., Vermant J. and Van den Mooter G. Formulate-ability of ten compounds with different physicochemical profiles in SMEDDS. *Eur J Pharm Sci.* 2009, 38: 479–88.

Thomas N., Holm R., Garmer M., Karlsson J.J., Mullertz A. and Rades T. Supersaturated Self-Nanoemulsifying Drug Delivery Systems (Super-SNEDDS) Enhance the Bioavailability of the Poorly Water-Soluble Drug Simvastatin in Dogs. *AAPS J.* 2013, 15: 219–27.

Tran T.H., Ramasamy T., Cho H.J., Kim Y.I., Poudel B.K., Choi H.G., Yong C.S. and Kim J.O. Formulation and optimization of raloxifene-loaded solid lipid nanoparticles to enhance oral bioavailability. *J Nanosci Nanotechnol.* 2014; 14: 4820–31.

Tran T.H., Ramasamy T., Truong D.H., Choi H.G., Yong C.S. and Kim J.O. Preparation and Characterization of Fenofibrate-Loaded Nanostructured Lipid Carriers for Oral Bioavailability Enhancement. *AAPS PharmSciTech.* 2014, 15: 1509–15.

Trevaskis N.L., Charman W.N. and Porter C.J.H. Lipid-based delivery systems and intestinal lymphatic drug transport: A mechanistic update. *Adv Drug Deliv Rev.* 2008; 60: 702–716.

Varshosaz J., Minayian M. and Moazen E. Enhancement of oral bioavailability of pentoxifylline by solid lipid nanoparticles. *J Liposome Res.* 2010, 20: 115–23.

Veerareddy P.R. and Bobbala S.K. Enhanced oral bioavailability of isradipine via proniosomal systems. *Drug Dev Ind Pharm.* 2012, 39: 909–17.

Wang Y., Sun J., Zhang T., Liu H., He F. and He Z. Enhanced oral bioavailability of tacrolimus in rats by self-microemulsifying drug delivery systems. *Drug Dev Ind Pharm.* 2011, 37: 1225–30.

Wei Y., Ye X., Shang X., Peng X., Bao Q., Liu M., Guo M. and Li F. Enhanced oral bioavailability of silybin by a supersaturatable self-emulsifying drug delivery system (S-SEDDS). Colloids and Surfaces A: *Physicochem. Eng. Aspects.* 2012, 396: 22–28.

Weisspapir M. and Schwarz J. (2000). Solid self-emulsifying controlled release drug delivery system composition for enhanced delivery of water insoluble phytosterols and other hydrophobic natural compounds for body weight and cholestrol level control. US 20020103139.

Yan Y.D., Kim J.A., Kwak M.K., Yoo B.K., Yong C.S. and Choi H.G. Enhanced oral bioavailability of curcumin via a solid lipid-based self-emulsifying drug delivery system using a spray-drying technique. *Biol Pharm Bull.* 2011; 34: 1179–86.

Yang L., Geng Y., Li H., Zhang Y., You J. and Chang Y. Enhancement the oral bioavailability of praziquantel by incorporation into solid lipid nanoparticles. *Pharmazie.* 2009, 64: 86–9.

Yao J., Lu Y. and Zhou J.P. Preparation of nobiletin in self-microemulsifying systems and its intestinal permeability in rats. *J Pharm Sci.* 2008, 11: 22–9.

Zhang T., Chen J., Zhang Y., Shen Q. and Pan W. Characterization and evaluation of nanostructured lipid carrier as a vehicle for oral delivery of etoposide. *Eur J Pharm Sci.* 2011, 43: 174–9.

Zhang X., Qiao H., Zhang T., Shi Y. and Nia J. Enhancement of gastrointestinal absorption of isoliquiritigenin by nanostructured lipid carrier. *Adv Powder Technol.* 2014; 25: 1060–1068.

Zhou L., Chen Y., Zhang Z., He J., Du M. and Wu Q. Preparation of tripterine nanostructured lipid carriers and their absorption in rat intestine. *Pharmazie.* 2012; 67: 304–10.

Zhou Y., Ning Q., Yu D., Li W. and Deng J. Improved oral bioavailability of breviscapine via a Pluronic P85-modified liposomal delivery system. *J Pharm Pharmacol.* 2014, 66: 903–911.

3

Soy protein-based nutraceuticals delivery systems

Himansu Bhusan Samal[1]*, Chinam Niranjan Patra[2], Suryakanta Swain[2] and Sriram Ashwath Narayan Setty Sreenivas[1]
[1]Department of Pharmaceutics, Guru Nanak Institutions Technical Campus- School of Pharmacy, Ibrahimpatnam, Hyderabad-501506, India.
[2]Department of Pharmaceutics, Roland Institute of Pharmaceutical Sciences, Berhampur, Odisha-760010, India.

3.1   Introduction

Nutraceutical delivery through oral route is considered to be the most acceptable and preferred route as it follows the same natural process of food and nutrient consumption in the body, is non-invasive, and involves neither special technique nor complex instructions. Functional foods and food supplements enriched with nutraceutical compounds are effective in preventing chronic diseases such as diabetes, obesity, cardio-vascular disease, inflammation, cancer. Functional foods are those that when consumed regularly exert a specific health-beneficial effect beyond their nutritional properties (a healthier status or a lower risk of disease). Dr Stephen DeFelice coined the term nutraceutical from "Nutrition" and "Pharmaceuticals" in 1989 and defined as any substance that may be considered a food or part of a food and provides medical or health benefits including the prevention and treatment of disease [1]. According to the International Food Information Council (IFIC), functional foods are "foods or dietary components that may provide a health benefit beyond basic nutrition" [13]. The International Life Sciences Institute of North America (ILSI) has defined functional foods as "Foods that by virtue of physiologically active food components provide health benefits beyond basic nutrition" [3]. Health Canada defines functional foods as "similar in appearance to a conventional food, consumed as part of the usual diet, with demonstrated physiological benefits, and/or to reduce the risk of chronic disease beyond basic nutritional functions." Nutraceuticals may range from isolated nutrients, dietary supplements and diets to genetically engineered "designer" foods, herbal products and processed products such as cereals, soups and beverages. Incorporation of nutraceutical compounds such as vitamins, probiotics, bioactive peptides and antioxidants, etc., into food systems provide a simple way to develop innovative functional foods. Effectiveness of nutraceutical products in preventing diseases depends on preserving the bioavailability of the active ingredients. This represents a formidable challenge, given that only a small proportion of molecules remain available following oral administration, due to insufficient gastric residence time, low permeability and/or solubility within the gut, as well as instability under conditions encountered in food processing (temperature, oxygen, light) or in the gastro-intestinal (GI) tract (pH, enzymes, presence of other nutrients), all of which limit the activity and potential health benefits of nutraceutical molecules. Delivery of these molecules will therefore require food formulators and manufacturers to provide protective mechanisms that (1) maintain the active molecular form until the time of consumption and (2) deliver this form to the physiological target within the organism [4].

As a vital macronutrient in food, protein possesses unique functional properties which allow them to be an ideal material for encapsulation of nutraceutical compounds. Food biopolymers, specifically food proteins, are widely used in formulated foods because they have high nutritional value and are generally recognized as safe (GRAS) and possess unique functional properties such as gelation, emulsification, foaming, water binding capacity and stabilize food texture. Among the food proteins, Soy proteins are used extensively in food manufacturing, because of their functional properties, low cost, availability, and high nutritional value and degradable by digestive enzyme [4].

Soybean (Glycine max L.) is currently one of the most abundant sources of plant proteins. The enriched form of soy protein, known as soy proteins isolate (SPI), has been reported to have high nutritional values and ingredient functionalities. Soy protein Isolate (SPI) is an important component of soybeans and provides an abundant source of dietary protein. Soy protein is unique among the plant-based protein because it is associated with isoflavone, a group of compounds with a variety of biological properties that may potentially benefit human health [5]. Soy protein is considered a complete protein that it contains most of the essential amino acids that are found in animal proteins. The nutritional value of soy protein is roughly equivalent to that of animal protein of high biological value. It is composed almost exclusively of two globular proteins called 7S (β-conglycinin, MW = 180,000, ~40%) & 11S (glycinin, MW = 360,000, ~60%) which comprise approximately 80% of the total protein. Glycinin has isoelectric point 4.90 and β-conglycinin has isoelectric point 4.64. Typical Composition of SPI is given in Table 3.1. SPI possesses a balanced composition of polar, nonpolar, and charged amino acids, allowing a variety of drugs to be incorporated. In an aqueous environment, glycinin and β-conglycinin exist as globular structures consisting of a hydrophilic shell and hydrophobic kernel, together with a certain amount of small water-soluble aggregates. Upon addition of dissolvent or crosslinking agents, SPI molecules continue to aggregate and form various structures such as microspheres, hydrogels and polymer blends. Soy protein nanoparticles can be prepared either from a freshly prepared SPI by desolvation or from the glycinin fraction of defatted soy flour extraction using a simple coacervation method. Gel property makes the Soy protein an ideal coating material for the encapsulation of bioactive compounds in the form of hydrogel, Tablet and Microspheres. Gels of diverse mechanical and microstructural properties can be formed by controlling the assembly of protein molecular chains, thus offering the possibility of developing GRAS biocompatible carriers for oral administration of sensitive nutraceuticals in a wide variety of foods. Protein hydrogels are undoubtedly the most convenient and widely used matrix in food applications. Soy protein hydrogel has the ability to protect the nutraceutical compounds from hostile environment and to deliver them in response to environmental stimuli such as pH and temperature. Soya protein acts as a substrate for the development of nutraceutical delivery systems. Various strategies have been developed to protect and deliver bioactive molecules such as Protein hydrogel, microparticles, nanoparticles, etc. [6].

3.2   Hydrogel

Biodegradable hydrogels have been widely researched to carry, protect and modify the delivery of a wide variety of pharmaceutical compounds including nutraceuticals. Over the past decade, hydrogels have been studied extensively in biomedical and pharmaceutical applications, due primarily to their ability to protect drugs from hostile environments and to deliver them in response to environmental stimuli such as pH and temperature. A hydrogel is an infinite water-swollen network of hydrophilic polymers that can swell in water and hold a large amount of water while maintaining a network structure. A three-dimensional network is formed by cross-linking polymer chains through covalent bonds, hydrogen bonding, van der Waals interactions, or physical entanglements. New synthetic methods have been used to prepare hydrogels for a wide range of drug delivery applications. Although successful as oral drug delivery systems, one of the inherent limitations of these hydrogels for food applications is that they contain components that are not generally recognized as safe for regular consumption by healthy individuals. In the food industry, the use of food proteins to develop environment-sensitive hydrogels for nutraceutical delivery constitutes an interesting strategy. A fundamental advantage of this approach is that nutraceutical carrier gels can stabilize food texture, which is a highly desirable characteristic in the manufacturing of food products. Moreover, the presence of acidic (e.g. carboxylic) and basic (e.g. ammonium) groups in polypeptide chains, which either accept or release protons in response to changes in the pH of the medium (acidic in the stomach and neutral in the intestine), allows the release rate of molecules from protein gels to be modulated by pH variations. Gelation of food proteins and particularly of globular proteins e.g. soybean has attracted much attention over the years because of its physicochemical and industrial significance. It is traditionally achieved through heat treatment. Thermal gels are produced by the unfolding of polypeptide chains with concomitant exposure of initially buried hydrophobic amino acid residues and subsequent self-aggregation of protein molecules into a three-dimensional network that entraps water by capillary forces. Forces involved in the aggregation process include hydrophobic effects, van der Waals, hydrogen bonding, and covalent interactions. Depending on preparation technique, gels can exhibit different microstructural properties, which are strongly related to aggregate molecular structure.

However, the heat needed to produce these gels limits their application to formulations that do not contain heat-sensitive ingredients. It has been shown that cold-induced gelation of globular soybean proteins can be achieved by adding Ca^{+2} ions to a preheated protein suspension. This method requires a heating step during which proteins are denatured and polymerized into soluble aggregates, followed by a cooling step and subsequent salt addition, which results in the formation of a network via Ca+2 mediated interactions of soluble aggregates. The formation of cold-set gels opens interesting opportunities for food proteins as carriers of sensitive nutraceutical compounds and in the development of innovative functional food ingredients. One advantage of this approach is exposure of multiple functional groups within the protein upon denaturing, which could be exploited to create different interactions between nutraceutical compounds and polypeptide chains such as (i) hydrogen bonding, (ii) hydrophobic interactions and (iii) electrostatic interactions, which may in turn be applied to target nutraceutical delivery [4].

Two types of gels could be obtained, depending on salt/ protein ratio, at a lower salt concentration, 'filamentous' gels composed of more or less flexible linear strands making up a regular network characterized by elastic behavior and high resistance to rupture are formed, while at a higher salt concentration, 'particulate' gels composed of large and almost spherical aggregates characterized by less elastic behavior and lower rupture resistance are obtained. These different microstructural properties are strongly linked to aggregate molecular structure, as revealed by Fourier transform infrared spectroscopy and rheological data. The filamentous form is created by linear aggregation of structural units maintained by hydrophobic interactions, whereas the aggregate form is produced by random aggregation of structural units essentially controlled by van der Waals forces. Modulation of gel microstructure and functional properties by cold gelation should allow tailoring of water-soluble delivery devices for nutraceutical and functional food system development [7, 8].

3.3   Microparticles

Protein-based Microparticles have found wide and rapidly increasing applications in the food industry because they can be precisely designed for use in many food formulations and virtually any ingredient can be encapsulated, whether hydrophobic, hydrophilic, or even microbial. A wide variety of processes have been developed to prepare protein-based micro (sub-micro) particles. The most common techniques are spray drying, emulsifying-cross linking or coacervation and cold-gelation. Soy protein isolate (SPI) use in microencapsulation has already been studied by various authors. SPI is generally used as an individual coating material, but can also be mixed with polysaccharides. The combination of proteins with carbohydrates as a carrier material favors better protection, oxidative stability and drying properties. Due to SPI hydrosolubility, microparticles are mainly produced using the spray-drying technique but coacervation and gelation have also been investigated [6]. For Soy protein isolate, microencapsulation is carried out by spray drying and coacervation technique [9–12]. The schematic representation of different absorption mechanisms of bioactive molecules are shown in Fig. 3.1.

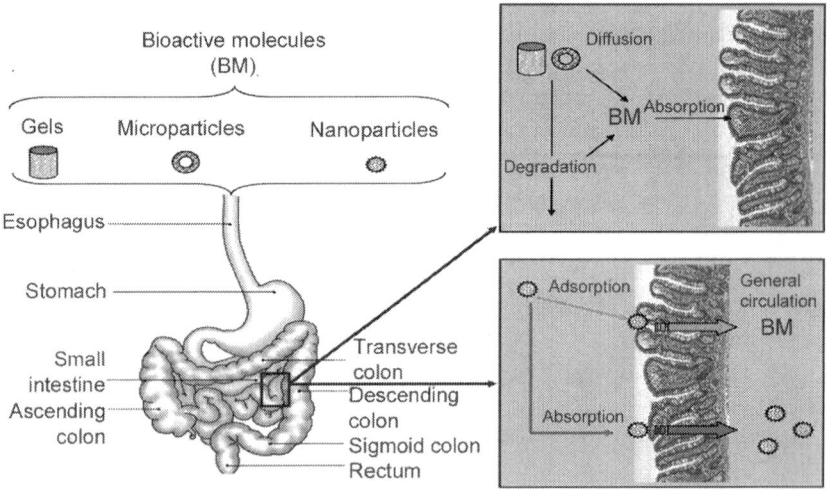

Figure 3.1 Schematic representation of different absorption mechanisms of bioactive molecules [4]

3.4   Nanoparticle

In the last decades, the growth of nanotechnology has opened several new possibilities in medical sciences, especially in the field of drug delivery. Different new drug carrier systems in the micro- and nanometer size range have been developed and the number of patents and products in the drug delivery field has increased tremendously. Various nanotechnology platforms are being investigated at either the developmental or clinical stages in order to obtain more effective and safer therapeutics for a myriad of clinical applications. Nanoparticles have increasingly been used for a variety of applications, most notably for the delivery of therapeutic and diagnostic agents. Natural biomolecules such as proteins are an attractive alternative to synthetic polymers which are commonly used in drug formulations because of their safety. In general, protein nanoparticles offer a number of advantages including biocompatibility and biodegradability. They can be prepared under mild conditions without the use of toxic chemicals or organic solvents. Moreover, due to their defined primary structure, protein-based nanoparticles offer various possibilities for surface modifications including covalent attachment of drugs and targeting ligands.

Nanoparticles are solid colloidal particles ranging in size from about 10 nm to 1000 nm. The major goal in designing nanoparticles as a delivery system is to control particle size, surface properties, and release of pharmacologically active agents in order to achieve the site-specific action of drugs at a therapeutically optimal rate and dosage regimen. Nanoparticle delivery systems offer certain distinct advantages for drug delivery. First, the particle size, particle morphology, and surface charge of nanoparticles can be controlled. Secondly, nanoscale drug delivery systems can carry or deliver a variety of therapeutic and diagnostic agents such as small molecules (hydrophilic or hydrophobic), peptides, proteins, and nucleic acids while releasing the active molecules in a controlled manner. The entrapped molecules can be released from the nanocarriers in a precise manner over time to maintain drug concentrations within a therapeutic window, or they can be triggered to be released by some stimuli unique to the delivery site. Thirdly, these nanocarriers can improve the solubility and stability of encapsulated drugs, providing an opportunity to reevaluate drug candidates that were previously ignored because of poor pharmacokinetics. Lastly, site-specific drug delivery can be achieved using nanoparticles delivered through various routes of administration. The nanocarriers can be engineered to have a prolonged circulation time or to have enhanced cellular uptake and targeting abilities.

The development of nanoparticle-based drug delivery systems is rapidly growing due to their great therapeutic potential. Various types of materials including polymers, lipids, polysaccharides, and proteins have been explored as drug delivery carriers. The selection of nanoparticle materials is dependent on many factors including (a) the size of nanoparticles needed, (b) inherent properties of the drug such as aqueous solubility and stability, (c) drug release profile desired, (d) surface charge and hydrophobicity of nanoparticles, (e) biocompatibility and biodegradability of nanomaterials, and (f) antigenicity and toxicity of the product. Biopolymer-based nanoparticles including protein nanoparticles have gained considerable interest in recent years due to their many desirable properties such as low toxicity and biodegradability. They are actively being developed for both pharmaceutical and nutraceutical delivery.

Proteins are a class of natural molecules that have unique functionalities and potential applications in both biomedical and material sciences. They are deemed as ideal materials for nanoparticle preparation because of their amphiphilicity which allows them to interact well with both the drug and solvent. Nanoparticles derived from natural proteins are biodegradable, metabolizable, and are easily amenable to surface modifications to allow attachment of drugs and targeting ligands. Preparation of protein nanoparticles is based on balancing the attractive and repulsive forces in the protein. It is generally accepted that increasing protein unfolding and decreasing intramolecular hydrophobic interactions are crucial to the formation of protein nanoparticles. During such particle formation, the protein undergoes conformational changes depending on its composition, concentration, crosslinking, and preparation conditions such as pH, ionic strength, and type of solvent. Unfolding of proteins during the preparation process exposes interactive groups such as disulfides and thiols. Subsequent thermal or chemical crosslinking leads to the formation of cross-linked nanoparticles with entrapped drug molecules. Coacervation/desolvation and emulsion-based methods are most commonly used for the preparation of protein nanoparticles. Coacervation or desolvation is based on the differential solubility of proteins in solvents as a function of solvent polarity, pH, ionic strength, and presence of electrolytes. The coacervation process reduces the solubility of the protein leading to phase separation. The addition of desolvating agent leads to conformation changes in protein structure resulting in coacervation or precipitation of the protein. By controlling processing variables, the size of nanoparticles in the coacervate can be controlled. After nanoparticles are formed, they are cross-linked by agents such as glutaraldehyde and glyoxal. Protein nanoparticles can be rigidized by crosslinking. An increase in the degree of crosslinking generally decreases the particle size due to the formation of denser particles. Organic solvents such as acetone and ethanol have been used as antisolvents for the preparation of protein nanoparticles. Crosslinking stabilizes the protein nanoparticles and reduces enzymatic degradation and drug release from the nanoparticles. Drugs can be loaded into particles by surface adsorption or by entrapping the drugs in the particles during the preparation process [4].

Table 3.1 Typical composition of Soya Protein Isolate (Moisture-free basis) [12]

Components	% Weight
Protein	90.0
Fat	0.5
Ash	4.5
Total Carbohydrate	0.3

Table 3.2 Soya Protein as carrier for nutraceutical compounds [6–9]

Wall material	Core material (Nutraceuticals)	Dosages form	References
Soy protein Isolate	Riboflavin	Tablets and Hydrogel	Anne Maltais, Gabriel E. Remondetto, Muriel Subirade (2009)
Soy protein Isolate	Riboflavin	Microspheres	Chen L, Subirade M. (2009)
Soy protein Isolate	Curcumin	Tablet	Arun Tapal, Purnima Kaul Tiku (2012)

Table 3.3 Microencapsulation with SPI based wall Materials [12]

Microencapsulation Process	Wall Materials	Core Materials	References
Spray-drying	SPI	Orange oil	Kim et al. (1996)
Spray-drying	SPI	Flavors	Chavre and Reineccius (2009)
Spray-drying	SPI	Casein hydrolysate	Ortiz et al. (2009)
Spray-drying	SPI	α- Tocopherol	Nesterenko et al. (2012)
Spray-drying	SPI	Fish oil	Gan et al. (2008)
Spray-drying	SPI	Riboflavin	Chen L, Subirade M. (2009)

3.5   Conclusions

Delivery of nutraceuticals through oral route is considered to be the most acceptable and preferred route. It follows the same natural process of food and nutrient consumption in the body. This processing of soy protein does not require any special technique. Soy protein can be used as nanoparticles, coating materials for the encapsulation of bioactive compounds, hydrogel in the formulation of tablet and microspheres. Hence soy protein based nutraceutical delivery system has excellent potential.

3.6　　References

[1] Brower V. Nutraceuticals: poised for a healthy slice of the healthcare market? *Nat Biotechnol.*, 1998, 16:728–731.

[2] Miller E.G., Gonzales-Sanders A.P., Couvillon A.M., Binnie W.H., Hasegawa S., Lam L.K.T., (1994).Citrus limonoids as inhibitors of oral carcinogenesis. seven citrus compounds were tested for their cancer-preventive activity. *Food Technol.* 48: 110–114.

[3] Fong C.H., Hasegawa S., Herman Z. and Ou P. (1990). Liminoid glucosides in commercial citrus juices, *J. Food Sci*, 54:1505–1506.

[4] Chen L., Remondetto G.E., Subirade M. (2006). Food protein-based materials as nutraceutical delivery systems. *Trends Food Sci Tech,* 17:272–83.

[5] Chen L., Remondetto G.E., Subirade M. (2006). Food protein-based materials as nutraceutical delivery systems. *Trends Food Sci Tech*, 17:272–283.

[6] Chen L. and Subirade M. (2009). Elaboration and characterization of soy/zein protein microspheres for controlled nutraceutical delivery, *Biomacromolecules*, 3327–3334.

[7] Maltais A., Subirade M. (2009). Soy protein cold-set hydrogels as controlled delivery devices for Nutraceutical compounds, Food Hydrocolloids, 1647–1653.

[8] Maltais A., Subirade M. (2010). Tabletted soy protein cold-set hydrogels as carriers of Nutraceuticals substances, Food Hydrocolloids, 518–524.

[9] Tapal A., Tiku P.K. (2012) Complexation of curcumin with soy protein isolate and its implications on solubility and stability of curcumin, *Food Chemistry*, 960–965.

[10] Liang L., Line V.L.S., Remondetto G.E., Subirade M. (2012). In vitro release of a-tocopherol from emulsion-loaded β-lactoglobulin gels. *International Dairy Journal* 20:176–181.

[11] Chen L., Subirade M. (2006). Alginate-whey protein granular microsphere as oral delivery vehicle for bioactive compounds, *Biomaterials*, 4646–4654.

[12] Nesterenko A., Alric I., Silvestre F., Durrieu V. (2013). Vegetable proteins in microencapsulation: a review of recent interventions and their effectiveness. *Industrial Crops Products,* 42: 469–471.

Spherical crystallization techniques

Chinam Niranjan Patra*, Suryakanta Swain, Kahnu Charan Panigrahi and Muddana Eswara Bhanoji Rao

Department of Pharamceutics, Roland Institute of Pharmaceutical sciences, Berhampur-760 010, Odisha, India.

4.1   Introduction

In the earlier period, the pharmaceutical industry did not realize the need to improve the manufacturing efficiency. The straightforward reason was pharmaceutical industry focused to bring new products to the market. Now-a-days, rising of the energy prices have made pharmaceutical companies to look for alternatives to cut expenditures. Direct compression of pharmaceutical materials is a modern method in tablet manufacturing. Such manufacturing of tablets involve simple mixing and compression of powders, which results in a number of overall benefits including time, cost and energy savings. However, the use of this technique depends on the following:

- Particle size and size distribution of materials.
- Flowability of the crystalline active pharmaceutical ingredient (APIs), consistent with the high-speed tablet press.
- Bulk density and compressibility of crystalline APIs.

The use of directly compressible vehicles may promote direct compression but this might not be favorable in terms of desired flowability. But drug crystals prepared by spherical crystallization technique may exhibit good flowability and compressibility. Kawashima developed spherical crystallization technique, a novel agglomeration technique to produce spherical crystals with controlled properties [1].

4.1.1   What is spherical crystallization technique?

Spherical crystallization is a novel particle engineering technique occurring in liquid system. In this system crystallization and agglomeration can be carried out simultaneously in one step. The production of spherical crystals has recently gained great attention and importance due to the fact that the crystal habit (form, surface, size and particle size distribution) can be modified during the crystallization process. In consequence of such modifications in the crystal habit, certain parameters of materials can also be changed: bulk density, flow property, compactibility, dissolution rate, stability, etc.

Advantages of spherical crystallization technique

1. Improvement in flowability and compressibility of crystalline drugs.
2. Enabling crystalline form of a drug to be converted into different polymorphic forms and thus attain better bioavailability.
3. Masking the bitter taste of drugs.
4. Improvement in solubility and bioavailability of drugs.
5. Improvement in wettability.

Disadvantages

1. Presence of residual solvent in the product.
2. Scaling up from laboratory scale to production scale.

4.2   Techniques of spherical crystallization

Spherical crystals can be obtained by two different techniques, either by typical spherical crystallization technique or non-typical spherical crystallization technique [2] as shown in Fig. 4.1.

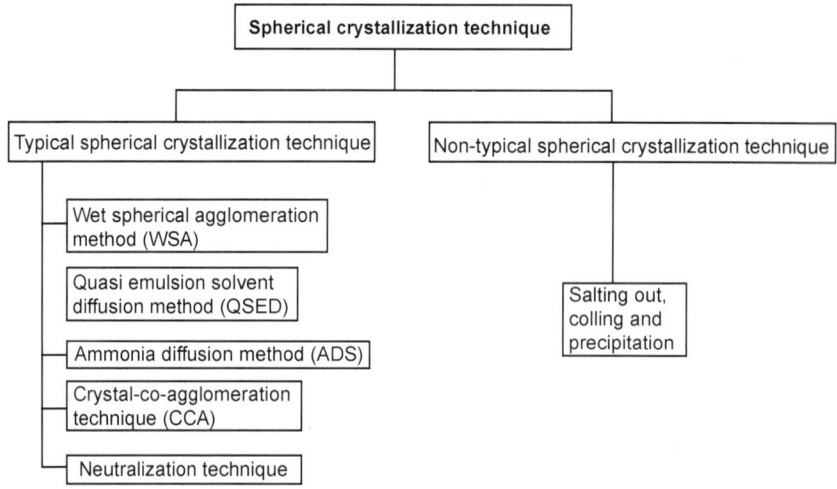

Figure 4.1 Techniques of spherical crystallization

The two most commonly used techniques of typical spherical crystallization are wet spherical agglomeration method (WSA) and quasi-emulsion solvent diffusion method (QESD, Transient emulsion) [2, 3]. But there are two extensions of these techniques: ammonia diffusion system (ADS) and crystal-co-agglomeration technique (CCA) [4, 5]. Another technique of this process is neutralization, where fine crystals are formed initially by neutralization then it will agglomerate by the help of a bridging liquid [6]. Non-typical spherical crystallization technique can also be considered as the traditional crystallization process (salting-out, cooling, precipitation, etc). This process is carried out by controlling the physical and chemical factors [7].

4.2.1   Typical spherical crystallization techniques

This system primarily requires three solvent systems, i.e. a good solvent dissolving the selected API, a poor solvent for precipitating the API and a bridging liquid which is immiscible with the solvent system but have an affinity with API crystals. The bridging liquid can be added to the poor solvent or to the good solvent.

Effect of bridging liquid

It has been found that the product properties are quite sensitive to the amount of the bridging liquid [8]. With decreasing amount of bridging liquid in the three-solvent system, the median diameter of agglomerated crystals increased, having a wider size distribution [9]. Less than the optimum amount of bridging liquid produces plenty of fines and more than optimum produces very coarse particles. So the amount of bridging is the important process parameters in crystallization process [10]. It is also reported that if the quantity of bridging liquid is too small, there is no significant agglomeration and when too much bridging liquid is used, the agglomerates become soft and pasty [11].

Effect of solvent

It was observed that the solvent has a significant effect on the properties of the spherical crystals. The median diameter of agglomerates decreased with increasing content of good solvent [9]. Whereas Zhang and coworkers have recently reported that the mean particle size of cefotaxime sodium spherical agglomerates increased with an increase of the poor solvent chloroform content in the crystallization system, and at the same time the particle size distribution became narrower, which can be ascribed to higher supersaturation and more effective crystallization and agglomeration [11].

Effect of stirring speed

It has been shown that increasing stirring speed makes the agglomeration process less efficient due to the shearing rate, disruptive forces, and higher probability of agglomerate collisions, which preferentially tear them apart. It has also been demonstrated that higher stirring speed at the same time results in decreased porosity and greater mechanical resistance of the agglomerates produced [12]. Less than optimum speed of rotation of stirrer produced agglomerates with higher diameter. Correspondingly more than optimum speed of rotation may produce irregularly shaped agglomerates of smaller diameter with rough surface. This could be due to high shear force of stirrer. Hence optimization of stirring speed also is a critical parameter in spherical crystallization technique to obtain agglomerates with good sphericity and flowability.

Effect of temperature

Temperature has shown variable effect in different cases. In one study, the temperature had no notable impact on the size of agglomerates; while in another study [13] higher temperature resulted in smaller particle size and increased density of agglomerates.

4.2.1.1   Wet spherical agglomeration (WSA) method

In this method, the good and the poor solvents are freely miscible and interaction (binding force) between the solvents is stronger than the interaction of API with the good solvent, which leads to precipitation of crystals immediately. Bridging liquid collects the crystals suspended in the system by forming liquid bridges between the crystals due to capillary negative pressure and interfacial tension between the interface of solid and liquid. WSA method proceeds in three steps as shown in Fig. 4.2A.

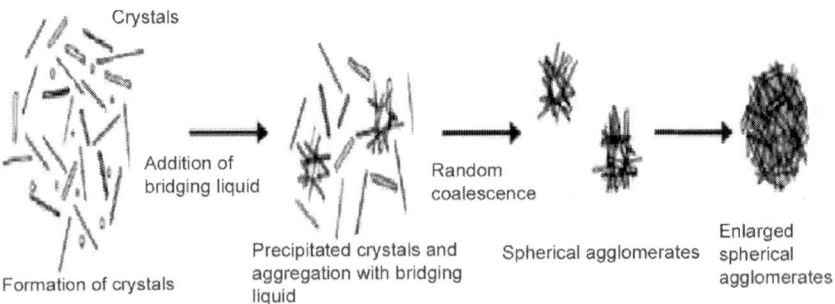

Figure 4.2A Wet Spherical Agglomeration method (Adopted from [14])

Step 1

The first one is the selection of crystallization method to precipitate crystals from solution, i.e., thermal method (temperature decrease or evaporation), physicochemical methods (addition of another solvent, salting out) and chemical reaction.

Step 2

The second step is the choice of the bridging liquid that will be immiscible with the solvent of crystallization.

Step 3

Finally, the third step is the hardening of the agglomerates.

Chow *et al.* postulated some general guidelines for the spherical agglomeration of drugs [15].

1. For compounds that are water soluble, a water immiscible organic solvent is used as the external medium and salt solutions of high concentration without common ions can be used as the bridging liquid.

2. For compounds that are soluble in one or more organic solvents, water is employed as the external phase and a water-immiscible organic solvent as the bridging liquid.

3. For compounds that are only soluble in water miscible organic solvents, a saturated aqueous solution of the compound can serve as the external phase and an organic solvent mixture as the bridging solvent.

4. For compounds that are insoluble in water or any organic solvents, a water-immiscible organic solvent can act as the external phase and a 20% calcium chloride solution as the bridging liquid.

5. In addition, a binding agent such as polyvinylpyrolidone (PVP) or polyethylene glycol (PEG) is required for agglomeration since the powders are not sufficiently soluble in the bridging liquids to allow binding through recrystallization and fusion.

Various drugs formulated as spherical crystals by WSA are summarized in Table 4.1.

The process of WSA can be explained with one practical example [16] as shown in Fig. 4.2B. For the model drug aceclofenac, acetone, distilled water and DCM were selected as good solvent, bad solvent and bridging liquid, respectively. In order to impart strength and sphericity besides increasing the solubility and drug release from aceclofenac agglomerates, hydroxypropyl methyl cellulose (HPMC) was selected.

Table 4.1 Reported literature on Wet Spherical Agglomeration (WSA) method

Drug	Solvent system			Results improvement in	References
	Good solvent	Bad solvent	Bridging liquid		
Acetyl salicylic acid	Ethanol	Water	Carbon tetrachloride	flowability, compressibility and cohesivity	[34]
Ascorbic acid	Purified water	Ethyl Acetate	Ethyl Acetate	micromeritic properties	[35]
Aceclofenac	Acetone	Water	DCM	micromeritic properties, solubility, dissolution and stability	[16]
Aceclofenac	DCM	Water	Acetone	micromeritic properties, solubility, dissolution and stability	[36]
Benzoic acid	Ethanol	Water	Chloroform	flowability and compressibility	[37]
Celecoxib	Acetone	Water	DCM	micromeritic properties, solubility and dissolution	[38]
Fenbufen	THF	Water	Isopropyl acetate	wettability and dissolution	[39]
Flurbiprofen	Acetone	Water	Hexane	micromeritic properties, wettability and dissolution.	[10]
Mebendazole	Acetone	Water	Hexane, Octanol, DCM	micromeritic properties, solubility and dissolution	[40]
Naproxane	Acetone-ethanol	Water	Chloroform	micromeritic properties, solubility, and dissolution	[41]
Propiphenazone	Ethyl alcohol	Water	Isopropyl acetate	better compressibility	[42]
Tranilast	Acetone	Water	DCM	micromeritic properties and dissolution	[43]
Simvastatin	DCM	Water	Isopropyl acetate	dissolution rate	[44]

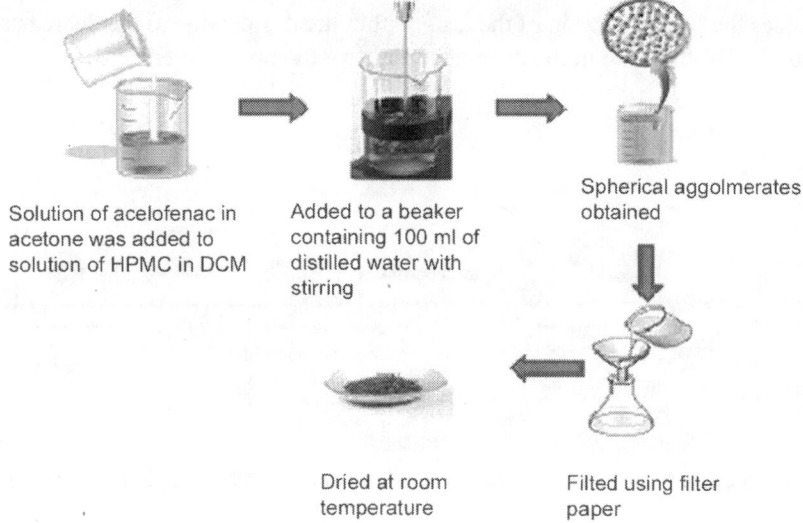

Solution of acelofenac in acetone was added to solution of HPMC in DCM

Added to a beaker containing 100 ml of distilled water with stirring

Spherical aggolmerates obtained

Dried at room temperature

Filted using filter paper

Figure 4.2B Spherical crystallization of aceclofenac by WSA method

A solution of aceclofenac in acetone (750 mg in 3 ml) was added to a solution of HPMC (50 mg) in DCM (1 mL). Drug was crystallized by adding the above solution to a 500 ml capacity beaker containing 100 mL of distilled water. The mixture was stirred continuously for a period of 0.5 h using a controlled speed stirrer (800 rpm) to obtain spherical agglomerates. The agglomerates were separated by filtration and dried at room temperature. The amount of DCM, speed of agitation and amount of polymer contributed significantly to get the agglomerates of desired properties. The authors reported that aceclofenac agglomerates and tablets exhibited excellent physicochemical and micromeritic properties, solubility, dissolution rate, stability and in vivo (preclinical and clinical) performance when compared with pure drug as well as marketed formulation besides exhibiting no preclinical toxicity.

4.2.1.2   Quasi-Emulsion Solvent Diffusion method (QSED)

This technique is usually applied for the preparation of microspheres [17] as shown in Fig. 4.3A. Here interaction between the drug and the good solvent is stronger than that of the good and poor solvents; hence, the good solvent drug solution is dispersed in the poor solvent, producing quasi emulsion droplets, even if the solvents are normally miscible [18]. This is because of an increase in the interfacial tension between good and poor solvent [19]. Then good solvent diffuses gradually out of the emulsion droplet into the outer poor solvent phase. The counter diffusion of the poor solvent into the droplet induces the crystallization of the drug within the droplet due to the decreasing solubility of the drug in the droplet containing the poor solvent.

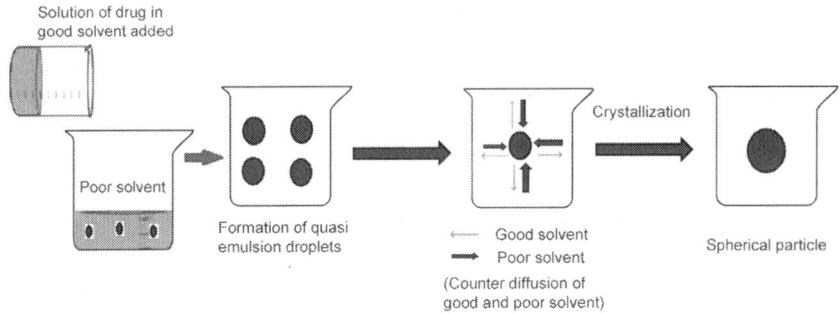

Figure 4.3A Steps involved in Quasi Emulsion Solvent Diffusion (QSED) method

The process of QSED can be explained with one simple practical example. Yang *et al.* [20] prepared spherical crystals of nitrendipine by QSED method employing acetone, DCM and water as good solvent, bridging liquid and poor solvent, respectively. Polymer (Eudragit RS) was used as a binding and release-retarding agent. Aerosil was used as an inert solid dispersing carrier to improve the dissolution rate of nitrendipine, and it also acts as an anti-adhesion agent to accelerate the solidification of droplets. Sodium deodecyl sulfate (SDS) assisted in the formation of stable droplets in the poor solvent. Nitrendipine (0.3 g) was dissolved with Eudragit RS (0.3–1.8 g) in a mixed solution of acetone (good solvent, 3.0–6.0 mL), and dichloromethane (bridging liquid, 2.0–4.0 ml). Then, aerosil (0.3–1.8 g) was suspended uniformly in the drug–polymer solution under vigorous agitation. The resultant drug–polymer–aerosil suspension was poured into 150 mL distilled water containing 0.02–0.15% of Sodium Deodecyl Sulfate (poor solvent) under agitation (400–700 rpm) and thermally controlled at 8–38°C. After agitating the system for 20 min, another 150 mL of poor solvent was added slowly and agitation was continued for another 40 min till the translucent quasi-emulsion droplets turned into opaque microspheres. The solidified microspheres were recovered by filtration, washed with water and dried in an oven at 50°C for 6 h. The resultant microspheres had the desired micromeritic properties. The above procedure is presented in Fig. 4.3B.

Various crystalline drugs reported by QSED method are enlisted in Table 4.2.

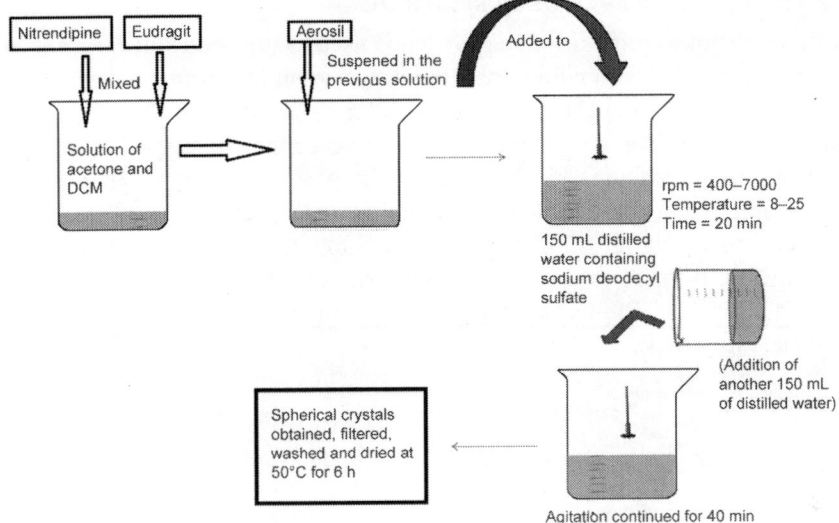

Figure 4.3B Spherical crystallization of Nitrendipine by QSED method

Table 4.2 Reported literature on Quasi-Emulsion Solvent Diffusion method

Drug	Solvent system			Improved properties	References
	Good solvent	**Bad solvent**	**Bridging liquid**		
Acebutalol HCl	Ethanol	Water	Isopropyl acetate	Micromeritics, solubility	[9]
Bucillamine	Ethanol	DCM	Water	compactibility	[45]
Carbamazepine	Ethanol	Water	Isopropyl acetate	Flowability and compactibility	[46]
Cefuroxime axetil	Acetone	Water	DCM	Solubility, wettability and dissolution	[3]
Griseofulvin	DCM	Water	DCM	Flowability, compressibility and solubility	[18]
Tolbutamide	Ethanol	Water	Isopropyl acetate	Particle shape and flowability	[19]
Nitrendipine	Acetone	Water	DCM	Compressibility, solubility, dissolution rate, bioavailability	[20]
Pioglitazone	Methanol	Water	chloroform	Micromeritics and bioavailability	[47]
Felodipine	Acetone	Water	DCM	Solubility	[48]
Carbamezapine	Ethanol	Water	Isopropyl acetate	Flowability, compressibility	[13]
Ketoprofen	Acetone	Water	Mowiol 8/88	Wetting and dissolution	[49]
Piroxicam	Ethanol	Water	DCM	Flowability, compressibility and dissolution	[50]

4.2.1.3   Ammonia diffusion system (ADS)

In this technique ammonia–water system is used as the good solvent and bad solvent is selected depending upon the drugs solubility in that solvent. The ammonia–water system also acts as a bridging liquid [21]. This technique usually intended for amphoteric drugs which cannot be agglomerated by conventional procedures [5, 22]. The whole process is completed in three stages. First, the drug dissolved in ammonia water is precipitated while the droplets collect the crystals (Fig. 4.4AI). Simultaneously, ammonia in the agglomerate diffuses to the outer organic solvent (Fig. 4AII). Its ability to act as a bridging liquid weakens and subsequently spherical agglomerates are formed (Fig. 4AIII).

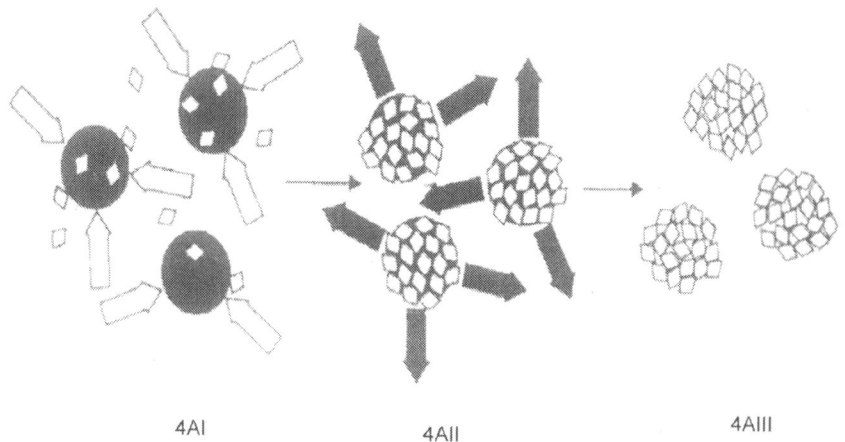

4AI   4AII   4AIII

Figure 4.4A Steps involved in Ammonia Diffusion (ADS) method (adopted from [14])

The process of ADS can be explained with one simple practical example. Bhadra *et al.* [23] prepared spherical crystals of mefenamic acid by ADS. Mefenamic acid was dissolved in a strong ammonia solution (25%) and mixed with 10 mL of acetone. Then chloroform was added with continuous stirring using a three-blade agitator in a cylindrical vessel. The agglomerated crystals were collected by filtration and washed with chloroform. Afterward, the crystals were dried at 50°C and stored in a tightly closed container. The effect of various parameters on the yield of spherical crystals was studied. These parameters included the amount of ammonia water used as a bridging liquid (10, 15, and 20 mL); the agitation speed (1000, 1500, and 2000 rpm); the agitation time (30, 60, and 90 min) and the concentration of ammonia in ammonia water (10%, 15%, 20%, and 25%). The effects were tested by changing one parameter at a time while keeping the other parameters constant. With 15 mL of ammonia water, uniform spherical agglomerates were formed. Similarly 1500 rpm produced uniform spherical agglomerates whereas lower and higher rpm produces larger and smaller agglomerates, respectively. They reported that the micromeritic properties of the agglomerated crystals such as flowability, packability, and compactibility were better than those of mefenamic acid powder. The above procedure is presented in the form of a flow chart in Fig. 4.4B. Various crystalline drugs reported by ADS method are enlisted in Table 4.3.

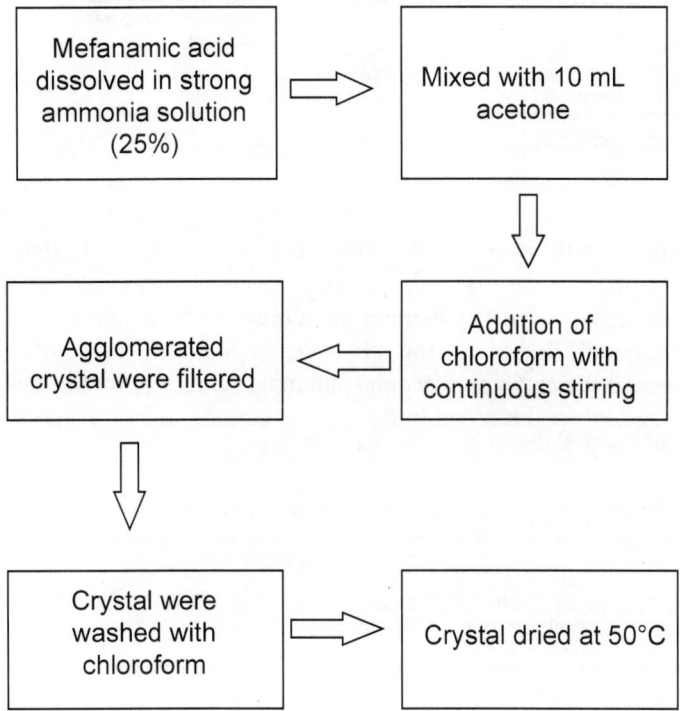

Figure 4.4B Spherical crystallization of Mefenamic acid by ADS method

4.2.1.4   Crystallo-co-agglomeration technique (CCA)

Applications of spherical crystallization to get directly compressible agglomerates without diluents are restricted to water insoluble large-dose drugs only. Most of the excipients, such as diluents and disintegrating agents, are hydrophilic in nature; hence, incorporation of these excipients in the agglomerates formed using organic bridging liquid is difficult. Because of this limitation, spherical crystallization could not be applied to obtain agglomerates of low-dose or poorly compressible materials [4].

Table 4.3 Reported literature on Ammonia diffusion method

Drug	Solvent system			Improved properties	References
	Good solvent	Bad solvent	Bridging Liquid		
Ampicilin trihydrate	Ammonia-water	Acetone	Ammonia-water	Flowability and compressibility	[21]
Enoxacin	Ammonia-water	Acetone and DCM	Ammonia-water	Preparation of agglomerates for the amphoteric drug was possible	[5]
Norfloxacin	Ammonia-water	Acetone	Ammonia-water	Particle shape and size	[51]

Crystallo-co-agglomeration (CCA) technique is a modification of the spherical crystallization technique in which a drug is crystallized and agglomerated with an excipient or with another drug, which may or may not be crystallized in the system. The agglomeration is performed using bridging liquid. The process enables design of agglomerates containing two drugs [4] or a low-dose [24] or poorly compressible drug in combination with diluents [25–26]. The difference in the physicochemical properties of the drug molecules and the excipient becomes the major challenge in the selection of a solvent system for the crystallo-co-agglomeration. Various crystalline drugs reported by CCA method are enlisted in Table 4.4.

Table 4.4 Reported literature on Crystallo-co-agglomeration (CCA) method

Drug	Solvent system			Improved properties	References
	Good solvent	Bad solvent	Bridging Liquid		
Bromohexin HCl-talc	DCM	Water	DCM	Flowability and compressibility	[24]
Ibuprofen-talc	DCM	Water	DCM	Flowability, compressibility and sustained release	[26]
Ibuprofen-Paracetamol	DCM	Water	DCM	Compressibility and dissolution	[4]
Ketoprofen-talc	DCM	Water	DCM	Sustained release with lower amount of polymers	[25]
Indomethacin-epirizole	Ethyl acetate	Water	Ethyl acetate	Flowability and compressibility	[1]
Racecadotril	DCM	Water	DCM	Flowability, compressibility	[52]

The process of crystallo-co-agglomeration technique can be explained by a practical example. Pawar *et al.* [26] prepared agglomerated ibuprofen with talc by crystallo-co-agglomeration technique. Ibuprofen (6 g) and HPMC were dissolved in DCM (6 mL), and talc was uniformly dispersed in it. An aqueous phase (45 mL) containing PEG and PVA were added, contents were stirred at 900 ± 25 rpm. The stirring was continued to obtain agglomerates, which were then filtered and dried overnight at room temperature. Needle shaped crystals with well-developed edges of ibuprofen were obtained. Presence of talc in formulation showed increased size of agglomerates with reduced surface deposit of polymers. Hence the CCA technique developed in the present study can be used for the design of sustained release ibuprofen-talc agglomerates containing lower amounts of polymers. The above procedure is presented in the form of a flow chart in Fig. 4.5.

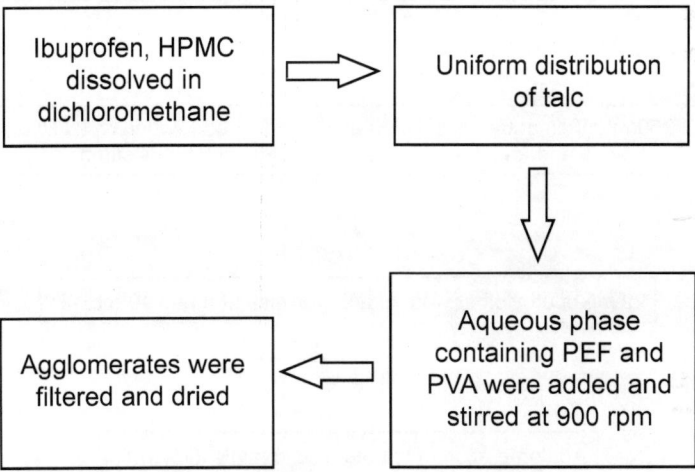

Figure 4.5 Spherical crystallization of ibuprofen with talc by crystallo-co-agglomeration (CCA) technique

4.2.1.5   *Neutralization technique (NT)*

This technique involves the formation of fine crystals by neutralization and consequently their agglomeration by a bridging liquid. Spherical crystallization of tolbutamide and phenytoin were reported by this technique [6, 27]. The drug was dissolved in alkaline solution and then poured into an acidic solution containing polymers and bridging liquid under constant agitation. The drug crystals are precipitated out by neutralization of the base with acid. Then the precipitated crystals were simultaneously agglomerated with the incorporated polymer through the wetting action of the bridging liquid [27].

Sano *et al.* [6] formulated spherical crystals of tolbutamide to improve dissolution and bioavailability of directly compressed tablets. Five gram of tolbutamide was dissolved in 25 mL 1 N sodium hydroxide in a cylindrical vessel. While stirring 250 mL of an aqueous solution of 2% HPMC and 25 mL of 1N HCl were added to neutralize the NaoH solution of tolbutamide and crystallize out the tolbutamide. Then 29 mL of ethylether was added dropwise at a rate of 10 mL/min, followed by agglomeration of tolbutamide crystals for 40 min. Agglomerated crystals showed excellent compressibility and solubility. The above procedure is presented in the form of a flow chart in Fig. 4.6. Various crystalline drugs reported by CCA method are enlisted in Table 4.5.

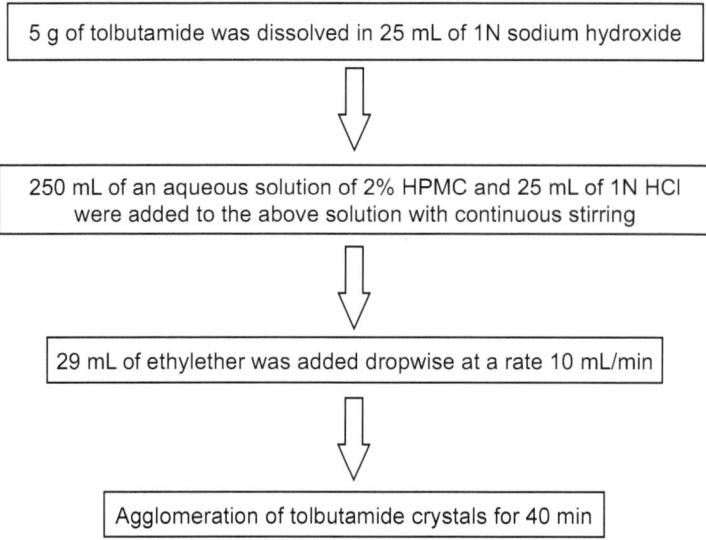

Figure 4.6 Spherical crystallization of tolbutamide by neutralization technique

Table 4.5 Reported literature on neutralization method

Drug	Solvent system			Improved properties	References
	Good solvent	Bad solvent	Bridging Liquid		
Phenytoin	1 N Sodium hydroxide	0.07 N hydrochloric acid	Isopropyl acetate	Mechanical strength, dissolution and bioavailability	[27]
Tolbutamide	1 N sodium hydroxide	Water and 1 N HCl	Ethyl ether	Wettability and dissolution	[6]

4.2.2   Non-typical spherical crystallization technique

These methods also can be used to produce spherically crystallized agglomerates, which are carried out by controlling the physical and chemical properties and can be called as the non-typical spherical crystallization process. These are known as salting out, cooling and precipitation. Various approaches have been attempted to get spherical crystals of drugs by non-typical spherical crystallization technique. Many scientists have attempted to prepare spherical crystals of paracetamol. DiMartino et al. [28] prepared a polymorph of paracetamol (orthorhombic form) by melting the drug at 170°C in an anhydrous nitrogen atmosphere. After slow cooling to room temperature, the solidified material was grounded. The obtained crystals exhibited better compressibility than untreated crystals. The disadvantage of this work was that a high temperature was required and therefore the atmosphere and time have to be strictly controlled. Garekani et al. [29] attempted crystallization of paracetamol by a 'watering-out' method from its ethanolic solution at low temperature caused marked modification to the crystal habit and produced thin plate-like crystals. However these crystals did not exhibit any improvement in their compaction properties compared to untreated paracetamol. Garekani et al. [30] investigated the effects of small amounts of different grades of PVP during crystallization of paracetamol, by the watering-out method. It was found that PVP is an effective additive during crystallization of paracetamol and significantly influenced the crystallization process and compressibility.

4.3   Characterization of spherical crystals

4.3.1   Yield of products

The percentage yield of spherical crystals can be calculated by using the following equation.

$$\% \text{ Yield} = \frac{\text{Total weight of drug}}{\text{Total weight of drug and polymers/excipients}} \times 100 \quad (4.1)$$

4.3.2   Particle size and shape analysis

This is determined by optical microscopy by suspending a small amount of powder in mineral oil. Place the above suspension on a microscope slide followed covering with a cover slip. Determine the diameter for hundred numbers of random particles and calculate the average diameter. Shape factor can be calculated using the following equation.

$$S = \frac{P^2}{4\pi A} \qquad (4.2)$$

Where A and P are the area and the perimeter of the particles, respectively. The shape factor (SF) is defined as the smoothness of the particle, if the value is close to 0 the particle is elongated and/or irregular, if it is close to 1, then it is smooth and rounded.

4.3.3   Scanning electron microscopy (SEM)

The shape and surface topography of spherical crystals are observed by scanning electron microscopy. The process is generally operated at 15 kV. The samples need to be sputter-coated with gold before examination. The photographs give information regarding spherical shape and rough or smoothness of the surface of the agglomerates.

4.3.4   Flowability and compressibility

4.3.4.1   Angle of repose

Pour the spherical crystals through the funnel until the apex of the conical pile just reached the tip of the funnel (H). Note the mean diameter (2R) of the powder cone. Determine the tangent of the angle of repose by using following equation

$$\theta = \tan^{(-1)} \frac{H}{R} \qquad (4.3)$$

4.3.4.2   Carr's compressibility index (C.I.) and Hausner's ratio (H.R.)

Determine tap and poured density of spherical crystals as per standard procedure. Calculate the Carr's index by using the following equation:

$$C.I. = \frac{(\text{Tapped density} - \text{Poured density})}{(\text{Tapped density})} \times 100 \qquad (4.4)$$

Determine Hausner's ratio by using the following equation:

$$H.R. = \frac{\text{Tapped density}}{\text{Poured density}} \qquad (4.5)$$

If the results of angle of repose, Carr's index and Hausner's ratio remain within the theoretical range [31]; then it is suitable for processing into tablet or capsule dosage forms.

4.3.4.3   Kawakita analysis

The compactibility and cohesiveness of the spherical crystals can be determined by the tapped density according to Kawakita equation as follows [32]:

$$N/C = N/a + 1/ab \qquad (4.6)$$

Where a and b are constants; a describes the degree of volume reduction at the limit of tapping and is called compactibility; 1/b is considered to be a constant related to cohesion and is called cohesiveness. C, the degree of volume reduction is calculated from the initial volume V0 and tapped volume V as:

$$C = (V_0 - V)/V \qquad (4.7)$$

Numerical values for constants a and 1/b are obtained from the slope, 1/a and the intercept, 1/ab, of plots of N/C against number of taps N. Lower value of "a" and "1/b" for spherical crystals is an indication for better flowability and less cohesiveness respectively [33].

4.3.4.4   Heckel Analysis

Compressibility of the spherical crystals can also be analyzed by using Heckel equation. Spherical crystals can be compressed using a hydraulic pellet press at loads of 10, 20, 30, 40, 50 kg/cm2 or more. Determine the dimensions (thickness, volume and diameter) of compacts at each compression pressure.

$$\ln \frac{1}{1 - \rho_r} = KP + A \qquad (4.8)$$

$$\rho_r = \rho_A/\rho_T \qquad (4.9)$$

$$K = 1/P_Y \qquad (4.10)$$

Where, ρ_r is the relative density of the compact, ρ_A is the apparent density and ρ_T is the true density, P is the applied pressure, K (the slope of the linear portion) is the reciprocal of the yield pressure, ρ_Y of the material. The yield pressure is inversely related to the ability of the material to deform plastically under pressure and A is a function of the original compact volume. The initial portions of Heckel plot curves are a general indication of the tendency for particle fracturing.

4.3.5   Polymorphism studies

4.3.5.1   Powder X-ray diffraction (P-XRD) study

To investigate polymorphic changes, P-XRD study is performed. PXRD patterns are generally obtained at room temperature with a Cu anode and a graphite monochromator, operated at a voltage of 35 kV and a current of 20 mA. The samples of spherical crystals can be analyzed in the 2θ angle range of 5–50°, and the process parameters were set as scan-step size of 0.02° (2θ) and scan-step time of 25 s.

4.3.6   Solubility and dissolution test

Solubility and dissolution test for the spherical crystals has to be performed as per the standard procedure for respective drugs as prescribed in official compendium.

4.4   Recent patents

Various patents related to development of spherical crystallization are enlisted in Table 4.6. Many United States and European patents are filed in the area of spherical crystallization which include preparation of drug-containing spherical fine particles, directly compressible spherical mannitol, agglomerate of spherical polyvalent-metal alginate particulates, diffusion layer modulated solids, method for controlled growth of particle size, etc.

Table 4.6 Patents on spherical crystallization method and products

S. no.	Title of the patent	Invention	Patent Number	Inventors	Publication date	References
1	Preparation method of drug-containing spherical fine particles.	Efficient preparation method of spherical fine particles containing a drug for an easily-swallowed, controlled-release preparation	US Patent 6692768	Ishibashi, T Nagao, K Ikegami, K Yoshino, H Mizobe, M	02/17/2004	[53]
2	Excipient for compressed tablets comprising novel spherical Mannitol	An excipient for direct compression	US Patent 20110135927	Satomi, J Imaoka, M	06/09/2011	[54]
3	Agglomerate of spherical polyvalent-metal alginate particulates, controlled-release preparation comprising difficultly soluble drug supported on said agglomerate and processes for producing the both.	A process for producing a controlled-release preparation capable of rapidly dissolving out at least 99% of a difficultly soluble drug having a low dissolution rate above the small intestine	European Patent EP0970705	Hom-ma, Myo, N Sato, T Nanbu, H	01/12/2000	[55]
4	Agglomerates by crystallization	novel agglomerates in crystalline form of β-lactum compounds	US Patent 20060079496	Booij, J Lefferts, GA	04/13/2006	[56]

Contd...

Contd...

S. no.	Title of the patent	Invention	Patent Number	Inventors	Publication date	References
5	Designed particle agglomeration	The present invention relates generally to the production of particle agglomerates, tailored to a specific size 6 and structure, from initially dispersed particles	US Patent 20050106310	Green, JH Sommer, MM Spahr, DE Wagner, NJ	05/19/2005	[57]
6	Controlled agglomeration	A method that enables a controlled growth in particle size	US Patent 7217431	Holm, P Buur, A Elema, MO Møllgaard, B Holm, JE Schultz, K	05/15/2007	[58]
7	Diffusion layer modulated solids	An excipient and a soluble salt of a poorly soluble, basic drug for improved delivery of drugs	US Patent 20050042291	Hawley, M Morozowich, W Bergren, MS. Skoug, JW Nixon, PR Heimlich, J Gao, P	02/24/2005	[59]

4.5   Conclusions

Spherical crystallization techniques can be widely used in pharmaceutical industry because they offer a solution for improving the tableting properties of crystalline drugs. This technique enables tableting by direct compression technique. The spherical crystallization process appears to be simple and inexpensive for scaling up to a commercial level. Appropriate solvents, excipients, and process parameters must be optimized for each drug to achieve improved flowability, compressibility, solubility, etc. This technique will shorten the time needed for production of tablets compared to wet granulation technique.

4.6   References

1. Kawashima Y. Development of spherical crystallization technique and its application to pharmaceutical systems. *Arch Pharm. Res.* 7(2): 145–151 (1984).

2. Nokhodchi A, Maghsoodi M, Hassanzadeh D. An Improvement of Physicomechanical Properties of Carbamazepine Crystals. *Iran. J. Pharm. Res.* 6(2): 83–93 (2007).

3. Yadav VB, Yadav AV. Polymeric Recrystallized Agglomerates of Cefuroxime Axetil Prepared by Emulsion Solvent Diffusion Technique. *Trop. J. Pharm. Res.* 8(4): 361–369 (2009).

4. Mahadik KR, Pawar AP, Paradkar AR, Kadam S. Crystallo-coagglomeration: A Novel Technique to Obtain Ibuprofen-Paracetamol Agglomerates. *AAPS Pharm SciTech.* 5 (3): 1–8 (2004).

5. Ueda M, Nakamura Y, Makita H, Imasato Y, Kawashima Y. Particle design of Enoxacin by spherical crystallization technique. I Principle of ammonia diffusion system (ADS). *Chem. Pharm. Bull.* 38 (9): 2537–2541 (1990).

6. Sano A, Kuriki T, Kawashima Y, Takeuchi H, Hino T, Niwa T. Particle design of Tolbutamide by the spherical crystallization technique IV, Improved of dissolution and bioavailability of direct compression tablets prepared using Tolbutamide agglomerated crystals, *Chem. Pharm. Bull.* 40: 3030–3035 (1992).

7. Szabo-Revesz P, Goczo H, Pintye-Hodi K, Kasa P, Eros IM, Hasznos-Nezdei M, Farkas B. Development of spherical crystal agglomerates of an aspartic acid salt for direct tablet making. *Powder Technol.* 114: 118–124 (2001).

8. Kawashima Y, Imai M, Takeuchi H, Yamamoto H, Kamiya K. Development of agglomerated crystals of Ascorbic acid by the spherical crystallization techniques. *Powder Technol.* 130: 283–289 (2003).

9. Kawashima Y, Cui F, Takeuchi H, Niwa T, Hino T, Kiuchi K. Parameters determining the agglomeration behavior and the micromeritic properties of spherically agglomerated crystals prepared by the spherical crystallization technique with miscible solvent systems. *Int. J. Pharm.* 119(2): 139–147 (1995).

10. Jain SK, Chourasia MK, Jain NK, Jain S. Preparation and characterization of agglomerates of flurbiprofen by spherical crystallization technique. *Ind. J. Pharm. Sci.* 65(3): 287–291 (2003).

11. Zhang H, Chen Y, Wang J, Gong J. Investigation on the spherical crystallization process of cefotaxime sodium. *Ind. Eng. Chem. Res.* 49: 1402–1411 (2010).

12. Blandin AF, Mangin D, Rivoire A, Klein JP, Bossoutrot JP. Agglomeration in suspension of salicylic acid fine particles: influences of some process parameters on kinetics and agglomerate final size, *Powder Technol.* 130: 316–323, (2003).

13. Maghsoodi M, Bakhsh AST. Evaluation of physico-mechanical properties of drug-excipients agglomerates obtained by crystallization. *Pharm. Dev. Technol.* 16(3): 243–249 (2011).

14. Mahanty S, Sruti J, Patra CN, Rao MEB. Particle Design of Drugs by Spherical Crystallization Techniques, *Int J Pharm Sci Nanotech.* 3(2): 912–918: (2010).

15. Chow AHL, Leung MWM. A study of the mechanisms of wet spherical agglomeration of pharmaceutical powders. *Drug Dev. Ind. Pharm.* 22(4): 357–371 (1996).

16. Usha AN, Mutalik S, Reddy MS, Ranjith AK, Kushtagi P, Udupa N. Preparation and, in vitro, preclinical and clinical studies of aceclofenac spherical agglomerates, *Eur. J. Pharm. Biopharm.*70: 674–683 (2008).

17. Cui F, Yang M, Jiang Y, Cun D, Lin W, Fan Y, Kawashima Y. Design of sustained-release nitrendipine microspheres having solid dispersion structure by quasi-emulsion solvent diffusion method. *J. Control. Rel.* 91: 375–384 (2003).

18. Yadav VB, Yadav AV. Effect of Different Stabilizers and Polymers on Spherical Agglomerates of Gresiofulvine by Emulsion Solvent Diffusion (ESD) System. *Int. J. Pharm. Tech. Res.* 1(2):149–150 (2009).

19. Sano A, Kuriki T, Kawashima Y, Hino T, Niwa T. Particle design of tolbutamide by the spherical crystallization technique. III. Micromeritic properties and dissolution

rate of tolbutamide spherical agglomerates prepared by the quasi-emulsion solvent diffusion method and the solvent change method, *Chem. Pharm. Bull.* 38: 733–739 (1990).

20. Yang MS, Cui FD, You BG, Fan YL, Wang L, Yue P, Yang H. Preparation of sustained-release nitrendipine microspheres with Eudragit RS and Aerosil using quasi-emulsion solvent diffusion method, *Int. J. Pharm.* 259: 103–113 (2003).

21. Gohel MC, Parikh RK, Shen H, Rubey RR. Improvement in flowability and compressibility of Ampicillin Trihydrate by spherical crystallization. *Ind. J. Pharm. Sci.* (65–96): 634–637 (2003).

22. Hector GP, Jorge B, Carlo A. Preparation of Norfolxacin spherical agglomerates using the ammonia diffusion system. *J. Pharm. Sci.* 87(4): 519–523 (1998).

23. Bhadra S, Kumar SM, Jain S, Agrawal S, Agrawal GP. Spherical crystallization of mefenamic acid, *Pharm. Technol.* Feb: 66–76 (2004).

24. Paradkar A, Jadhav N, Pawar A. Design and Evaluation of Deformable Talc Agglomerates Prepared by Crystallo-Co-Agglomeration Technique for Generating Heterogeneous Matrix. *AAPS Pharm Sci Tech.* 8(3): E1-E7 (2007).

25. Chavda V, Maheshwari RK. Tailoring of ketoprofen particle morphology via novel crystallo-coagglomeration technique to obtain a directly compressible material. *Asian J. Pharm.* 2(1): 61–67 (2008).

26. Pawar A, Paradkar A, Kadam S, Mahadik K. Agglomeration of Ibuprofen with Talc by Novel Crystallo-Co-Agglomeration Technique. *AAPS PharmSciTech.* 5(4): 1–6, (2004).

27. Kawashima Y, Handa T, Takeuchi H, Okumura M, Katou H, Nagata O. Crystal modification of phenytoin with polyethylene glycol for improving mechanical strength, dissolution rate and bioavailability by a spherical crystallization technique. *Chem. Pharm. Bull.* 34(8): 3376–3383 (1986).

28. Di Martino P, Guyot-Hermann AM, Conflant P, Drache M, Guyot JC. A new pure paracetamol for direct compression: the orthorhombic form. *Int. J. Pharm.* 128: 1–8, (1996).

29. Garekani HA, Ford, J.L., Rubinstein, M.H., Rajabi-Siahboomi, A.R. Formation and compression characteristics of polyhedral and thin plate-like crystals of paracetamol. *Int. J. Pharm.* 187: 77–89 (1999).

30. Garekani HA, Ford JL, Rubinstein MH, Ali RRS. Highly compressible paracetamol: I: crystallization and Characterization, *Int. J. Pharm,* 208: 87–99 (2000).

31. Aulton ME. Pharmaceutical Preformulation. In: Taylor K, editor. Aulton's Pharmaceutics. The design and manufacture of medicines. Hungary: Elsevier; 2007: 355–357.

32. Yamashiro M, Yuasa Y, Kawakita K. An experimental study on the relationships between compressibility, fluidity and cohesion of powder solids at small tapping numbers. *Powder Technol* 34, 225–231 (1983).

33. Patra CN, Pandit HK, Singh SP, Devi MV. Applicability and Comparative Evaluation of Wet Granulation and Direct Compression Technology to Rauwolfia serpentina Root Powder: A Technical Note. *AAPS PharmSciTech,* 9(1), 100–104 (2008).

34. Goczo H, Szabo-Revesz P, Farkas B, Hasznos-Nezdei M, Serwanis FS, Pintye-Hodik, Kasa P, Eros I, Antal I, Marton S. Development of Spherical Crystals of Acetylsalicylic Acid for Direct Tablet-making, *Chem. Pharm. Bull.* 48(12): 1877–1881 (2000).

35. Kawashima Y, Imai M, Takeuchi H, Yamamoto H, Kamiya K. Development of Agglomerated Crystals of Ascorbic Acid by the Spherical Crystallization Technique for Direct Tableting, and Evaluation of Their Compactibilities, KONA. 20: 251–262 (2002).

36. Pandey S, Patil AT. Preparation, evaluation and need of spherical crystallization in case of high speed direct tabletting. *Curr Drug Deliv.* 11(2): 179–190 (2014).

37. Rasmuson CA, Katta J. Spherical crystallization of benzoic acid. *Int. J. Pharm.* 348: 61–69 (2008).

38. Gupta VR, Mutalik S, Patel M, Girish K. Spherical crystals of celecoxib to improve solubility, dissolution rate and micromeritic properties. *Acta Pharm.* 57: 173–184 (2007).

39. Di Martino P, Barthélémy C, Piva F, Joiris E, Palmieri GF, Martelli S. Improved dissolution behavior of fenbufen by spherical crystallization. *Drug Dev. Ind. Pharm.* 25(10): 1073–1081 (1999).

40. Kumar S, Chawla G, Bansal A. Spherical Crystallization of Mebendazole to Improve Processability. *Pharm. Dev. Technol.* 13(6): 559–568 (2008).

41. Maghsoodi M, Hassan-Zadeh D, Barzegar-Jalali M, Nokhodchi A and Martin G. Improved Compaction and Packing Properties of Naproxen Agglomerated Crystals Obtained by Spherical Crystallization Technique. *Drug Dev. Ind. Pharm* 33: 1216–1224 (2007).

42. Di Martino P, Di Cristofaro R, Barthélémy C, Joiris E, Palmieri FG, Sante M. Improved compression properties of propyphenazone spherical crystals. *Int. J. Pharm.* 197 (1–2): 95–106 (2000).

43 Kawashima Y, Niwa T, Takeuchi H, Hino T, Itoh Y, Furuyama S. Characterization of polymorphs of Tranilast anhydrate and Tranilast monohydrate when crystallization by two solvents changes spherical crystallization technique, *J. Pharm. Sci.* 80(5): 472–78 (1991).

44. Varshosaz J, Tavakoli N, Salamat FA. Enhanced dissolution rate of simvastatin using spherical crystallization technique, *Pharm Dev Technol.* 16(5): 529–35 (2011).

45. Morshima K, Kawashima Y, Takeuchi H., Niwa, T, Hino T. Tabletting properties of Bucillamine agglomerates prepared by the spherical crystallization technique, *Int. J. Pharm.* 105: 11–18 (1994).

46. Nokhodchi A, Maghsoodi M, Hassanzadeh D. An Improvement of Physicomechanical Properties of Carbamazepine Crystals. Iran. *J. Pharm. Res.* 6(2): 83–93 (2007).

47. Patil SV, Pawar AP, Sahoo SK. Improved compressibility, flowability, dissolution and bioavailability of pioglitazone hydrochloride by emulsion solvent diffusion with additives, *Pharmazie.* 67(3): 215–223 (2012).

48. Tapas AR, Kawtikwar PS, Sakarkar DM. Enhanced dissolution rate of felodipine using spherical agglomeration with Inutec SP1 by quasi emulsion solvent diffusion method. Res Pharm Sci. 4(2): 77–84 (2009).

49. Ribardire A, Tchoreloff P, Couarraze G, Puisieux F. Modification of ketoprofen bead structure produced by the spherical crystallization technique with a two-solvent system, *International Journal of Pharmaceutics* 144: 195–207 (1996).

50. Maghsoodi M, Sadeghpoor F. Preparation and evaluation of solid dispersions of piroxicam and Eudragit S100 by spherical crystallization technique, *Drug Dev. Ind. Pharm.* 36(8): 917–925 (2010).

51. Puechagut HG, Bianchotti J, Chiale CA. Preparation of norfloxacin spherical agglomerates using the ammonia diffusion system, *J. Pharm. Sci.* 87: 519–523 (1998).

52. Garala K, Patel J, Patel A, Raval M, Dharamsi A. Influence of excipients and processing conditions on the development of agglomerates of racecadotril by crystallo-co-agglomeration. *Int J Pharm Investig.* 2(4): 189–200 (2012).

53. Ishibashi T, Nagao K, Kengo YH, Mizobe M. Preparation method of drug-containing spherical fine particles. US Patent 6692768, February 17, 2004.

54. Satomi J, Imaoka M. Excipient for compressed tablets comprising novel spherical mannitol. US Patent 20110135927, June 9, 2011.

55. Hom-ma M, NagayoshiSato T, Nanbu H. Agglomerate of spherical polyvalent-metal alginate particulates, controlled-release preparation comprising difficultly soluble drug supported on said agglomerate, and processes for producing the both. European Patent EP0970705, January 12, 2000.

56. Booij J, Lefferts GA. Agglomerates by crystallization. US Patent 20060079496, April 13, 2006.

57. Green JH, Sommer MM, Spahr DE, Wagner NJ. Designed particle agglomeration. United States Patent 20050106310, May19, 2005.

58. Holm P, Buur A, Elema MO, Møllgaard B, Holm JE, Schultz K. Controlled agglomeration. United States Patent 7217431, May 15, 2007.

59. Hawley M, Morozowich W, Bergren M, Skoug JW, Nixon PR, Heimlich JM, Gao P. Diffusion layer modulated solids, United States Patent 20050042291, February 24, 2005.

5

An overview of liquisolid technology

Chinam Niranjan Patra*, Suryakanta Swain, Kahnu Charan Panigrahi
and Muddana Eswara Bhanoji Rao

*Department of Pharamceutics, Roland Institute of Pharmaceutical sciences,
Berhampur-760010, Odisha, India.*

5.1 Introduction

The biopharmaceutical classification system (BCS) of classifying active pharmaceutical ingredients (APIs) is based on their aqueous solubility and gastrointestinal permeability. According to BCS, API is classified into four categories as shown in Table 5.1. Depending on the properties of API as well as therapeutic need, at times solubility enhancement, quick release, delayed release, sustain release, modified release formulations etc may be preferred. Generally enhancement of solubility by micronization, nanonization, solid dispersions, use of surfactants, co-solvents, complexation with cyclodextrin etc is attempted for APIs belonging to BCS class II and IV. Similarly in order to sustain release many approaches such as matrix tablet, microencapsulation, osmotic tablets etc. are available for API belonging to BCS class I and III. But one of the most simple, promising and alternative techniques for achieving both quick as well as sustain release is liquisolid technique [1]. In this technique drug release can be manipulated by judicious selection of non-volatile solvent, carrier and coating materials.

Table 5.1 Bio-pharmaceutical classification system

BCS class	Solubility	Permeability	Absorption
I	High	High	Well absorbed
II	Low	High	Variable
III	High	Low	Variable
IV	Low	Low	Poorly absorbed

5.1.1 What is liquisolid technique?

Formulations prepared using liquisolid techniques are considered as free-flowing and compressible powders containing a nonvolatile liquid vehicle and solid drug particles. The solid drug particles can be dissolved completely or partly by the non-volatile liquid vehicle, which do not evaporate during the processing and thus the drug is carried within the liquid system. This liquid system is transformed into a dry looking, free-flowing, non-adherant and compressible powder by mixing with suitable excipients termed as carrier and coating materials. In liquisolid formulations, the excipient with a capability to adsorb liquid on its surfaces is called the carrier which usually has the highest contribution in the formulation. The excipient with very high surface area which usually covers the carrier surfaces containing liquid to improve flowability of liquisolid powders is known as coating materials.

Advantages
- Poorly water-soluble solid drugs can be formulated into a liquisolid system.
- As the drug is in liquid form, it exhibits improved dissolution and bioavailability.
- Liquisolid systems can also be used for sustain release of water-soluble drugs.
- Simple technique and low production cost.
- Large-scale manufacturing feasibility.
- Lower cost of the formulation than soft gelatin capsules.

Disadvantages
- Drug should have good solubility in non-volatile liquid vehicle for enhancing dissolution rate of drugs.
- Drug with high dose poses problems in formulation of liquisolid systems. These drugs require a large amount of liquid vehicle and also carrier and coating material to prepare dry powder with desired flowability and compressibility. This could increase the mass of each formulation leading to large size tablet or large size capsule difficult to swallow orally.
- Compressibility of liquisolid formulations may be poor. Sometimes liquid may squeeze out of the compact during compression.

Applications
- Liquisolid powders can be filled in capsules or compressed into tablets.
- Improvement in dissolution rate and bioavailability of poorly soluble drugs.
- Sustained release of water-soluble drugs.

- Liquisolid technique a promising alternative to conventional coating for improvement of drug photostability in solid dosage forms [2].

5.2   Theory

A powder can retain only certain amounts of liquid while maintaining adequate flow and compression properties. In order to determine the required amounts of carrier and coating materials, a mathematical approach for liquisolid formulations has been developed by Spireas [3]. This approach is based on the flowable (Φ-value) and compressible (ψ-value) liquid retention potential introducing constants for each powder-liquid combination.

The flowable liquid retention potential (Φ-value) of a powder represents the highest amount of a given non-volatile liquid that can be retained inside its bulk (w/w) while maintaining an acceptable flowability. The flowability may be determined by the angle of repose. The compressible liquid retention potential (ψ-value) of a powder is defined as the maximum amount of liquid the powder can retain inside its bulk (w/w) while maintaining acceptable compactability. The compact must have desired hardness without any liquid squeezing out during compression [4]. The compactability may be determined by the so-called 'pactisity', which describes the maximum (plateau) crushing strength of a 1 g tablet compacted at sufficiently high compression forces.

Basing on the excipient ratio (R) of the powder substrate, an acceptably flowing and compressible liquisolid system can be obtained only if a maximum liquid load on the carrier material is not exceeded. This liquid:carrier ratio is termed 'liquid load factor L_f (w/w) and is defined as the weight ratio of the liquid formulation (W) and the carrier material (Q) in the system.

$$L_f = W/Q \tag{5.1}$$

Excipient ratio (R) represents the ratio between the weights of the carrier (Q) and the coating (q) material present in the formulation

$$R = Q/q \tag{5.2}$$

The liquid load factor that ensures acceptable flowability (Φ) can be determined by

$$\Phi L_f = \Phi + \text{ø}\,(1/R) \tag{5.3}$$

Where Φ and ø are the Φ values of the carrier and coating material, respectively.

Similarly, the liquid load factor for production of liquisolid systems with acceptable compactability ($^{\Psi}L_f$) can be determined by

$$^{\Psi}L_f = \psi + \psi\,(1/R) \tag{5.4}$$

Where ψ and ψ are the ψ numbers of the carrier and coating material, respectively.

Therefore, the optimum liquid load factor (Lo) required obtaining acceptably flowing and compressible liquisolid systems is equal to either $^{\Phi}L_f$ or $^{\Psi}L_f$, whichever represents the lower value. As soon as the optimum liquid load factor is determined, the appropriate quantities of carrier (Qo) and of liquid formulation (W) into an acceptably flowing and compressible liquisolid system may be calculated as follows: coating (qo) material required for converting a given amount of liquid formulation (W) into an acceptably flowing and compressible liquisolid system may be calculated as follows:

$$Qo = W/Lo \qquad (5.5)$$

and
$$qo = Qo/R \qquad (5.6)$$

The validity and applicability of the above-mentioned principles have been tested and verified by producing liquisolid compacts possessing acceptable flow [5] and compaction properties [6].

5.3   Formulation of liquisolid systems

This technique can be extended for the formulation of both quick release and sustain release formulations.

5.3.1   Liquisolid formulations to enhance drug release

Preparation of liquisolid systems is based on the principles of conversion of the drug in the liquid state into a free flowing, readily compressible and apparently dry powder by simple physical blending with selected excipients, carriers and coating materials. General method of preparation of liquisolid formulations are shown in Fig. 5.1. Liquid drug could be also sprayed into the carrier material in fluid-bed equipment for homogenous distribution of the active substance. Liquid drug is incorporated into the porous structure of a carrier material due to adsorption and absorption. Literature on liquisolid formulations with enhanced dissolution rate are summarized in Table 5.2.

5.3.1.1   Mechanisms of enhancement of drug release

As the drug is present in the form of liquid medication, it is either in a solubilized or a molecularly dispersed state. This results in following changes:

- Increased surface area of drug available for drug release
- Increased solubility of the drug
- Increased wettability of the drug

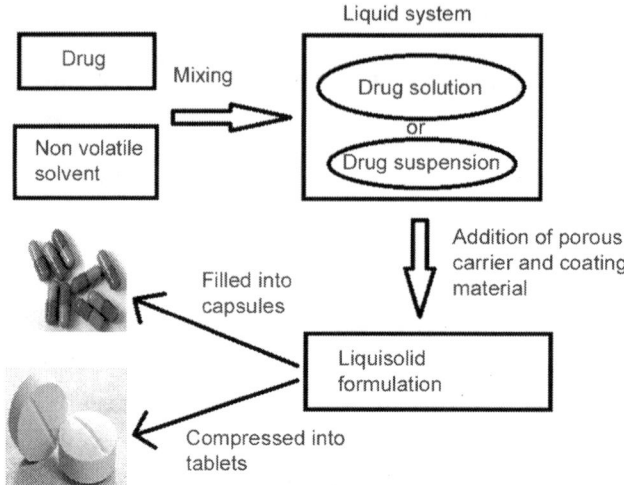

Figure 5.1 General method of preparation of liquisolid formulations

5.3.1.2   Concentration of drug in liquisolid systems

The liquisolid technique has been applied successfully to low dose water-insoluble drugs like rosuvastatin [21], olmesartan medoxonil [20], mesapride citrate [18] etc. It is difficult to design a liquisolid formulation for high dose drugs as it requires a higher of amount of solvent and further this will also need high amount of carrier to convert to free flowing solid. But the use of modern carriers and coating materials with a large specific surface area and high absorption capacity (e.g., Neusilin®, Sylysia) is another way of incorporation of higher doses of water insoluble drugs into liquisolid systems.

Table 5.2 Literature on liquisolid systems with improved dissolution rate

S. no.	Name of the drug	Liquid vehicle	Carrier	Coating	Dosage form	References
1	Clonazepam	Propylene glycol (PG)	Avicel PH 102	Aerosil200	Tablet	[7]
2	Ezetimibe	Labrasol PEG 400 Transcutol HP Tween 80	Avicel PH 101 Avicel PH 200	Aerosil 200	Tablet	[8]
3	Famotidine	Propylene glycol	Avicel PH 102	Aerosil 200	Tablet	[9]
4	Furosemide	Synperonic® PE/L 81 Caprol® PGE-860 PEG 400	Avicel PH 101	Cab-O-Sil® M-5	Tablet	[10]

Contd...

Contd...

S. no.	Name of the drug	Liquid vehicle	Carrier	Coating	Dosage form	References
5	Glyburide	PEG 400	MCC PH 102	Aerosil 200	Tablet	[11]
6	Griseofulvin	PEG 300	Neusilin US2	*	Tablet	[12]
7	Hydrochlorothiazide	PEG200	Avicel PH 101, PH 102, Light magnesium carbonate	Talc, Magnesium stearate	Tablet	[13]
8	Hydrocortisone	Propylene glycol	Avicel PH 200	Cab-O-Sil M5	Tablet	[14]
9	Indomethacin	PEG 200 Glycerin	MCC	Silica	Tablet	[15]
10	Indomethacin	Propylene glycol	MCC	Silica	Tablet	[16]
11	Irbesartan	PEG400	Avicel PH 102	Cab-O-Sil M5	Tablet	[17]
12	Mesapride citrate	Propylene glycol	Avicel PH102, Mannitol, Lactose DC	Aerosil 200	Tablet	[18]
13	Naproxen	Cremophor EL, Synperonic PE/L61, PEG400	Avicel PH102,	Cab-O-Sil M5	Tablet	[19]
14	Olmesartan Medoxomil	Acrysol El 135	Avicel PH 101, Fujicalin Neusilin	Aerosil	Tablet	[20]
15	Prednisolone	PG, glycerin, PEG400, Polysorbate 80.	Avicel PH 102	Cab-O-Sil M5	Tablet	[3]
16	Rosuvastatin	PEG 200 PEG 400	Avicel PH 102	Aerosil200	Tablet	[21]
17	Spironolactone	Capryol 90, Synperonic - PE/L61, Solutol HS-15, Kollicoat SR 30 D	Avicel PH 101	Cab-O-Sil M5	Tablet, powder	[22]
18	Valsartan	Propylene glycol	Avicel PH 101	AEROSIL 200	Tablet	[23]
19	Piroxicam	Tween 80	Avicel PH 101	Cab-O-Sil M5	Tablet	[24]
20	Piroxicam	Propylene glycol	Avicel PH 101	Cab-O-Sil M5	Tablet	[25]
21	Piroxicam	Polysorbate 80	Avicel PH 101	Cab-O-Sil M5	Tablet	[26]

5.3.1.3   Non-volatile solvent

Various non-volatile, high-boiling point, preferably water miscible and not highly viscous solvents are selected in the formulation of liquisolid systems. It was demonstrated in several studies that the solvent had a significant effect on drug release from liquisolid systems. For enhanced drug release from liquisolid preparations, a liquid vehicle in which the active ingredient is most soluble is usually selected. The most widely used non-volatile liquids used in the formulation of liquisolid formulations include polyethylene glycol 200 [13], polyethylene glycol 400 [21], propylene glycol [18], polysorbate 80 [26], glycerin [3, 15]. Apart from these many researchers have also used novel liquid vehicles like caproyl 90 [22], acrysol El 135 [20], labrasol [8] etc. One of the prerequisite in the formulation of liquisolid systems is to determine the solubility of drug in different non-volatile solvents. Solvent with highest solubility is selected for formulation. Low solubility of drug candidates in solvents leads to insignificant improvement in dissolution rate.

Apart from solubility enhancement, solvents can also impart compactness to liquisolid formulations. This was ascribed to hydrogen bonding owing to the presence of hydroxyl groups [27]. In addition, non-volatile solvent can act as a binder in low concentration, and shows a negative effect on compaction properties of liquisolid compacts in higher concentrations. Excessive non-volatile solvent causes generation of the capillary state of powder aggregation, and hence the surface tension effect becomes less significant in bringing the particles together, leading to poor bonding between powder particles. Also, decrease in tensile strength at high levels of solvent results in the formation of multilayer of solvent around the surface of the particles. These layers disturb or reduce intermolecular attraction forces and hence decrease tablet strength. Therefore, in higher concentrations, non-volatile solvent covers contact points between particles and acts as a lubricant and reduces the binding of particles [25]. List of various non-volatile solvents for enhancement of dissolution rate are enlisted in Table 5.2.

5.3.1.4   Method of determining solubility of drug in non-volatile solvent

Solubility studies are carried out by preparing a saturated solution of the drug by adding an excess amount of drug into non-volatile solvents by using rotary shakers till equilibrium is attained. After this step, the saturated drug solution is filtered, dissolved in a specific solvent and evaluated by a suitable analytical technique. Solvents with greater ability to solubilize the drug are selected for the formulation of liquisolid systems for enhancement of drug release [28]. Similarly solvents with lowest ability to solubilize the drug are selected for the formulations of liquisolid systems for sustain release [29].

5.3.1.5 Carrier materials

In liquisolid formulations, role of carriers is vital in obtaining the dry form of powder from the drug in liquid state. Carriers should be a porous material possessing high liquid absorption capacity. Specific surface area of the carrier is an important factor in the formulation of liquisolid systems. In general an ideal carrier should have the following properties:

- High loading capacity
- Facilitates easy processing with standard processes of solid dosage forms like tablets and capsules
- High storage stability
- No negative impact on drug stability
- Complete drug release in the body
- No toxicity

Usually various carriers like microcrystalline cellulose [MCC (PH 102, PH 101 and PH 200)], lactose, mannitol are used. It was observed that Avicel is a better carrier because of high specific surface area. Apart from these other porous carriers with high specific surface area such as Neusilin, sylysia, anhydrous fujacalin are also used in liquisolid systems. It was reported that large size particles of carrier (MCC 200) showed lower dissolution rate, poor flowability and lower tensile strength of tablets compared to smaller size particles of MCC 101 [25].

5.3.1.6 Coating materials

Materials used as coating material must be of fine size (0.01–5.0 μm) with good absorbing capacity. Coating material should cover the wet carrier particles adsorbing excess liquid to ensure good flowability [30]. In liquisolid formulations, this role is played by materials with a large specific surface area and high absorption capacity. At present, the coating materials used in liquisolid formulations are colloidal silicon dioxide (Aerosil®, Cab-O-Sil® M5), amorphous silica gel (Syloid®, Sylysia®), granulated silicon dioxide (Aeroperl®), silica aerogel, magnesium alumino metasilicates (Neusilin®), calcium silicate (Florite®) and ordered mesoporous silicates can be also used to prepare liquisolid system with suitable flowability and compressibility.

5.3.2 Liquisolid formulations to sustain release of drug

Liquisolid technique is a relatively new and potential technique to provide sustained drug release. In such preparations the surface area of liquisolid powders should be reduced by converting the powder into a tablet dosage form. These formulations should have the following characteristics:

- Free flowing
- Capable of forming stable compacts at low compaction pressures (good compactibility)
- Should not stick to the punches

No literature was reported on liquisolid technique for sustaining the release of drug till 2008. The first research article was reported by Javadzadeh et al. [32] who formulated sustained release liquisolid compacts for propranolol HCl by using polysorbate 80 as non-volatile liquid vehicle, Eudragit RS and RL as carrier material and silica as coating material. Literature on sustain release liquisolid compacts are shown in Table 5.3.

5.3.2.1   Mechansims of sustain release of drug

There are many mechanisms proposed for the sustain release of drugs from liquisolid formulations such as:
- Reduction of glass transition temperature (Tg) of the polymer.
- Low surface area of carriers like Eudragit RS and RL.
- Hydrophobic coat of silica around the carriers.

5.3.2.2   Non-volatile liquid

Liquid vehicle with less capability to dissolve the drug can be selected as a non-volatile solvent to obtain a better retardation effect. This is in disparity to fast release liquisolid formulations where a liquid vehicle with a high potential to dissolve the drug should be selected. For example, propranolol HCl (a highly water-soluble drug) and theophylline (low solubility in water) both showed the lowest solubility in polysorbate 80 and this liquid vehicle was incorporated in liquisolid sustained release formulations of propranolol HCl and theophylline. This does not mean that the other liquid vehicles cannot be used in liquisolid formulations as other liquid vehicles like PEG 200, 400, PG and glycerin can also be used to obtain sustained release liquisolid formulations [33]. It was shown that the liquid vehicle polysorbate may act as a plasticizer and thus decrease the glass transition temperature (Tg) of the used polymer (Eudragit® RS). Plasticizers affect the intermolecular bonding between polymer chains, thereby increasing their flexibility. At temperature above Tg, a better coalescence of polymer particles occurs, and a fine network and matrix with low porosity forms. This results in decreased penetration of dissolution media into the compact structure. Therefore, reduction of Tg of the polymer might be the reason for prolongation of drug release from liquisolid tablets [31].

5.3.2.3   Carrier and coating materials

The hypothesis behind the liquisolid technology to sustain the release of drug is that if hydrophobic carriers such as Eudragit RL, RS, ethyl cellulose etc. are used instead of hydrophilic carriers, sustained release systems can be obtained. The presence of hydrophilic HPMC polymer grades further increase the retardation effect of liquisolid technique [33]. Additionally in liquisolid sustained release formulation, generally the surface area of carrier (Eudragit RS or RL) is low compared to the carriers used in fast release liquisolid formulation (microcrystalline cellulose); therefore, in order to increase loading factor and obtain a better flowability the amount of coating material (silica) is usually high [31]. This in turn helps to impart more sustaining properties to the liquisolid formulations as silica forms a hydrophobic coat around the material.

5.3.2.4   Adjuvants

Liquisolid compacts for sustained release should be robust and not disintegrate in the dissolution medium or else, the sustaining effect might be reduced. These limitations can be resolved by increasing the percentage of adjuvant in the formulations. However, this may lead to a final weight of the tablets >1 g which makes them difficult to swallow. Thus, in practice it is relatively difficult to prepare liquisolid compacts for drugs with high dose. A recent approach was incorporation of hydrophilic polymers such as PVP, HPMC and PEG 35000 which results in increased loading factor. In case of carbamezapine (high dose drug) the use of hydrophilic polymers reduced the weight of the tablet. By addition of such materials to the liquid system, a low amount of carrier is required to obtain a dry powder with free flowability and good compactibility. Addition of additives such as such as PVP, HPMC or PEG 35000 to the liquid system may increase the loading factor above 0.25. This could be ascribed to the granulating effect of the co-solvents used in the formulation of the liquisolids.

5.4   Evaluation of liquisolid systems

5.4.1   Angle of repose

Pour the liquisolid powders through the funnel until the apex of the conical pile just reached the tip of the funnel (H). Note the mean diameter (2R) of the powder cone. Determine the tangent of the angle of repose by using following equation

$$Q \ = \ \tan^{-1} \frac{H}{R} \tag{5.7}$$

Table 5.3 Literature on liquisolid systems with sustain release

S. no.	Name of the drug	Liquid vehicle	Carrier	coating	Dosage form	References
1	Griseofulvin	Cremophor EL, Synperonic PE/ L61 Capryol™ 90, Kollicoat® SR 30D	Avicel PH102	Cab-O-Sil® M5	Tablet	[34]
2	Propranolol HCl	Polysorbate 80	Eudragit RL or RS	Silica	Tablet	[31]
3	Trimetazidine HCl	Polysorbate-80	Ethyl cellulose Eudragit L-100, Eudragit RS-100	Aerosil 200	Tablet	[34]
4	Venlafaxine HCl	Tween-80, PEG-400, PG	Avicel PH101, Eudragit L-100, Eudragit RS-100	Aerosil 200	Tablet	[8]
5	Tramadol	Tween 80 PEG 400, PG	Avicel 101	Aerosil 200	Tablet	[35]
6	Indomethacin	PEG400	Avicel PH 102, Eudragit RL100	Aerosil 200	Tablet	[36]

5.4.2   Carr's compressibility index (C.I) and Hausner's ratio (H.R)

Determine tap and poured density of liquisolid powders as per standard procedure. Calculate the Carr's index by using the following equation:

$$C.I. = \frac{(\text{Tapped density} - \text{Poured density})}{(\text{Tapped density})} \times 100 \quad (5.8)$$

Determine Hausner's ratio by using the following equation:

$$H.R. = \frac{\text{Tapped density}}{\text{Poured density}} \quad (5.9)$$

If the results of angle of repose, Carr's index and Hausner's ratio remain within the theoretical range [37], then it is suitable for processing into tablet or capsule dosage forms.

5.4.3   Wetting time determination

The time needed for complete wetting of the prepared liquisolid powders can be measured by taking 2 g of powder in a sintered glass funnel. Tap the poured powder 20 times before beginning of the test. Bring the sintered glass disk into contact with saturated drug solutions of the dissolution media and measure the time for the powder to be completely wetted [38].

5.4.4   Powder x-ray diffraction study (P-XRD)

To investigate polymorphic changes P-XRD study is performed. P-XRD patterns are generally obtained at room temperature with a Cu anode and a graphite monochromator, operated at a voltage of 35 kV and a current of 20 mA. The samples of liquisolid formulations can be analyzed in the 2θ angle range of 5–50°, and the process parameters were set as scan-step size of 0.02° (2θ) and scan-step time of 25 s. Disappearance of characteristic peaks or crystals of the drug generally indicates that the drug is converted into the amorphous form or is solubilized in the liquisolid formulation [39].

5.4.5   Differential scanning calorimetry

Differential scanning calorimetry (DSC) is conducted to assess the thermal behaviors of the drug, excipients used in the formulation, as well as the liquisolid system prepared. The following parameters like melting point, glass transition temperature, change of crystalline structure are analyzed. Samples are sealed in aluminium pans and evaluated at a constant heating rate the scanning temperature range according to the drug used. DSC is also used to determine compatibility between the drug and excipients detectable by shifting of the characteristic peak, used in the formulation. The drug has a characteristic peak and the absence of this peak in the DSC thermogram indicates that the drug is in the form of solution in the liquisolid formulation and hence it is molecularly dispersed within the system [40].

5.4.6   Quality control tests for liquisoild compact or tablets

Various quality control tests for liquisolid compacts as per USP. The liquisolid compacts must pass all the tests such as hardness, friability, disintegration test and dissolution test.

5.5   Conclusions

Poor water solubility of newly developed drugs is a challenge to formulation scientists. Conventional formulation approaches often fail to achieve the desired dissolution and bioavailability for such drugs. The liquisolid technology has demonstrated improvement in solubility and bioavailability. It is an efficient technology in terms of large-scale production capability similar to that of tablets with low cost of formulations. The present chapter also showed that the liquisolid technology can be used for the purpose of sustaining the release of highly water-soluble drugs if the suitable non-volatile solvent, carrier, coating and adjuvant materials are selected.

5.6   References

1. Vranikova B., Gajdziok J. Liquisolid systems and aspects influencing their research and development. *Acta Pharm.* 63: 447–465 (2013).

2. Ahmed K. Liquisolid technique: a promising alternative to conventional coating for improvement of drug photostability in solid dosage forms. *Expert Opin. Drug Deliv.* 10(10): 1335–1343 (2013).

3. Spireas S., Sadu S. Enhancement of prednisolone dissolution properties using liquisolid compacts. *Int. J Pharm.* 166: 1771–1788 (1998).

4. Grover R., Spireas S., Lau-Cam C. Development of a simple spectrophotometric method for propylene glycol detection in tablets. *J. Pharm. Biomed. Anal.* 16: 931–938 (1998).

5. Spireas S.S., Jarowski C.I., Rohera B.D. Powdered solution technology: principles and mechanism. *Pharm Res.* 9:1351–1358 (1992).

6. Spireas S. Liquisolid systems and methods of preparing same. US6423339B1: (2002).

7. Sanka K., Poienti S., Abdul B.M., Diwan P.V. Improved oral delivery of clonazepam through liquisolid powder compact formulations: In-vitro and ex-vivo characterization. *Powder Technology.* 256: 336–344 (2014).

8. Khanfar M., Salem M.S., Hawari R. Formulation factors affecting the release of ezetimibe from different liquisolid compacts. *Pharm. Dev. Technol.* 18(2): 417–427 (2013).

9. Rania H.F., Mohammed A.K. Enhancement of famotidine dissolution rate through liquisolid tablets formulation: In vitro and in vivo evaluation. *Eur. J. Pharma. Biopharm.* 69: 993–1003 (2008).

10. Akinlade B., Elkordy A.A., Essa E.A., Elhagar S. Liquisolid systems to improve the dissolution of furosemide. *Sci Pharm.* 78: 325–344 (2010).

11. Singh S.K., Srinivasan K.K., Gowthamarajan K., Prakash D., Gaikwad N.B., Singare D.S. Influence of formulation parameters on dissolution rate enhancement of glyburide using liquisolid technique. *Drug Dev. Ind. Pharm.* 38(8): 961–970 (2012).

12. Hentzschel C.M., Alnaief M., Smirnova I., Sakmann A., Leopold C.S. Enhancement of griseofulvin release from liquisolid compacts. *Eur. J. Pharma. Biopharm.* 80: 130–135 (2012).

13. Khaled K.A., Asiri Y.A., Yousry M., Sayed E. In vivo evaluation of hydrochlorothiazide liquisolid tablets in beagle dogs. *Int. J. Pharm.* 222: 1–6 (2001).

14. Spireas S., Sadu S., Grover R. In Vitro Release Evaluation of Hydrocortisone Liquisolid Tablets. *J Pharm Sci.* 87(7): 867–872 (1998).

15. Saeedia M., Akbaria J., Morteza-Semnani K., Enayati-Fard R., Dara S.S.R. Soleymania A. Enhancement of dissolution rate of indomethacin using liquisolid compacts. *Iran J Pharm Res.* 10 (1): 25–34(2011).

16. Javadzadeh Y., Mohammad R.S.S., Mohammad B.J. The effect of type and concentration of vehicles on the dissolution rate of a poorly soluble drug (indomethacin) from liquisolid compacts. *J Pharm Pharmaceut Sci.* 8(1):18–25 (2005).

17. Boghra R., Patel A., Desai H., Jadhav A. Formulation and evaluation of irbesartan liquisolid tablets. *Int J Pharm Sci Rev Res.* 9 (2): 32–37 (2011).

18. Mahmoud A.B., Amany O.K., Omaima A.S. Use of biorelevant media for assessment of a poorly soluble weakly basic drug in the form of liquisolid compacts: in vitro and in vivo study. *Drug Deliv,* Early Online: 1–10.

19. Tiong N., Elkordy A.A. Effects of liquisolid formulations on dissolution of naproxen. *Eur J Pharm Biopharm.* 73: 373–384 (2009).

20. Prajapati S.T., Bulchandani H.H., Patel D.M., Dumaniya S.K., Patel C.N. Formulation and Evaluation of Liquisolid Compacts for Olmesartan Medoxomil. *J Drug Delivery.* Volume 2013, Article ID 870579, 9 pages.

21. Kamble P.R., Shaikh K.S., Chaudhari P.D. Application of Liquisolid Technology for Enhancing Solubility and Dissolution of Rosuvastatin. *Adv Pharm Bulletin.* 4(2): 197–204 (2014).

22. Elkordy A.L., Tan X.N., Essa E.A. Spironolactone release from liquisolid formulations prepared with Capryol™ 90, Solutol HS-15 and Kollicoat SR 30 D as non-volatile liquid vehicles. *Eur J Pharm Biopharm.* 83: 203–223 (2013).

23. Chella N., Shastri N., Tadikonda R.R. Use of the liquisolid compact technique for improvement of the dissolution rate of valsartan. *Acta Pharmaceutica Sinica* B. 2(5): 502–508 (2012).

24. Javadzadeh Y., Siahi-Shadbad M.R., Barzegar-Jalali M., Nokhodchi A. Enhancement of dissolution rate of piroxicam using liquisolid compacts. *IlFarmaco.* 60: 361–365 (2005).

25. Javadzadeh Y., Shariati H., Movahhed-Danesh E. Effect of some commercial grades of microcrystalline cellulose on flowability, compressibility, and dissolution profile of piroxicam liquisolid compacts. *Drug Dev Ind Pharm.* 35:243–251 (2009).

26. Javadzadeh Y., Siahi M.R., Asnaashari S., Nokhodchi A. An Investigation of Physicochemical Properties of Piroxicam Liquisolid Compacts. *Pharm Dev Technol.* 12: 337–343 (2007).

27. Karmarkar A.B., Gonjari I.D., Hosmani A.H. Dissolution rate enhancement of Fenofibrate using liquisolid tablet technique. *Lat Am J Pharm.* 28:219-25 (2009).

28. Karmarkar A.B., Gonjari I.D., Hosmani A.H. Liquisolid technology for dissolution rate enhancement or sustained release, *Expert Opin. Drug Deliv.* 7: 1227–1234 (2010).

29. Nokhodchi A., Hentzschel C.M., Leopold C.S. Drug release from liquisolid systems: speed it up, slow it down, *Expert Opin. Drug Deliv.* 8(2): 191–205 (2011).

30. Kulkarni A.S., Aloorkar N.H., Mane M.S., Gaja J.B. Liquisolid systems: a review, *Int. J.Pharm. Sci. Nanotechnol.* 3: 795–802 (2010).

31. Javadzadeh Y., Musaalrezaei L., Nokhodchi A. Liquisolid technique as a new approach to sustain propranolol hydrochloride release from tablet matrices. *Int. J. Pharm.* 362: 102–108 (2008).

32. Nokhodchi A., Aliakbar R., Desai S., Javadzadeh Y. Liquisolid compacts: the effect of cosolvent and HPMC on theophylline release. Colloids Surf B Biointerfaces.79: 262–269 (2010).

33. Elkordy A.A., Essa E.A., Dhuppada S., Jammigumpula P. Liquisolid technique to enhance and to sustain griseofulvin dissolution: Effect of choice of non-volatile liquid vehicles. *Int. J. Pharm.* 434: 122–132 (2012).

34. Pavani E., Noman S., Syed I.A. Liquisolid Technique Based Sustained Release Tablet of Trimetazidine Dihydrochloride. *Drug Invention Today.* 5: 302–310 (2013).

35. Karmarkar A.B., Gonjari I.D., Hosmani A.H., Dhabale P.N. Evaluation of in vitro dissolution profile comparison methods of sustained release tramadol hydrochloride liquisolid compact formulations with marketed sustained release tablets. *Drug Discoveries & Therapeutics.* 4:26–32 (2010).

36. Elkhodairy K.A., Elsaghir H.A., Amal M.A. Formulation of indomethacin colon targeted delivery systems using polysaccharides as carriers by applying liquisolid technique. *BioMed Res. Int.* Article ID 704362, 1–17 (2014).

37. Aulton M.E. Pharmaceutical Preformulation. In: Taylor K, editor. Aulton's Pharmaceutics. The design and manufacture of medicines. Hungary: Elsevier; 2007, p. 355–357.

38. Yadav V.B. Enhancement of solubility and dissolution rate of bcs class II pharmaceuticals by non-aqueous granulation technique. *Int. J. Pharm. Res. Dev.* 1: 1–12 (2010).

39. Rajesh K., Rajalakshmi R., Umamaheswari J., Kumar C.K.A. Liquisolid technique: a novel approach to enhance solubility and bioavailability, *Int. J. Biopharm.* 2: 8–13 (2011).

40. Gavali S.M., Pacharane S.S., Sankpal S.V., Jadhav K.R., Kadam V.J. Liquisolid compact: A new technique for enhancement of drug dissolution, *Int. J. Res. Pharm. Chem.* 1: 705–713 (2011).

6

Cyclodextrin as carrier for drug delivery systems

Jnyanaranjan Panda*, Biswa Mohan Sahoo and Suryakanta Swain

*Faculty of Pharmacy, Roland Institute of Pharmaceutical Sciences, Khodasingi,
Berhampur (Ganjam), Odisha-760 010, India.*

6.1   Introduction

Cyclodextrins are cyclic oligomers of glucopyranose units discovered in
1891 by Villiers and studied later by Schardinger who identified α, β and
γ-cyclodextrin known today as the parents of cyclodextrins. The natural
cyclodextrins are obtained from starch due to degradation by cycloglycosyl
transferase amylase which is produced by the bacteria Bacillus macerans.
When the above enzyme degrades starch, the primary product of chain
splitting undergoes an intramolecular reaction without the participation
of water molecule and α-1→4-linked cyclic products are formed known as
cyclodextrins. Depending on the reaction conditions, there are three natural
types of cyclodextrins produced such as α-, β-, and γ-cyclodextrin with 6,
7 and 8 glucopyranose units joined through 1,4 alpha glucosidic linkage.
Each glucose unit posseses two secondary alcohols and a primary alcohol
that provide 18–24 sites for chemical modification and derivatization [1].
Cyclodextrins with lipophilic inner cavities and hydrophilic outer surfaces
are capable of interacting with a large variety of guest molecules to form
non-covalent inclusion complexes. Owing to lack of free rotation about the
bonds connecting the glucopyranose units, the cyclodextrins are not perfectly
cylindrical molecules but are toroidal or cone shaped (Fig. 6.1). Based on this
structural design, the primary hydroxyl groups are located on the narrow side
of the torus while the secondary hydroxyl groups are located on the wider
edge. Cyclodextrins have been found as potential candidates due to their
ability to alter physical, chemical and biological properties of guest molecules
through the formation of inclusion complexes [2]. Cyclodextrin molecules
are relatively large with a number of hydrogen donors and acceptors and
thus they do not permeate lipophilic membranes. In pharmaceutical industry,
cyclodextrins have been used primarily as complexing agents to increase
the aqueous solubility of poorly soluble drugs, and also to increase their

bioavailability and stability [3]. In addition, cyclodextrins can be used to reduce gastrointestinal drug irritation, convert liquid drugs into microcrystalline or amorphous powder, and prevent drug–drug and drug–excipients interactions.

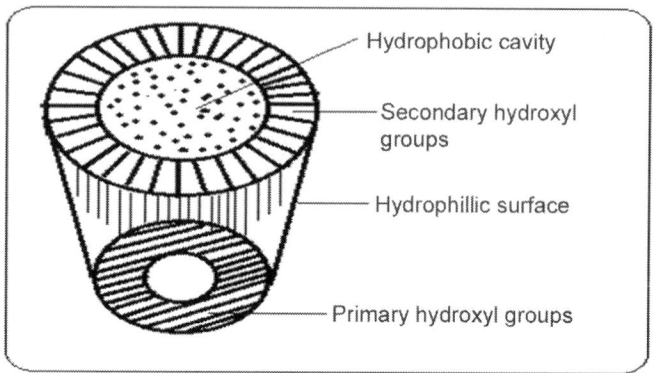

Figure 6.1 Planner representation and the molecular shape of Cyclodextrins

6.1.1   History of development of cyclodextrins

The history of cyclodextrins can be divided into three distinct periods such as (a) discovery from 1891 to 1930, (b) development from 1930 to 1970, and (c) industrial use from 1970 onwards (Fig. 6.2) [4].

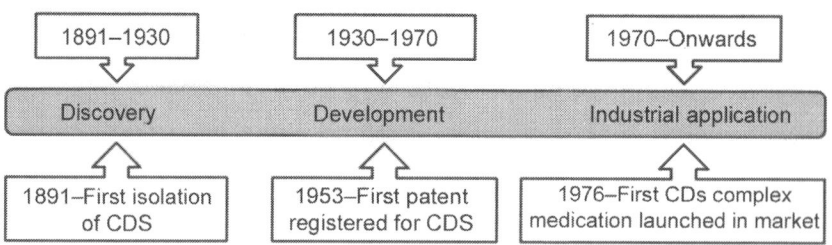

Figure 6.2 History of development of cyclodextrins

6.1.2   Types of cyclodextrin

Cyclodextrins are classified according to their number of glucopyranose units. The natural and most employed cyclodextrins are crystalline, homogeneous, non-hygroscopic substances which includes α-cyclodextrin (α-CD, cyclohexaamylose, 6 units of glucopyranose), β-cyclodextrin (β-CD, cycloheptaamylose, 7 units of glucopyranose) and γ-cyclodextrin (γ-CD, cyclooctaamylose, 8 units of glucopyranose) (Fig. 6.3) [5].

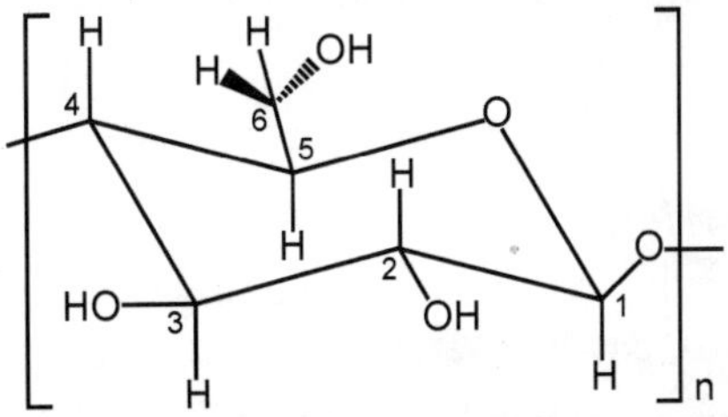

Figure 6.3 Structural representations of various cyclodextrins

6.1.3   Cyclodextrin derivatives

Considering that cyclodextrins contain 18 (α-CD), 21 (β-CD), or 24 (γ-CD) substitutable hydroxyl groups, the number of possible derivatives is unlimited. The β- cyclodextrin derivatives are normally distributed based on their interaction with the water molecules. So β-cyclodextrin is categorized into three derivatives such as hydrophilic, hydrophobic or ionisable derivatives. The first group (hydrophilic) has better solubility in water and is suitable for inclusion formation with poor water-soluble "guest" molecules. The dimethyl-β-cyclodextrin (DM-β-CD), trimethyl-β-cyclodextrin (TM-β-CD), hydroxyalkylated CDs such as hydroxypropyl-β-cyclodextrin (HP-β-CD) and branched CDs like G-β-CD are some examples of hydrophilic CD derivatives. The hydrophobic derivative such as DE-β-CD is capable of decreasing and modulating the release rate of water-soluble drug molecules. The ionisable cyclodextrins, such as CM-β-CD, CME-β-CD, and sulphobutyl ether-β-cyclodextrin (SBE-β-CD), can enhance the dissolution rate, the inclusion capacity and also the decrease of the side effects of some drug molecules [6].

6.1.3.1   Ideal properties of cyclodextrin derivatives

The optimum cyclodextrin derivatives should posses following ideal properties:

- Very soluble in water
- High purity
- Nontoxic even in high doses or in chronic treatment
- Characterized by high solubilizing power for various drugs
- Stable during heat sterilization and storing in aqueous solution

- Non-reacting with cholesterol and phospholipids
- Free from any intrinsic pharmacological effect
- Biodegradable in the blood circulation and eliminated as small molecular metabolites
- Cheaply available

6.1.4   Synthesis of cyclodextrins

The production of cyclodextrins is relatively simple and involves treatment of ordinary starch with a set of easily available enzymes. Commonly cyclodextrin glycosyltransferase (CGTase) is employed along with α-amylase. CGTase(s) are predominantly extracellular enzymes which are produced by various bacteria such as *Bacillus macerans, Bacillus circulance, Bacillus megaterium, Klebsiella pneumonia, Bacillus stearothermophillus, Bacillus amyloliquefaceins, Alkolophilic Bacillus sp., Bacillus lentus, Thermoanaerobacter sp. and Micrococcus.* Commercially available CGTases have been produced from *Bacillus macerans.* Recently attempts were made to isolate thermo tolerant CGTase producing bacteria, i.e. *B. stearothermophilus.*

Generally two types of cyclodextrin production processes are used such as solvent process and non-solvent process. Solvent process requires an organic solvent mainly toluene, ethyl alcohol or acetone that acts as a complexing agent which extracts one type of cyclodextrin selectively and thus directs the enzyme reaction to produce particular type of cyclodextrin of interest. Non-solvent process does not require complexing agents and produces a mixture of cyclodextrins. The preparation of cyclodextrins consists of following four major phases.

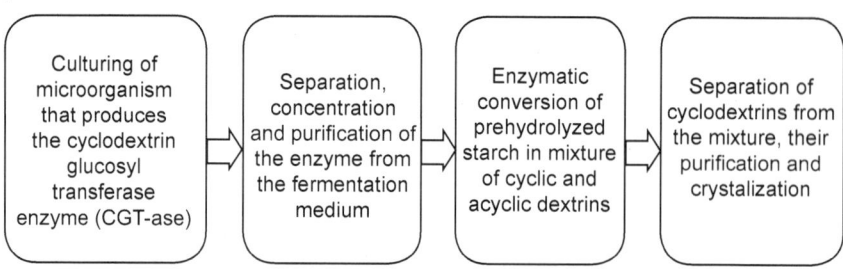

First starch is liquified either by heat treatment or using α-amylase followed by the addition of CGTase for enzymatic conversion. CGTases can synthesize all forms of cyclodextrins, thus the product of the conversion results in a mixture of the three main types of cyclic molecules, in ratios that are strictly dependent on the enzyme used: each CGTase has its own characteristic α:β:γ synthesis ratio. Purification of the three types of cyclodextrins takes advantage of the different water solubility of the molecules: β-cyclodextrin which is very poorly water soluble can be easily retrieved through crystallization while the more soluble α- and γ-CDs (145 and 232 g/l respectively) are usually purified by means of expensive and time-consuming chromatography techniques. As an alternative a "complexing agent" can be added during the enzymatic conversion step: such agents (usually organic solvents like toluene, acetone or ethanol) form a complex with the desired cyclodextrin which subsequently precipitates. The complex formation drives the conversion of starch towards the synthesis of the precipitated cyclodextrin, thus enriching its content in the final mixture of the products [7].

6.1.5   Properties of cyclodextrin

6.1.5.1   Physical properties

All the three cyclodextrins are white crystalline powder. They have no definite melting point but start decompose from temperature of 200°C. The solubility of cyclodextrins depends mainly on the temperature. Cyclodextrins are fairly soluble in water. However, β-cyclodextrin shows remarkably lower solubility than α-CD and γ-CD. The aqueous solubility of all cyclodextrins increases at an elevated temperature. Cyclodextrins are insoluble in most of the organic solvents but they are soluble in some polar and aprotic solvents. Temperature-dependent solubility of cyclodextrin may change due to its complexation with the guest molecule. The inclusion complex is more soluble than cyclodextrin itself when the guest molecule is highly soluble in water. In contrast, inclusion complex with the guest molecule of poor water solubility generally results in a decrease in solubility of the cyclodextrin, although solubility of inclusion complex is generally less than that of the cyclodextrin, it is greater than that of the guest molecule [8].

6.1.5.2   Chemical properties

Cyclodextrins have no reducing end groups. Periodate oxidation of α-, β-, and γ-CD will open the glucopyranose ring, but no formic acid or formaldehyde is formed, consistent with the fact that cyclodextrins do not contain free end groups. The glycosidic bonds of cyclodextrins are fairly stable towards alkali and even at elevated temperature. Cyclodextrins are more resistant to acid hydrolysis than is starch. Strong acid such as hydrochloric acid hydrolyzes the cyclodextrins to produce a mixture of oligosaccharides. The rate of acid hydrolysis increases as functions of both increased temperature and the concentration of acid. The hydrolysis of cyclodextrins is minimum in the presence of weak organic acids. Cyclodextrins are more resistant to acid catalyzed hydrolysis as compared to that of linear sugars. The rate of ring opening of cyclodextrins increases with increase in cavity size in the order of α- cyclodextrin < β- cyclodextrin < γ-cyclodextrin (Table 6.1) [9].

Table 6.1 Properties of cyclodextrins

Properties	α- cyclodextrin	β-cyclodextrin	γ-cyclodextrin
No. of glucopyranose unit	6	7	9
Molecular Weight (g/mol)	972	1135	1297
Inner cavity diameter (Å)	4.7–5.3	6.0–6.5	7.5–8.3
Outer cavity diameter (Å)	14.6	16.4	17.5
Cavity height (Å)	7.9	7.9	7.9
Cavity volume (A³)	174	262	427
Water solubility at 25°C (g/100 mL)	14.5	1.85	23.2
Specific rotation	150.5 ± 0.5	162.5 ± 0.5	177.4 ± 0.5
pKa by potentiometry	12.33	12.20	12.08
Hydrolysis by α-amylase	Negligible	Slow	Fast
Shape of the crystals	Hexagonal lattice	Monocyclic parallelogram	Quadratic prism

6.1.6   Inclusion complexes

Inclusion complexes are entities comprising two or more molecules in which one of the molecule, the "host" molecule and the second one is a "guest" molecule (Fig. 6.4). Molecules or part of molecules which are hydrophobic and can fit into the cavity of host in the presence of water are included into the host cavity [10].

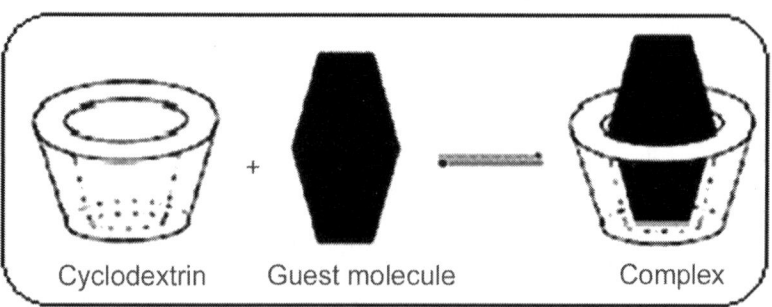

Cyclodextrin   Guest molecule   Complex

Figure 6.4 Formation of inclusion complexes

6.1.6.1   Mechanism of cyclodextrin inclusion complex

The complexes are formed when a "guest" molecule is partially or fully included inside a "host" molecule, like cyclodextrin, with no covalent bonding. Most frequently, CD/substrate inclusion complexes exist with a host: guest ratio of 1:1, which signifies the essence of molecular encapsulation, can be represented as

Drug + CD $\rightleftharpoons$ Drug : CD Complex

The inclusion of a guest in a cyclodextrin cavity consists basically of a substitution of the included water molecules by the less polar guest. The process is energetically favored by the interactions of the guest molecule with the solvated hydrophobic cavity of the host. Generally there are various types of interactions take place which help during the formation of inclusion complex.

1. The displacement of polar water molecules from the apolar cyclodextrin cavity.
2. The increased number of hydrogen bonds formed as the displaced water returns to the large pool.
3. A reduction of the repulsive interaction between the hydrophobic guest and the aqueous environment.
4. An increase in the hydrophobic interactions as the guest inserts itself into the apolar cyclodextrin cavity.
5. Van der Waals interactions

The energy of covalent chemical bonding is 400 kJ/mol, hydrogen bond is about 40 kJ/mol, and the Van der Waals force is about 4 kJ/mol. In the case of inclusion complexes, the species may achieve a stability which is proportional to covalent bonding due to the spatial arrangement produced. The energy of Van der Waals forces is proportional to molecular polarizability and molecular refraction. The stability of the complex grows with the increase in the electron-donor character of the substituents of the included molecule. The most widely used approach to study inclusion complexation is the phase solubility method described by Higuchi and Connors, which examines the effect of a solubilizer, i.e. cyclodextrin or ligand on the drug being solubilized, i.e. the substrate. Various factors are considered to select the type of cyclodextrin for a given study: physico-chemical properties of the guest, size of the cyclodextrin cavity, solubility, the preparation method and possibility of co-encapsulation. Basically, the inclusion complexes involve interactions of the molecules of both the host and the guest, and it is a combination of different non-covalent interactions such as ionic, dipolar, electronic, van der Waals and the hydrophobic effect, besides the size and the shape of the molecules [11].

6.1.6.2   Techniques involved in preparation of inclusion complex

The various techniques used for the preparation of inclusion complex are discussed below [12].

Grinding

Inclusion complexes can be prepared by simply grinding the guest with cyclodextrin. This method is suitable for oils or liquid guests. The degree of complexation achieved by this process is very slow.

Solid dispersion/co-evaporated dispersion

The drug is dissolved in ethanol and cyclodextrin is either dissolved in alcoholic solution or dissolved separately in water, or other suitable medium. The cyclodextrin solution is then added to the drug solution or vice versa, and stirred to attain equilibrium. The resulting solution is evaporated to dryness preferably under vacuum.

Neutralization method

In this method equimolar concentration of drug and cyclodextrin are separately dissolved in 0.1 N NaOH, mixed and stirred for about half an hour. pH is recorded and 0.1 N HCl is added drop wise with stirring until pH reaches 7.5, where upon complexes precipitates. The residue is filtered and washed until free from chlorine; it is dried at 250°C for 24 h and stored in a desiccators.

Kneading method

In this method, cyclodextrin is not dissolved but kneading like a paste with small amount of water to which the guest component has been added. Guest component can be added without a solvent or in a small amount of ethanol in which guest has been suspended. After grinding, paste solvent gets evaporated and powder like complex is formed.

Precipitation method

In this method, drug and cyclodextrin are dispersed in water, and solution is heated to obtain concentrate, viscous and translucent liquid. The solution is left to give precipitate of inclusion complex. The precipitate can be collected by decanting, centrifugation or filtration. The precipitate may be washed with a small amount of water or other water-miscible solvent such as ethyl alcohol, methanol or acetone. Solvent washing may be detrimental with some complexes, so this should be tested before scaling up. On the other hand, additives such as ethanol can promote complex formation in the solid or semisolid state. Un-ionised drugs usually form a more stable cyclodextrin complex than their ionic counterparts and, thus, complexation efficiency of basic drugs can be enhanced by addition of ammonia to the aqueous complexation media. For example, solubilisation of pancratistatin with hydroxypropyl-cyclodextrins was optimized upon addition of ammonium hydroxide.

Spray drying (Atomization)

In spray drying, cyclodextrin is dissolved in a solution previously alkalinized with 25% aqueous ammonia (final pH 9.5). The guest is dissolved in 100 ml of 96% ethyl alcohol. Both solutions are mixed and sonicated; the final solution is spray-dried to get the complexes.

Freeze-drying (lyophylization)

In freeze-drying, physical mixture of guest and cyclodextrin is wetted with a small amount of buffer solution and is kneaded forming a homogeneous suspension which is then freeze-dried. The final complexes are pulverized and sieved through appropriate sieve. Freeze-drying is an industrially applicable method for heat labile guests, but large amount of water, if it is used as solvent, and excessive cyclodextrin would be required because of the low solubility of hydrophobic guest in aqueous solution and makes the process time consuming.

Melting

Complexes can be prepared by simply melting the guest, mixed with finely powdered cyclodextrin. In such cases there should be a large excess of guest, and after cooling this excess is remove by careful washing with a weak complex, forming solvent or by vacuum sublimation. The later is preferred method and is used to sublimate guests such as menthol.

Extrusion

It is a variation of the heating and mixing method. Cyclodextrin, guest and water can be mixed as added to the extruder. Degree of mixing, amount of heating and time can be controlled in the barrel of the extruder. Depending upon the amount of water, the extruded complex may dry as it cools or the complex may be placed in an oven to dry. Extrusion has the advantages of being a continuous process and using very little water. Because of the generation of heat, some thermo-labile guests decompose by this method.

Co-precipitation

Cyclodextrin is dissolved in water and the guest is added while stirring the cyclodextrin solution. By heating, more cyclodextrin can be dissolved (20%) if the guest can tolerate the higher temperature. The cyclodextrin and guest solution must be cooled under stirring before a precipitate is formed. The precipitate can be collected by decanting, centrifugation or filtration and washed. The main disadvantage of this method lies in the scale-up.

Slurry complexation

It is not necessary to dissolve the cyclodextrin completely to form a complex. Cyclodextrin can be added to water as high as 50–60% solids and stirred. The aqueous phase will be saturated with cyclodextrin in solution. Guest molecules will complex with the cyclodextrin in solution and, as the cyclodextrin complex saturates the water phase, the complex will crystallize or precipitate out of the aqueous phase. The cyclodextrin crystals will dissolve and continue to saturate the aqueous phase to form the complex and precipitate or crystallize out of the aqueous phase, and the complex can be collected in the same manner as with the co-precipitation method. The time required to complete the complexation is variable, and depends on the guest. Generally slurry complexation is performed at ambient temperatures. With many guests, some heat may be applied to increase the rate of complexation, but care must be applied since too much heat can destabilize the complex and the complexation reaction may not be able to take place completely. The main advantage of this method is the reduction of the amount of water needed and the size of the reactor.

Damp mixing

The guest and cyclodextrin are thoroughly mixed and placed in a sealed container with a small amount of water. The contents are heated to about 100°C and then removed and dried.

Microwave irradiation method

This technique involves the microwave irradiation reaction between drug and cyclodextrin using a microwave oven. The drug and cyclodextrin are dissolved in a mixture of water and organic solvent in a specified proportion. Then the prepared mixture is reacted for few minutes in the microwave oven. After completion of the reaction, sufficient quantity of solvent mixture is added to the above reaction mixture to remove the uncomplexed free drug and cyclodextrin. The precipitate is filtered and dried under vacuum. Microwave irradiation technique is a novel method for industrial scale preparation due to its major advantage of shorter reaction time and higher yield of the product.

Supercritical antisolvent technique

In this technique, supercritical carbon dioxide is used as anti-solvent due to its low critical temperature and pressure which makes it useful for preparation of thermo labile substances. It is also non-toxic, nonflammable, in-expensive and is much easier to remove from the polymeric materials when the process is completed. Supercritical carbon dioxide is suggested as a new complexation medium due to its properties of improved mass transfer and increased solvating power. This method constitutes one of the most innovative methods to prepare the inclusion complex of drug with cyclodextrin in solid state. In this method, first drug and cyclodextrin are dissolved in a suitable solvent and then the solution is fed into a pressure vessel under supercritical conditions through a nozzle in which supercritical fluid is sprayed as an anti-solvent. When the solution is sprayed into supercritical fluid anti-solvent, the anti-solvent rapidly diffuses into that liquid solvent as the carrier liquid solvent. Because of the expansion of supercritical fluid solvent, the mixture becomes supersaturated which results in the precipitation of the complex. Thus, supercritical particle generation processes is an efficient route for improving bioavailability of pharmaceutically active compounds.

6.1.6.3   Factors affecting inclusion complex formation

The various factors which affect the formation of cyclodextrin inclusion complex are type of cyclodextrin, method of preparation, cavity size, pH, ionization state and temperature etc. [13].

Type of cyclodextrin

Type of cyclodextrin can influence the formation as well as the performance of drug-cyclodextrin complexes.

Method of preparation

Method of preparation, viz. co-grinding, kneading, solid dispersion, solvent evaporation, co-precipitation, spray drying, or freeze drying can affect drug/ cyclodextrin complexation. The effectiveness of a method depends on the nature of the drug and cyclodextrin. In many cases, spray drying, and freeze drying were found to be most effective for drug complexation.

Cavity size of cyclodextrin

For complexation, the cavity size of cyclodextrin should be suitable to accommodate a drug molecule of particular size. Compared with neutral cyclodextrins, complexation can be better when the cyclodextrin and the drug carry opposite charge but may decrease when they carry the same charge. Cavity size of β-cyclodextrin was suitable for complexation while that of α-cyclodextrin was insufficient for some drugs such as Gliclazide. The effective enhancement of dissolution rate with only with β- and γ- cyclodextrins but the cavity of α-cyclodextrin is less suitable.

pH and ionization state

The strong drug/ cyclodextrin interaction takes place in acidic region of pH 4 whereas the solubility of the cationic drug is increased at pH 1. The stability constants of complex are low with the highly polar drug at pH 5 due to its lesser ability to enter the cyclodextrin cavity but were high with anionic less polar form at pH 10. In case of ionizable drugs the presence of charge plays an important role in drug-CD complexation. In general ionic forms of the drugs are weaker complex forming agents than their non-ionized forms. In case of mebendazole, the unionized form was less included in HP-β-CD than the cationic form.

Temperature

The increase in temperature decreases the apparent stability constant where as the increase in the temperature decreases the association constant for binding.

Co-solvents or additives

Some additives may compete with the drug molecules for cyclodextrin cavities and thus decrease the apparent complex stability constant, for example additives with positive and negative hydrotropic effect. Co-solvents can improve the solubilizing and stabilizing effects of cyclodextrins, for example the use of 10% propylene glycol in development of an oral Itraconazole preparation containing 40% HP-β-CD.

6.1.6.4   Advantages of Complexation

The formation of cyclodextrin-drug complex provides advantageous changes in their properties in the following ways [14].

(a) Enhance solubility, i.e. increased aqueous solubility of water-insoluble drugs

(b) Enhance bioavailability

(c) Enhance stability, i.e. stabilization of light or oxygen-sensitive substances and protection of drug from ultraviolet radiation, oxidation, and hydrolysis

(d) Simplest to formulate

(e) Convert liquids and oils to free flowing powders

(f) Reduce evaporation and stabilize flavours

(g) Reduction of unpleasant odor, tastes and irritations of drugs

(h) Masking pigments or the color of substances

(i) Reduce haemolysis

(j) Modification of the chemical reactivity of guest molecules

(k) Fixation of very volatile substances

(l) Protection against degradation of substances by microorganisms

(m) Prevent admixture incompatibilities

(n) Emulsification of hydrocarbons, steroids, and fats

(o) Prevention of drug–drug interactions

(p) Enable mixing of incompatible drugs

(q) Controlled release of drugs

(r) Provision of a constrained medium for chemical synthesis

(s) Used as antidote for metal poisoning

6.1.6.5   Disadvantages of complexation

Due to formation of inclusion complex between drug and cyclodextrins, the following limitations are observed [15].

(a) Laborious and expensive methods of preparation

(b) Reproducibility of physicochemical characteristics

(c) Difficulty in incorporating into formulation of dosage forms

(d) Scale-up of manufacturing process

(e) Stability issues

(f) Only small dose drugs are complexed

6.1.6.6   Characterization of inclusion Complex

Spectroscopic methods

The quantitative determination can be performed by spectroscopic methods such as UV-Visible spectroscopy, FT-IR, NMR and Mass [16–21].

(a) UV-Visible Spectroscopy – UV-Visible spectroscopy is useful for the detection of inclusion complex by change in the absorption spectrum and molar extinction coefficient. During the spectral changes, the chromophore of the guest is transferred from an aqueous medium to the non-polar cyclodextrin. Hypsochromic or bathochromic shift or increases in the absorption intensity without change in the λmax have been considered as evidence for interaction between cyclodextrin and the drug molecule. Hydrogen bonding is considered as the main force behind the formation of inclusion complex. As hydrogen bonding lowers the energy of 'n' orbitals, a hypsochromic shift (blue shift) is observed. The cleavage of the existing hydrogen bonds in the compound can lead to a bathochromic shift due to complexation.

(b) Fourier Transform Infra Red Spectroscopy (FT-IR) – FT-IR is widely used to determine the functional groups of a compound. It is also used to prove the presence of component of inclusion complex which belongs to the host and guest molecules. FT-IR is a very useful tool to prove the existence of both guest and host molecules in their inclusion complexes. The characteristic bands of cyclodextrin, representing the overwhelming part of the complex, are scarcely influenced by the complex formation. Bands due to the included part of the guest molecule are generally shifted or their intensities are altered, but since the mass of the guest molecule does not exceed 5–15% of the mass of the complex, these alterations are usually obscured by the spectrum of the host. In literature, most often the IR spectroscopic studies of such cyclodextrin complexes are reported which have a carbonyl group-bearing guest. This is due to the adequate and well-separated bands of the carbonyls (1680–1700 cm^{-1}), which is significantly covered and shifted by complexation with cyclodextrin. The IR spectrum of β=cyclodextrin shows prominent peaks at 3392 cm^{-1} (O-H stretch), 2927 cm^{-1} (C-H stretch), and 1630 cm^{-1} (C=O stretch). The broad peak of β-cyclodextrin at 3392.14 cm^{-1} was gradually reduces its intensity in physical mixture and in inclusion complex (Fig. 6.5).

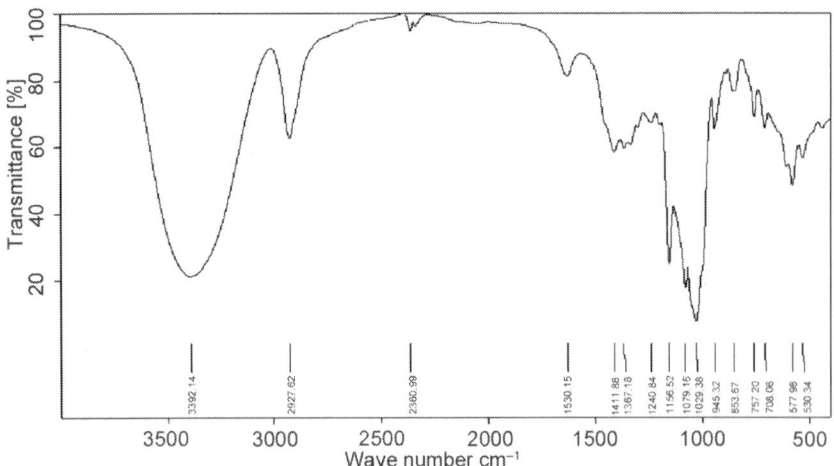

Figure 6.5 FTIR spectrum of β-cyclodextrin

(c) Nuclear Magnetic Resonance (NMR) Spectroscopy – NMR is a powerful technique which provides the most evidence for the inclusion of a guest molecule into the hydrophobic cyclodextrin cavity. It is also useful for the investigation of interactions between guest and cyclodextrins molecules because it gives valuable information regarding physico-chemical parameters, orientation of the guest molecule inside the cavity and stability of the inclusion complex. The insertion of guest molecule into the hydrophobic cavity of the β-cyclodextrin results in the chemical shift of guest and host molecules in the NMR spectra. Generally, the large chemical shifts are observed at H3 and H5, which are placed in the inner cavity of cyclodextrin due to the inclusion complex (Fig. 6.6). The chemical shift (δ) is defined as the difference in chemical shift change, positive sign means a downfield shift and negative sign means an upfield shift.

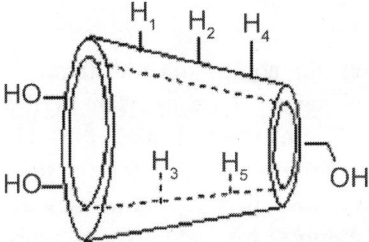

Figure 6.6 Chemical shifts observed in β-cyclodextrin complexes

(d) Mass Spectrometry (MS) – Cyclodextrin guest-host complexes have been identified with several classes of molecules, including aromatics, barbiturates and dicarboxylic acids in both negative and positive ion mode mass spectrometry. Mass spectrometry can used to determine the orientation and stoichiometry for such complexes. Currently the LC-MS analysis is carried out on a triple quadrupole mass spectrometer with an electrospray ionization source in the positive ion full scan mode.

(e) Thermo Analytical methods – Thermal analysis of inclusion complex is useful to differentiate between inclusion complexes and adsorbates along with characterization of the thermal affects due to molecular entrapment. The various thermo analytical methods used are differential scanning calorimetry, thermo derivatography and thermal evolution analysis. Differential scanning calorimetry determines the rate of heat evolved or absorbed by the sample during a temperature programme. After melting, a small exothermic peak is recorded which suggests the complex formation. The broadening, shifting, and appearance of new peaks or disappearance of certain peaks in thermo gram can be characterized with the help of differential scanning colorimetric and differential thermal analysis.

Chromatographic method

The various chromatographic methods available for characterization of cyclodextrins are thin layer chromatography, gas chromatography, high performance liquid chromatography. In thin layer chromatography, the R_f values of a guest molecule decreases to large extent which helps in identifying the formation of complex between guest and host molecule. Inclusion complexation between guest and host molecules is a reversible process. As a result, the complex is separated completely in guest and host molecules during the chromatographic process and only the spots of the guest and host molecules are found on the TLC plate [22].

X-ray diffraction

Liquid guest molecules do not produce diffraction pattern. When guest molecule is a solid substance, a comparison of the diffractogram of the complex with a mechanical mixture of the guest and cyclodextrin has to be made. When the diffractograms are different, i.e. the characteristic peaks of one or other components disappear and new ones appear as a result of the complex experiments, complex formation is very probable [23].

Scanning Electron Microscopy

The scanning electron microscope produce images of the sample surface by scanning it with a high-energy beam of electrons in a raster scan pattern. The surface morphology of raw materials such as drug and cyclodextrin is examined by means of a scanning electron microscope. The samples are fixed on a brass stub using double-sided tape and then made electrically conductive by coating in a vacuum with thin layer of copper and the photographs are taken [20].

6.1.6.7   *Release of drugs from the drug-cyclodextrin complex*

Various mechanisms play major role in drug release from the drug-cyclodextrin complex. Complexation of the drug (D) to cyclodextrin occurs through a non-covalent interaction between the molecule and the cyclodextrin cavity [24]. This is a dynamic process whereby the drug molecule continuously associates and dissociates from the host cyclodextrin. Assuming a 1:1 complexation, the interaction will be as follows.

Drug + cyclodextrin $\overset{K}{\rightleftarrows}$ Drug – Cyclodextrin complex

There are two important parameters involved in the drug release mechanism such as the complexation constant (K) and the lifetime of the complex. The various mechanisms involved in the release of drugs from the complex are discussed below.

Dilution

Dissociation due to dilution appears to be a major release mechanism. Dilution is minimal when a drug-CD complex is administered ophthalmically. Efficient corneal absorption is further exacerbated by contact time.

Competitive displacement

Competitive displacement of drugs from their cyclodextrin complexes probably plays a significant role in vivo. Addition of parabens to parenterals not only leads to decreased antimicrobial activities of the parabens, due to complexation, but also decreases the drug solubility due to its displacement from complexes. Oral administration of the complex showed less bioavailability than expected, based on the in vitro dissolution experiments. It was suggested that cinnarizine was too strongly bound to the cyclodextrin so that complex dissociation was limiting oral bioavailability. Co-administration of phenylalanine, a displacing agent, improved the bioavailability of cinnarizine from the complex but not from conventional cinnarizine tablets.

Protein binding

Drug binding to plasma proteins may be an important mechanism by which the drug may be released from a drug-CD complex. It is evident that proteins may effectively compete with cyclodextrins for drug binding and thus facilitate the in vivo release of drugs from drug-CD complexes. Dilution alone may be effective in releasing free drugs from weak drug-CD complexes but when the strength of the binding between the drug and cyclodextrin is increased, a mechanism such as competitive displacement is at work. Plasma and tissue protein binding may also play a significant role.

Drug uptake by tissue

A potential contributing mechanism for drug release from cyclodextrin is preferential drug uptake by tissues. When the drug is lipophilic and has access to tissue, and is not available to the cyclodextrin or the complex, the tissue then acts as a sink, causing dissociation of the complex based on simple mass action principles. This mechanism is more relevant for strongly bound drugs or when the complex is administered at a site where dilution is minimal, e.g. ocular, nasal, sublingual, pulmonary, dermal or rectal sites. For example, cyclodextrin has been used in ophthalmic delivery of poorly water-soluble drugs to increase their solubility and/or stability in the tear fluid, and in some cases to decrease irritation.

Change in ionic strength and temperature

In the case of a weak electrolyte, the strength of binding to cyclodextrin is dependent on the charged state of the drug, which is dependent on dissociation constant(s) of the drug and the pH of environment. For most molecules, the ionized or charged form of the molecule has poorer binding to cyclodextrin compared to the non-ionized or neutral form of the drug, especially when bound to a neutral cyclodextrin. Binding of substrate to cyclodextrin is an exothermic process. Hence any increase in temperature results in a weakening of the complex and thus increases the free fraction of substrate. Drug-CD complexes are usually prepared and stored at/or below room temperature. Since the normal body tissue temperature can be as high as 37°C, this difference in temperature may be another contributing factor to drug dissociation in vivo.

6.2   Applications of cyclodextrins in drug delivery

6.2.1   Oral drug delivery

6.2.2.1   Immediate release

The rate of dissolution of poorly water-soluble drugs is generally responsible for rate and extent of oral bioavailability of the drugs. Highly hydrophilic cyclodextrin derivatives such as HP-β-CD, maltosyl-β-CD and SBE-β-CD have been used to achieve the immediate release formulation which is readily dissolved in GIT thereby enhancing the oral bioavailability of poorly water-soluble drugs through the formation of inclusion complexes. The hydrophilic cyclodextrins have been widely applied to increase the oral bioavailability of cardiac glycosides, vasodilators, barbiturates, anti-epileptics, benzodiazepines, antidiabetics, steroidal and non-steroidal anti-inflammatory drugs etc. [25].

6.2.2.2   Prolonged release

The slow-release formulations are designed to achieve zero order or pH-independent release of drugs to provide a constant blood level for a longer period of time. They have several advantages such as reducing the frequency of dose, prolongation of the drug efficacy and avoiding the toxicity associated with the administration of a simple plain tablet. For this reason, hydrophobic cyclodextrins such as alkylated and acylated derivatives are used as slow-release carriers for water-soluble drugs. Hydrophobic alkylated cyclodextrins such as DE-β-CD, TE-β-CD were the first slow-release carriers to be used in conjugation with Diltiazem [26].

6.2.2.3   Modified release

Modified release involves the release of drug in a different physical state. The conventional formulation of nifedipine, a typical calcium-channel antagonist, must be dosed either twice or three times daily, because of the short elimination half-life due to the considerable first-pass metabolism. Moreover, it has some pharmaceutical problems such as low oral bioavailability due to poor aqueous solubility and a decrease in dissolution rate during the storage due to the crystal growth. Therefore, the release rate of nifedipine must be modified in order to obtain a more balanced oral bioavailability with prolonged therapeutic effect [27].

6.2.2.4   Delayed release

CME-β-CD was developed to exhibit pH-dependent solubility for use in selective dissolution of the drug CD-complex. Molsidomine absorption from tablets containing CME-β-CD was studied in gastric acidity-controlled dogs in fasted and fed states. Under high gastric acidity, molsidomine absorption was significantly retarded compared to that found under low gastric acidity conditions [27].

6.2.2.5   Cyclodextrin in osmotically controlled oral drug delivery

Osmotically controlled oral drug delivery systems are available to control the drug release based on the principle of osmosis. Osmotic tablets provide various advantages such as zero order delivery rate, improvement of patient compliance and high degree of in vivo-in vitro correlation. Oral osmotic pump tablets generally consist of a core including the active agent, an osmogent and other common excipients coated with semi-permeable membrane. A delivery orifice drilled through the coating which provides a passage way for drug release by hydrostatic pressure created from the core osmogent when exposed to an aqueous environment. The rate of drug release from osmotic pump depends on the drug solubility and osmotic pressure of the core. Hence, these systems are suitable for delivery of drugs with moderate water solubility. The push–pull osmotic tablets were developed in the 1980s and were used to deliver drugs having low to high water solubility. Drugs such as oxybutynin chloride, nifedipine and glipizide are based on this technology. These pumps are in the form of a two-layer tablet with a drug and a push layer. When the system comes in contact with an aqueous environment, both layers absorb water. The lower part, which does not have an orifice swells and pushes the drug through the orifice as a solution or suspension in the upper chamber. However, this system has some disadvantages such as laser drilling technology should be required to drill the orifice next to the drug compartment and lag time for drug release from osmotic pumps after coming in contact with the aqueous media is long. So, it has been reported that Sulfo butyl ether-7m-β-CD can act as solubilizer as well as osmotic agent. Therefore, the elementary osmotic pump (EOP) tablets were prepared to improve the solubility and dissolution rate of lovastatin by complexation with β-CD by kneading method. Hence, the dissolution rate of lovastatin β-CD is significantly enhanced and this system has better solubility performance in EOP tablet formulations [28].

6.2.3   Sublingual drug delivery

Sublingual drug delivery is one of the most effective routes of administration to bypass hepatic first pass metabolism. In this method, the drug enters into the systemic circulation by dissolving in the mucosa. In the sublingual formulations, the complexation of poorly water soluble drugs with cyclodextrin increase the bioavailability of various lipophilic drugs. For example, 2-hydroxy propyl-β-cyclodextrin is used to increase the bioavailability of 17β-oestradiol, androstenediol, clomipramine and danazol. The increased bioavailabilities achieved by cyclodextrins are due to the increased aqueous solubility and drug resolution rate. Along with this, they also act as penetration enhancers. The drug should be released from the inclusion complex before it can be absorbed. This can be a problem for sublingual application due to the small volume of aqueous saliva and the relatively short residence time. The dissolved drug is removed from the buccal cavity within few minutes after administration, thereby not allowing enough time for the drug to be released from cyclodextrin complex [29].

6.2.4   Parenteral drug delivery

Cyclodextrin derivatives have been widely investigated; particularly HP-β-CD and SBE-β-CDs have been widely investigated for parenteral use because of their high aqueous solubility and minimal toxicity. Applications of cyclodextrins in parenteral delivery are solubilization of drugs, reduction of drug irritation at the site of administration and stabilization of drugs unstable in aqueous environment etc. An IM dosage form of Ziprasidone Mesylate with targeted concentration of 20–40 mg/mL was developed by inclusion complexation of the drug with SBE-β-CD. Formation of a stable, water-soluble Dexamethsone complex with sugar branched β-CDs suggested the potential of these cyclodextrins as excellent carriers in steroidal injectable formulations. SBE-β-CD was found to be useful in the preparation of parenteral solutions of poorly water-soluble drugs with positive charge [30].

6.2.5   Ophthalmic drug delivery

Topically applied drug formulations for instance suspensions, oily drops, gels, ointments and solid inserts have been used, but most of these show unwanted side effects such as irritation and blurred vision of eye. Cyclodextrins act as an anti-irritant by formation of inclusion complex and thereby masking the irritating property of drugs or by replacing the irritating additives from the formulation. The irritation due to Pilocarpine is because of rapid absorption of the lipophilic prodrug into the lipophilic corneal epithelium or by precipitation of prodrug in the pre-corneal area. Pilocarpine/SBE7 β-CD complexes can act as depot which limits the free prodrug concentration in the precorneal area to a non-irritating level. Similarly, the anti-allergic drug like Cetirizine causes strong irritation after ophthalmic administration but co-administration of α-CD, β-CD and γ-CD reduces this irritation. Eye drops are generally prepared in a multi-dose container which contains an antimicrobial preservative. The antimicrobial activity of preservative can be reduced by complexation of cyclodextrin with the preservatives. The lipophilic preservatives such as p-hydroxy benzoic acid esters form complexes with HP-β-CD and their antimicrobial effect decreases. Contrary to this, the antimicrobial effect of hydrophilic preservatives like Thimerosal, Bronopol, Benzalkonium chloride and Chlorohexidine gluconate are not much affected due to less effective binding to cyclodextrin [30].

6.2.6 Nasal drug delivery

The administration of drugs by nasal route eliminates the first pass metabolism. The lipophilic drugs are difficult to deliver through this route because of their poor water solubility. Hydrophilic drugs like proteins and peptides show inadequate absorption by nasal route. This route shows large interspecies differences which exist in the nasal absorption of drugs and also the nasal mucosa possesses enzymatic activity as a protective mechanism against exogenous chemicals. Due to these above limitations, the complexation of drugs with cyclodextrin is one the best approach for making the nasal drug delivery for poorly water-soluble drug. The bioavailability of the drugs can be increased by using methylated cyclodextrins. Recent study of cyclodextrin complexes with lipophilic drugs in the nasal drug delivery system provides information regarding decrease in dose of Estradiol, enhanced nasal absorption of Morphine HCl and improved stability of Dihydroergotamine [31].

6.2.7 Pulmonary drug delivery

Pulmonary drug delivery is a suitable route for systemic drug delivery and used for the treatment of diseases like asthma and chronic obstructive pulmonary disease. The first pass drug metabolism and degradation in gastrointestinal tract can be eliminated by pulmonary drug delivery system. Due to large surface area in pulmonary system, it involves better flow of blood, low enzymatic activity and there by effective absorption of drugs takes place in the lungs. However the pulmonary drug delivery can be limited by low aqueous solubility and slow dissolution. Thus, the solubility, stability, dissolution rate of water insoluble and chemical unstable drugs can be improved by the use of cyclodextrins complex, which leads to decrease in clearance, increase of drug absorption and faster onset of action. Due to complexation of the drug with cyclodextrin, the liquid drugs can be converted into solid form, two incompatible drugs can be mixed in dry powder formulation, local irritation in lungs can be reduced, bad smell and taste can be reduced [30].

6.2.8   Dermal and transdermal drug delivery

Cyclodextrins have a significant role in dermal application and can be used to optimize the transdermal drug delivery intended either for local or systemic use. They also improves the stability and solubility of drugs in the topical preparations, enhances the transdermal absorption of drugs, sustains release of the drug from the vehicle and avoids undesirable side effects associated with dermally applied drugs. The main barrier for dermal drug absorption through the skin is the outer most layer stratum corneum. Penetration enhancers like alcohols, fatty acids etc. are used to decrease its barrier properties. The cyclodextrins enhance drug delivery through aqueous diffusion barriers, but not through lipophilic barriers like stratum corneum. If the drug release is from an aqueous based vehicle or if an aqueous diffusion layer at outer surface of skin is a rate determining factor, then cyclodextrins can acts as penetration enhancers. But if drug penetration through the lipophilic stratum corneum is the main rate-determining factor then cyclodextrins are unable to enhance the drug delivery. For example, the in vitro release rate of corticosteroids from water containing ointments is markedly increased by using hydrophilic cyclodextrins. The enhancement of drug release can be achieved by increase in solubility, diffusibility and concentration of the drug in the aqueous phase of the ointment through water-soluble complex formation. Hydroxy propyl beta cyclodextrin increases drug bioavailability in dermal formulations. There is no physical and chemical changes takes place in the skin by using HP-β-CD [32].

6.2.9   Cyclodextrin in site-specific drug delivery

6.2.9.1   Rectal drug delivery

Rectal delivery is an effective delivery system for drugs which have bitter or nauseous taste, first pass metabolism and degraded in the stomach pH. It is an ideal route of administration of drugs for unconscious patients, children and infants. The use of cyclodextrins in rectal delivery system include improvement of drug absorption from a suppository base either by enhancing drug release from the base or by increasing drug mucosal permeability and also increase drug stability in the base or at the absorption site along with sustained release of drugs. The drug release from the suppository base is essential in rectal absorption because of the high viscosity of rectal fluids. The effect of cyclodextrins on rectal drug absorption can be influenced by partition coefficient of the drug and its cyclodextrin complex and nature of the suppository base such as oleaginous or hydrophilic. The hydrophilic cyclodextrins particularly methylated and hydroxylpropyl cyclodextrins enhance the absorption of lipophilic drugs either by improving the drug release from oleaginous vehicles or by increasing the dissolution rate of drugs in rectal fluids. The rectal absorptions of Flurbiprofen and Biphenylacetic acid were improved by DM-β-CD and HP-β-CD respectively. Similarly, the rectal absorption of non-absorbable, hydrophilic drugs such as antibiotics, proteins and peptides are improved by complex formation with cyclodextrins [30].

6.2.9.2   Colon-specific drug delivery

Cyclodextrins are not properly hydrolysed but absorbed in stomach and small intestine, where as they are absorbed in large intestine after fermentation into small saccharides by colonic microbial flora. The above properties of cyclodextrins make them valuable for colon drug targeting. Biphenyl acetic acid prodrugs for colon-specific delivery were developed by complexation of the drugs with cyclodextrins [30].

6.2.9.3   Brain drug delivery or brain targeting

The blood brain barrier is a highly selective permeability barrier that separates the circulating blood from the brain extracellular fluid in the central nervous system. This barrier is formed by capillary endothelial cells which are connected by tight junctions. It allows the passage of water, some gases and lipid soluble molecules by passive diffusion as well as the selective transport of molecules such as glucose and amino acids. It restricts the passage of polar drugs to the brain and thus blocks the specific delivery of neuro-pharmaceuticals to the brain. It is reported that P-glycoprotein mediated peptide transport may play an important role in greatly reducing the peptide delivery to central nervous system *in vivo*. Cyclodextrins such as DM-β-CD, due to their inhibitory effect on P-glycoprotein efflux function, may enhance drug delivery to brain [30].

6.2.10   Peptide and protein drug delivery system

The various problems associated with the use of proteins and peptides are their poor absorption through biological membranes, rapid plasma clearance, peculiar dose response curves, chemical and enzymatic instability. The enhancement of absorption can be achieved by cyclodextrins due to their ability to reduce the physical and metabolic barriers to proteins and peptides. The limited systemic bioavailability of proteins and peptides is due to the existence of enzymatic barrier in the epithelial cells. Cyclodextrins can protect proteins and peptides against enzymatic as well as chemical degradation [33].

6.2.11   Gene and oligonucleotide delivery

Cyclodextrins are considered as carrier for the delivery of therapeutically used nucleic acids. They can eliminate the problems associated with *in vivo* delivery of oligonucleotides such as high degree of susceptibility to endonucleases with potential toxicity of their breakdown products and potential immunogenicity. Cyclodextrins can improve cellular uptake of oligonucleotides and also delay their degradation by increasing their stability against endonucleases. Oligonucleotide-adamatane conjugates associated with HP-β-CD affords considerable increased cellular uptake of oligonucleotides. The conjugated oligonucleotides-cyclodextrin complex delivered to the colon provides better absorption site to get adequate therapeutic levels of oligonucleotides. Neutral cyclodextrins like β-, DM-β-, and HP-β-CDs were reported to increase DNA cellular uptake by increasing its permeability. The permeability of DNA is increased due to interaction of the cyclodextrins with membrane components such as cholesterol. Cyclodextrins is also enhanced the physical stability of viral vector formulations for gene therapy [34].

6.2.12   Applications of cyclodextrins in novel delivery systems

6.2.12.1   Liposomes

The main purpose of cyclodextrins complex with liposome in novel drug delivery systems provides increased drug solubility by cyclodextrin and drug targeting by liposomes. The problems associated with cyclodextrin can be eliminated by incorporating it into liposome. The complexation with cyclodextrin can also improve the stability of liposomes, for e.g. most stable liposomal formulations of metronidazole and verapamil were obtained by direct spray drying of lipid, drug and HP-β-CD mixture. Complexation with cyclodextrin increases the liposomal entrapment of nifedipine by reducing its interaction with lipid bilayers and also improved the liposomal stability in plasma. Liposomal entrapment of prednisolone was higher when incorporated as HP-β-CD complex than as free drug [35].

6.2.12.2   Microspheres

The role of cyclodextrins in microsphere preparation was first studied by Loftsson. HP-β-CD acted as a promising agent for stabilizing lysozyme and bovine serum albumin (BSA) during primary emulsification of poly (d, l-lactide-co-glycolide) (PLGA) microsphere preparation. The stabilizing effect was reported to be due to increased hydrophilicity of the proteins caused by shielding of their hydrophobic residues by HP-β-CD; this also reduces their aggregation and denaturation by keeping them away from methylene chloride water interface. HP-β-CD enhanced BSA conformational stability and also increased its recovery from w/o emulsion by preventing the adsorption of the protein to PLGA. Cyclodextrins were also used to modulate peptide release rate from microspheres. The amount of cyclodextrin linked in microspheres was in the order β> γ> α-CD and the dimensions of the microspheres with γ-CD were much higher than those with α- or β-CDs [36].

6.2.12.3   Microcapsules

It was suggested that crosslinked β-CD microcapsules, because of their ability to retard the release of water-soluble drugs through semipermeable membranes, can act as release modulators to provide efficiently controlled release of drugs. Terephthaloyl chloride (TC) crosslinked β-CD microcapsules were found to complex with p-nitrophenol rapidly and the amount of complex increased as the size of the microcapsules decreased. TC crosslinked β-CD microcapsules retarded the diffusion of propranolol hydrochloride through dialysis membrane. Double microcapsules, prepared by encapsulating methylene blue with different amounts of β-CD microcapsules inside a crosslinked human serum albumin (HSA), showed decreasing release rate of methylene blue with increasing amount of β-CD microcapsules. Dissociation of methylene blue complex with β-CD microcapsules was found to serve as an additional mechanism in controlling the release kinetics of HAS double microcapsules. In the case of HSA microcapsules with parent β-CD, the hydrating property of the cyclodextrin, by promoting the diffusion of water into the microcapsules, caused an increased release rate of methylene blue compared with those without the cyclodextrin. However, in the case of HSA double microcapsules (i.e., with β-CD microcapsules), the hydrophobic groups introduced during crosslinking suppressed the cyclodextrin hydration and provided controlled release without enhancing the diffusion of water that can impair the complexation of methylene blue [37].

6.2.12.4   Nanoparticles

Nanoparticles are considered to be more stable than liposomal delivery systems. However, a major drawback is associated with the drug-loading capacity of polymeric nanoparticles. Cyclodextrins are used for this reason to improve water solubility and sometimes the hydrolytic or photolytic stability of drugs for better loading properties. D/CD complexes act to solubilize or stabilize active ingredients within the nanoparticles, resulting in increased drug concentration in the polymerization medium and increased hydrophobic sites in the nanosphere structure when large amounts of cyclodextrin are associated to the nanoparticles. The antiviral agent saguinavir was complexed to HP-β-CD to increase saquinavir loading into poly alkyl cyano acrylate nanoparticles by providing a soluble drug reservoir in the polymerization medium that is the basis of nanoparticle formation. Cyclodextrins increasing the loading capacity of nanoparticles and the spontaneous formation of either nanocapsules or nanospheres is achieved by nanoprecipitation of amphiphilic cyclodextrins diesters. Incorporation of the steroidal drugs hydrocortisone and progesterone in complex with β-CD and HP-β-CD reduced the particle size for solid lipid nanoparticles (SLNs) below 100 nm. HP-β-CD addition in the polymerization medium of poly ethyl cyano acrylate (PECA) nanospheres improved the subcutaneous absorption of metoclopramide in rats. PECA nanospheres with HP-β-CD provided the highest drug concentration and enhanced drug absorption compared with those with dextran or with drug solution. However, in addition to drug absorption from subcutaneous sites, HP-β-CD also enhanced the drug elimination by enhancing the drug absorption to reticulo endothelial tissues. Progesterone complexed to HP-β-CD or DM-β-CD was loaded into bovine serum albumin (BSA) nanospheres. Dissolution rates of progesterone were significantly enhanced by complexation to cyclodextrin with respect to free drug. In an approach, cyclodextrin properties of complexation were combined with those of chitosan. Complexation with cyclodextrin was believed to permit solubilization as well as protection for labile drugs while entrapment in the chitosan network was expected to facilitate absorption. Chitosan nanoparticles including complexes of HP-β-CD with the hydrophobic model drugs triclosan and furosemide, were prepared by ionic crosslinking of chitosan with sodium tripolyphosphate (TPP) in the presence of cyclodextrin. Nanoparticles were then prepared by ionotropic gelation using the obtained drug HP-β-CD inclusion complexes and chitosan. On the other hand, drug entrapment increased up to 4 and 10 times by triclosan and furosemide, respectively. The release profile of nanoparticles indicated an initial burst release followed by a delayed release profile lasting up to 4 h. The HP-β-CD inclusion complex with insulin was encapsulated into the nanoparticles resulting in a pH dependent release profile. The biological activity of insulin was demonstrated with enzyme linked immunosorbent assay (ELISA). Cyclodextrin complexed to insulin encapsulated into mucoadhesive nanoparticles was believed to be a promising candidate for oral insulin delivery [38].

6.2.12.5   Nanosponges

Cyclodextrin-based nanosponges, which are proposed as a new nanosized delivery system, are innovative cross-linked cyclodextrin polymers nanostructured within a three-dimensional network. This type of cyclodextrin polymer can form porous insoluble nanoparticles with a crystalline or amorphous structure and spherical shape or swelling properties. Nanosponge functionalisation for site-specific targeting can be achieved by conjugating various ligands on their surface. They are a safe and biodegradable material with negligible toxicity on cell cultures and are well tolerated after injection in mice. Cyclodextrin based nanosponges can form complexes with different types of lipophilic or hydrophilic molecules. The release of the entrapped molecules can be varied by modifying the structure to achieve prolonged release kinetics or a faster release. The nanosponges could be used to improve the aqueous solubility of poorly water-soluble molecules, protect degradable substances, obtain sustained delivery systems or design innovative drug carriers for nanomedicine. Cyclodextrin nanosponges possess particular properties in terms of their encapsulation ability, biocompatibility and solubilisation capacity with regard to different types of molecules [39].

6.2.12.6   Hydrogels

Hydrogels have been gaining increased relevance as drug delivery systems, medical devices, scaffolds for tissue regeneration and substitution, and in several chemical applications as well. These hydrogels may offer interesting possibilities as dosage forms, administered by almost any route, if their limited ability for the direct loading of poorly water-soluble drugs is overcome. Thus crosslinked cyclodextrins that enable the combination of the hydrogel versatility with the complexation capability of cyclodextrins could be particularly useful. Polymerized cyclodextrins maintain or promote the complexation ability of free cyclodextrins in solution. When solutions of drug–cyclodextrin complexes are diluted in the physiological fluids, the release of the drug is practically instantaneous. By contrast, in the case of the cyclodextrin hydrogels, the cyclodextrin units are covalently attached to each other and the volume of water which can enter the hydrogel is limited by the own network. This provides a microenvironment rich in cavities available to interact with the surrounding drug molecules. Consequently, delivery systems comprising chemically linked cyclodextrins offer considerable possibilities of achieving sustained release. Cyclodextrin hydrogels are obtained by copolymerization of cyclodextrin monomeric derivatives with other acrylic or vinyl monomers.

HP-β-CD hydrogels using diglycidylethers as cross-linking agents can be developed directly in order to avoid the important drawbacks of the chemical modification of cyclodextrins (low reproducibility of the synthesis or residual toxic monomers). Hydrogels of different cyclodextrin varieties, crosslinked with ethyleneglycol diglycidylether (EGDE) as sustained delivery systems for estradiol demonstrated increase solubility in physiological environments [40].

6.2.12.7   Beads

Beads can be prepared by using soft conditions (no organic solvent, no cross-linking or surface-active agents, moderate heating). Morphologically, these beads appear as minispheres consisting of a partial crystalline matrix of cyclodextrins surrounding micro-domains of oil. Beads can be prepared by continuous external orbital shaking of a mixture of an α-cyclodextrin aqueous solution and soybean oil at room temperature. α-CD is employed for its ability to interact with components of vegetable oil and more especially with triglycerides whereas the high oil content offers interesting prospects for the microencapsulation of lipophilic drugs. Freeze-drying advantageously transforms beads into dry powder in which the oil content reaches 80% weight and also facilitates ease of handling and use for oral administration. Encapsulation of isotretinoin (poorly stable and lipophilic molecule) in "beads, for oral delivery demonstrated high drug loading/encapsulation efficiency which can be attributed to inner structure (micro-domains of oil) and increase oral bioavailability in rats. Thus beads may open up new prospects for oral delivery of lipophilic drugs [41].

6.2.12.8   Nanogels/nanoassemblies

In the past few decades, submicronic polymeric particles have attracted considerable attention as potential drug delivery devices for the controlled release of active molecules and targeting. So the technical roadblock in their use is the fact that their preparation needs to employ large amount of potentially toxic organic solvents and surfactants, which is often not acceptable, at least for parenteral administration. Therefore to overcome these technological issues new self-assembling nanogels/nanoassemblies were developed avoiding the use of organic solvents and surfactants. They consist of a hydrophilic polymer backbone, on which hydrophobic moieties are grafted. Among the associative polymers, hydrophobized polysaccharides, such as cellulose derivatives, dextran, chitosan or pullulan are particularly attractive due to their biocompatibility, biodegradability and low toxicity, which are advantageous for biological and pharmaceutical applications. The development of supramolecular assemblies, in which cyclodextrins were associated to macromolecules, attracted much attention. More recently, supramolecular gel-like networks were obtained by mixing a cyclodextrin-bearing host polymer and a hydrophobically modified guest polymer. Spherical supra molecular nano assemblies (nanogels) may be obtained in pure water just by mixing two neutral polymers which instantaneously associate together. Colloidal systems generally result from the association of amphiphilic polymers in water, from the complexation of oppositely charged poly ions or from hydrogen-bonding interactions. Thus the new supra molecular nano assembly concept avoids some of the inconveniences of the currently employed nano technologies [42].

6.2.13   Cyclodextrin containing polymers

Cyclodextrin-containing polymers are now being explored as vehicles for delivering nucleic acids into cells. The structures of the cyclodextrin-containing polycations affect the nucleic acid delivery efficiencies and their toxicities. The cyclodextrin-containing polymers reveal lower toxicities than polymers that lack the cyclodextrins. The cyclodextrins endow the nucleic acid delivery vehicles with the ability to be modified by compounds that form inclusion complexes with the cyclodextrins, and these modifications can be performed without disruption of the polymer–nucleic acid interactions. The development of polyplexes (cationic polymer + nucleic acid) for use as DNA delivery agents is based on hypothesis that it may be possible to prepare low toxicity polycations from cyclodextrins because numerous individual cyclodextrins were known to reveal low toxicity and to not elicit immune responses in animals. These new families of cyclodextrin-containing cationic polymers were able to provide effective DNA delivery to cultured cells with low toxicity. Numerous cyclodextrin-containing, cationic polymers currently exist. For example, within the class of cyclodextrin pendent polymers, several are polycations, e.g. PEI, poly (allylamine), dendrimers. Polyplex formulations (cationic polymer + nucleic acid) optimized for in vitro delivery are typically not appropriate for in vivo use because successful systemic delivery requires different particle properties. After intravenous injection, cationic polyplexes interact with serum proteins and are quickly eliminated from the bloodstream by phagocytic cells. But use of cyclodextrin-containing polycations for polyplex formation provides the means to create modified particles in an entirely new manner. Pun and Davis recently developed methodologies to modify the surface of polyplexes formed with cyclodextrin-containing polymers whether they are of the CDP-type or not. This concept exploits the use of cyclodextrin/guest compound complexation to provide modified polyplexes appropriate for systemic application as gene delivery vehicles. As an example of this methodology, adamantane was conjugated to PEG and the resulting compound exposed to CDP-based polyplexes for self-assembly between the adamantane and the cyclodextrins. This methodology can provide CDP-based particles that are appropriate for systemic gene delivery [43].

6.2.14   Cyclodextrin use as excipients in drug formulation

As excipients, cyclodextrins have been finding different applications in the formulation and processing of drugs. β-CD, due to its excellent compactability and minimal lubrication requirements, showed considerable promise as a filler binder in tablet manufacturing but its fluidity was insufficient for routine direct compression. β-CD was also found to be useful as a solubility enhancer in tablets. The ability of β-CD to complex progesterone by wet granulation was found to be dependent on both binder solution and mixture type. Complexation can cause subtle changes in the tabletting properties of drugs or CDs that can substantially affect the stability and tabletting performance of tablet formulations containing drug/CD complexes. Cyclodextrins also affect the tabletting properties of other excipients, e.g. microcrystalline cellulose codried with β-CD showed improved flowability, compactability and disintegration properties suitable for direct compression. In the case of high-swelling wheat starches, β-CD (1%) increased the peak viscosity (PV) but decreased the cool paste viscosity (CPV) and in the case of low swelling starches, the same cyclodextrin slightly decreased the PV but increased the CPV. However, β-CD reduced the heat paste viscosity of both the starches. Avicel/β-CD codried product showed improved flowability and disintegration properties but its rounder particles, because of their sensitivity to lubrication, gave tablets weaker than those with avicel. But on addition of magnesium stearate, the codried excipient with improved powder flowability served as a better excipient in wet granulation. Cyclodextrins can be used to mask the taste of drugs in solutions, e.g. suppression of bitter taste of oxyphenonium bromide by cyclodextrins. With the assumption that only the free drug molecule exhibits bitter taste regardless of the kind and concentration of cyclodextrin, the suppression of drug bitter taste by cyclodextrins was reported to be in the order of α-CD > G γ-CD > G β-CD, reflecting the stability constants of the complexes. Cyclodextrins were also indicated to stabilize protein and peptide pharmaceuticals during spray drying, e.g. inhibition of spray drying induced inactivation of β-galactoside by HP-β-CD.

Cyclodextrins were found to inhibit adsorption or absorption of drugs to container walls. SBE-β-CD and HP-β-CD reduced the adsorption of DY-9760e to PVC tubes but the effect was more significant with SBE-β-CD reflecting the stability constants of the cyclodextrin complexes. Compared with HP-β-CD, SBE-β-CD was found to exhibit a greater masking effect against the hydrophobic interaction between the surface of PVC tubes and the drug. Hydrophilic cyclodextrins, including maltosyl-β-CD, inhibited the adsorption of bovine insulin to containers and also inhibited insulin aggregation by interacting with the hydrophilic regions of the peptide. β-CD inhibited the adsorption of FK 906, a surface-active drug, from aqueous solution onto container walls by shifting the critical micellar concentration of the drug to a higher value [44].

6.2.15   Other applications of cyclodextrins

6.2.15.1   Applications in the food industry

Cyclodextrins is taken into consideration in applications for food processing and also as food additives in the following ways [45].

(a) To protect lipophilic food components that are sensitive to oxygen and light- or heat-induced degradation

(b) To solubilise food colourings and vitamins

(c) To stabilize fragrances, flavours, vitamins and essential oils against unwanted changes

(d) To suppress unpleasant odours or tastes

(e) To achieve a controlled release of certain food constituents

6.2.15.2   Cosmetics, personal care and toiletry

Cosmetic preparation is another area which demands cyclodexytrin use, mainly in volatility suppression of perfumes, room fresheners and detergents by controlled release of fragrances from inclusion compounds. The major benefits of cyclodextrins in this area are stabilization, odour control and process improvement upon conversion of a liquid ingredient to a solid form. The various applications include toothpaste, skin creams, liquid and solid fabric softeners, paper towels, tissues and underarm shields. The interaction of the guest with cyclodextrins produces a higher energy barrier to overcome to volatilise, thus producing long-lasting fragrances. Fragrance is enclosed with the cyclodextrin and the resulting inclusion compound is complexed with calcium phosphate to stabilise the fragrance in manufacturing bathing preparations. Cyclodextrin-based compositions are also used in various cosmetic products to reduce body odours. The use of CD-complexed fragrances in skin preparations such as talcum powder stabilizes the fragrance against the loss by evaporation and oxidation over a long period. The antimicrobial efficacy of the product is also improved. Dry cyclodextrin powders of size less than 12 mm are used for odour control in diapers, menstrual products, paper towels, etc., and are also used in hair care preparations for the reduction of volatility of odorous mercaptans. The hydoxypropyl-cyclodextrin surfactant, either alone or in combination with other ingredients, provides improved antimicrobial activity. Dishwashing and laundry detergent compositions with cyclodextrins can mask odours in washed items. Cyclodextrins are used in the preparation of sunscreen lotions in 1:1 proportion (sunscreen/hydroxypropyl-CD) as the CD's cavity limits the interaction between the UV filter and the skin, reducing the side effects of the formulation. Similarly, by incorporating cyclodextrin in self-tanning emulsions or creams, the performance and shelf life are improved [46, 47].

6.2.15.3   Agricultural and chemical industries

Cyclodextrins form complexes with a wide variety of agricultural chemicals including herbicides, insecticides, fungicides, repellents, pheromones and growth regulators. Cyclodextrins can be applied to delay germination of seed. In grain treated with β-cyclodextrins, some of the amylases that degrade the starch supplies of the seeds are inhibited. Initially the plant grows more slowly, but later on this is largely compensated by an improved plant growth yielding a 20–45% larger harvest. In the chemical industry, cyclodextrins are widely used to separate isomers and enantiomers, to catalyse reactions, to aid in various processes and to remove or detoxify waste materials. The low cost, biocompatible and effective degradation makes cyclodextrins a useful tool for bioremediation process [48].

6.2.15.4   Adhesives, coatings and other polymers

Cyclodextrins increase the tackiness and adhesion of some hot melts and adhesives. They also make additives and blowing agents compatible with hot melt systems. The interaction between polymer molecules in associative thickening emulsion-type coatings such as paints tends to increase viscosity and cyclodextrins can be used to counteract this undesirable effect [49].

6.2.15.5   Application of cyclodextrins in chromatographic techniques

The chromatographic techniques are utilized for the separation of large number of organic and inorganic compounds. The chromatographic separation can be increased by modification of both stationary and mobile phase of the system. Due to adsorption property of cyclodextrins and their derivatives, they are applied for the improvement of the separation parameters of various chromatographic methods. Cyclodextrins are used as additives of stationary phase and modifier of the mobile phase. Because of their versatility, reproducibility and sensitivity, cyclodextrins are used in various chromatographic methods such as liquid chromatography (HPLC), gas chromatography (GC), size exclusion chromatography (SEC), gel permeation chromatography (GPC), supercritical fluid chromatography (SFC) and ultra performance liquid chromatography [50].

6.3   Conclusions

Cyclodextrins can form inclusion complexes with several guest molecules by taking up a whole molecule or some part of it, into its cavity. These complex formations modify solubility of the molecule, increase the stability and decrease the volatility of compounds. These versatility properties of cyclodextrins and modified cyclodextrins have demonstrated wide range of applications of cyclodextrins in various fields such as pharmaceutical, cosmetics, food, agriculture, chemical industry and chromatographic techniques. However, it is also necessary to find out any possible interaction between these agents and other formulation additives because the interaction can adversely affect the performance of both. It is also important to have knowledge of different factors that can influence complex formation in order to prepare economically drug-cyclodextrin complexes with required properties. Recent developments in the field of biotechnology have gained lot of improvements in the efficient manufacture of cyclodextrins lowering the cost of these materials making highly purified cyclodextrins and cyclodextrins derivates.

6.4   Future prospects of cyclodextrins

From the above discussion, it is evident that cyclodextrins will play important role in the pharmaceutical industry starting from drug delivery system to role in the treatment of viral infections and cosmetics in future. The application of cyclodextrins is liable to be explored as the properties of cyclodextrins are expanding with good number of commercialized and FDA-approved variants. Recently, conventional formulations such as tablets, capsules, solutions, ointment and intravenous solutions have been commercialized using cyclodextrins. Similarly in case of novel drug delivery system, cyclodextrins are extensively studied for their application in novel drug delivery such as nanoparticles, liposomes, microspheres, microcapsules, hydrogels and targeted drug delivery system. They can become commercially available in future and plays tremendous role to solve various problems associated with the delivery of different novel drugs through different routes of administration.

6.5   References

1. Szetjli, J. Introduction and general overview of cyclodextrin chemistry. *Chem Rev*; 98: 1743–1753, 1998.

2. Challa, R., Ahuja, A., Ali, J., Khar, R.K. Cyclodextrins in drug delivery: an updated review. *AAPS Pharm Sci Tech*; 6(2): 329–357, 2005.

3. Loftsson, T., Brewester, M. Pharmaceutical applications of cyclodextrins. Drug solubilization and stabilization. *J Pharm Sci*; 85: 1017–1025, 1996.

4. Rajewski, R.A., Stella, V.J. Pharmaceutical applications of cyclodextrins. In vivo drug delivery. *J Pharm Sci*. 85: 1142–1168, 1996.

5. Hashimoto H. Preparation, structure, property and application of branched cyclodextrins. In: Duchene, D. (Ed.), New Trends in Cyclodextrins and Derivatives. *Editions de Sante, Paris*. 97–156; 1991.

6. Szente L., Szejtli J. Highly soluble cyclodextrin derivatives: chemistry, properties, and trends in development. *Adv Drug Deliv Rev.* 36: 17–38, 1999.

7. Biwer, A., Antranikian, G., Heinzle, E. Enzymatic production of cyclodextrins. Appl *Microbiol Biotechnol.* 59 (6): 609–617, 2002.

8. Sanjoy, K.D., Rajan, R., Sheba, D., Nasimul, G., Jasmina, K., Arunabha, N. Cyclodextrins – The Molecular Container, Research Journal of Pharmaceutical, *Biological and Chemical Sciences*, 4(2): 1694–1720, 2013.

9. Bender, M.L., Komiyama, M. Cyclodextrin chemistry. Berlin, Springer-Verlag, 1978, 1–94.

10. Chen, G., Jiang, M. Cyclodextrin-based inclusion complexation bridging supramolecular chemistry and macromolecular self-assembly. *Chemical Society Review*. 40: 2254–2266, 2011.

11. Liu, L., Guo, Q.X. The driving forces in the inclusion complexation of cyclodextrins. *J Incl Phenom*. 42: 1–14, 2002.

12. Nisar, A.K., Mohi D. Cyclodextrin: An overview. *International Journal of Bioassays*. 2 (6): 858–865, 2013.

13. Davis, M.E., Brewster, M.E. Cyclodextrin-based pharmaceutics: past, present and future. *Nature reviews*. 3: 1023–1035, 2004.

14. Martin, D.V.E.M. Cyclodextrins and their uses: a review. *Process Biochemistry*. 39 (9): 1033–1046, 2004.

15. Rasheed, A., Kumar, A., Sravanthi, V.V.N.S.S. Cyclodextrins as Drug Carrier Molecule: A Review. *Sci Pharm*. 76: 567–598, 2008.

16. Kralj, B., Smidovnik, A., Kobe, J. Mass spectrometric investigations of alpha and beta-cyclodextrin complexes with ortho-, meta- and para-coumaric acids by negative mode electrospray ionization. *Rapid Commun Mass Spectrom*. 23(1): 171–180, 2009.

17. Scheneider, H.J., Hacket, F., Rudiger, V. NMR Studies of Cyclodextrins and Cyclodextrins Complexes, *Chem. Rev*. 98: 1755–1785, 1998.

18. Camelia, N., Corina, A., Crina-maria M. Preparation and characterization of inclusion complexes between repaglinide and β-cyclodextrin, 2-hydroxypropyl-β-cyclodextrin and randomly methylated β-cyclodextrin. *Farmacia*, 58: 78–88, 2010.

19. Cannava, C., Crupi, V., Ficarra, P., Guardo, M., Majolino, D., Stancanelli, R., Venuti, V. Physicochemical characterization of coumestrol/beta-cyclodextrins inclusion

complexes by UV-Vis and FTIR-ATR spectroscopies, *Vibrational Spectroscopy*, 48: 172–178, 2008.

20. Olaru, A., Borodi, G., Kacso, I., Vasilescu, M., Bratu, I., Cozar, O. Spectroscopic studies of the inclusion compound of lisinopril in β-cyclodextrin, *Journal of optoelectronics and advanced materials-symposia*, 2: 1–5, 2010.

21. Giordano, F., Novak, C., Moyano, J.R. Thermal analysis of cyclodextrins and their inclusion compounds. *Thermochimica Acta*, 380: 123–151, 2001.

22. Singh, R., Bharti, N., Madan, J., Hiremath, S.N. Characterization of cyclodextrin inclusion Complexes-A review. *J. Pharm. Sci.* Technol. 2: 171–183, 2010.

23. N.V. Roik, L.A. Belyakova, IR Spectroscopy, X-Ray Diffraction and Thermal Analysis Studies of Solid β-Cyclodextrin-Para-Aminobenzoic Acid Inclusion Complex, Physics and Chemistry of solid state, 12(1): 168–173, 2011.

24. Stella, V.J., Rao, V.M., Zannou, E.A., Zia, V. Mechanisms of drug release from cyclodextrin complexes, *Adv Drug Del Rev.* 36: 3–16, 1999.

25. Singh, M., Sharma, R., Banerjee, U.C. Biotechnological applications of cyclodextrins. *Biotechnology Advances*, 20: 341–359, 2002.

26. Rasheed, A., Ashok, K.C.K., Sravanthi, V. Cyclodextrins as Drug Carrier Molecule. *Sci Pharm.* 76: 567–598, 2008.

27. Pagington, J.S. Cyclodextrin: the success of molecular inclusion. *Chem. Brit.* 23: 455–458, 1987.

28. Mehramizi, A., Monfared, A.E., Pourfarzib, M., Bayati, K.H., Dorkoosh, F.A. Rafiee-Tehrani. Influence of β-cyclodextrin complexation on lovastatin release from osmotic pump tablets (OPT). *Daru.* 15 (2): 71–78, 2007.

29. Badawy, S.I.F., Ghorab, M.M., Adeyeye, C.M. Bioavailability of danazolhydroxypropyl-β-cyclodextrin complex by different routes of administration. *Int J Pharm.* 145: 137–143, 1996.

30. Loftsson, T. Cyclodextrins and the biopharmaceutical classification system of drugs. *J. Incl. Phenom. Macrocycl. Chem.* 44, 63–67, 2002.

31. Merkus, F.W., Verhoef, J.C., Marttin, E., Romeijn, S.G., Vander, K.P.H., Hermens, W.A., Shipper, N.G. Cyclodextrins in nasal drug delivery. *Adv Drug Deliv Rev.* 36 (1): 41–57, 1999.

32. Cal, K., Centkowska, K. Use of cyclodextrins in topical formulations: Practical aspects. *Eur J Pharm Biopharm.* 68: 467–478, 2008.

33. Hirayama, F., Uekama, K., Cyclodextrin-based controlled drug release system. *Drug Delivery Rev.* 36 (1): 125–141, 1999.

34. Del Valle E.M.M. Cyclodextrins and their uses: a review. Process Biochemistry. 39: 1033–1046, 2004.

35. Brenda McCormack,Gregory Gregoriadis, Drugs-in-cyclodextrins-in liposomes: A novel concept in drug delivery, *International Journal of Pharmaceutics,* 112(3): 249–258, 1994.

36. Skiba M.,Bounoure F.,Barbot C.,Arnaud P.,Skiba M. Development of cyclodextrin microspheres for pulmonary drug delivery, *J Pharm Pharm Sci.*8(3): 409–418, 2005.

37. Pariot N.,Edwards-Lévy F.,Andry M.C.,Lévy M.C. Cross-linked beta-cyclodextrin microcapsules: preparation and properties, *Int J Pharm*. 211(1–2): 19–27, 2000.

38. Kanwar J.R.,Long B.M.,Kanwar R.K. The use of cyclodextrins nanoparticles for oral delivery, *Curr Med Chem*. 18(14): 2079–2085, 2011.

39. Francesco, T., Marco, Z., Roberta, C. Cyclodextrin-based nanosponges as drug carriers, Beilstein *J Org Chem*. 8: 2091–2099, 2012.

40. Rubén Machín,José Ramón Isasi,Itziar Vélaz. β-Cyclodextrin hydrogels as potential drug delivery systems, Carbohydrate Polymers, 87(3): 2024–2030, 2012.

41. Trichard L.,Fattal E.,Le Bas G.,Duchêne D.,Grossiord J.L.,Bochot A. Formulation and characterisation of beads prepared from natural cyclodextrins and vegetable, mineral or synthetic oils. *Int J Pharm*. 354(1–2): 88–94, 2008.

42. Maria D. Moya-Ortega,Carmen Alvarez-Lorenzo,Angel Concheiro,Thorsteinn Loftsson. Cyclodextrin-based nanogels for pharmaceutical and biomedical applications, *International Journal of Pharmaceutics*, 428(1–2): 152–163, 2012.

43. Akira Harada,Masaoki Furue,Shun-ichi Nozakura. Cyclodextrin-Containing Polymers. 1. Preparation of Polymers, *Macromolecules*, 9(5): 701–704, 1976.

44. Thompson, D.O. Cyclodextrins-enabling excipients: their present and future use in pharmaceuticals. *Crit Rev Ther Drug Carrier Syst*. 14: 1–104, 1997.

45. Prasad, N., Strauss, D., Reichart, G. Cyclodextrins inclusion for food, cosmetics and pharmaceuticals. European Patent. 84: 625, 1999.

46. Tatsuya, S. Stabilisation of fragrance in bathing preparations. Japanese Patent. 11: 209, 1999.

47. Foley, P.R., Kaiser, C.E., Sadler, J.D., Burckhardt, E.E., Liu, Z. Detergent composition with cyclodextrin perfume complexes to mask malodours. *PCT Int Appl WO*. 23: 516, 2000.

48. Hedges, R.A. Industrial applications of cyclodextrins. *Chem Rev*. 98: 2035–44, 1998.

49. Gao, S., Wang, L. Application of cyclodextrin in environmental science. *Huanjing Kexue Jinzhan*. 6: 80–6, 1998.

50. Gyula, O., Tibor, C., Maria, S. Cyclodextrins in Chromatography. Part 1. Liquid chromatographic methods, *Eur Chem Bull*. 2(11): 920–926, 2013.

Porous carriers in drug delivery systems

Chinam Niranjan Patra*, Suryakanta Swain,
Kahnu Charan Panigrahi and Muddana Eswara Bhanoji Rao

*Department of Pharmaceutics, Roland Institute of Pharmaceutical Sciences,
Khodasingi, Berhampur (Ganjam), Odisha-7600 10, India.*

7.1   Introduction

Drug is a chemical entity used for the diagnosis, prevention, treatment or cure of diseases. Novel drug delivery systems (NDDS) have many benefits, which include improved therapy, increased patient compliance through decreased dosing frequency, convenient routes of administration and improved targeting for a specific site to reduce unwanted side effects. The sustained or prolong release can be achieved by use of many natural, semi-synthetic or synthetic polymers which are coated on surface of drug to provide the release for longer duration [1]. Various drug delivery systems such as liposome, micelles, emulsion, polymeric nanoparticles have been showing great promise in controlled and targeted drug delivery [2–4]. Owing to wide range of useful properties, porous carriers have been used in pharmaceuticals for many purposes including development of novel drug delivery system, sustained drug delivery system and improvement in solubility of poorly soluble drugs. Liquid penetration into and its subsequent flow through such porous materials depends on both the molecular and the bulk property of the liquid and the geometric and surface property of the porous carrier. The required displacement depends upon the pore size, surface tension of liquid and the contact between surface and liquid [5]. When a porous hydrophobic polymeric drug delivery system is placed in contact with the appropriate dissolution medium, release of drug to medium must be preceded by the drug dissolution in the water-filled pores or from surface and by diffusion through the water filled channels [6]. Drug release from the porous carrier may be complete within 10 min or incomplete after several hours or days. Solvent polarity and surface properties play an important role in the adsorption and release from the porous carriers [7]. Various properties of porous carriers include stable uniform porous structure, high surface area, and tuneable pore size with narrow distribution and readily adaptable for thermo labile drugs. Due to the these properties, it is used in development of novel drug delivery system, floating drug delivery system, sustained drug delivery system. Entrapment into carrier molecules leads to increase in physicochemical stability because of hydrogen groups that form inter and intra molecular bonded structure of loaded drug [8–11].

7.2   Types of pores

Physical picture of a porous solid particle depicts surface roughness, bink bottle (blind pores), closed pores, transport pores (pores through the solid) and cylindrical blind [12]. The different types of pores are depicted in Fig. 7.1 and Fig. 7.2, respectively. The various types of pores are briefly explained as follows. Open pores are connected to the external surface of a solid and allows the passage of an adsorbate through the solid. Closed pores are void within the solid which is not connected to the external surface. Transport pores are connected parts of the external surface of the solid to the inner micro porosity. Blind pores are connected to the transport pores but do not lead to any other pore or surface.

Figure 7.1 Physical picture of a porous solid particle showing A – Surface roughness, B – Bink bottle (blind pores), C – Closed pores, D – Transport pores (pores through the solid) and E – Cylindrical blind

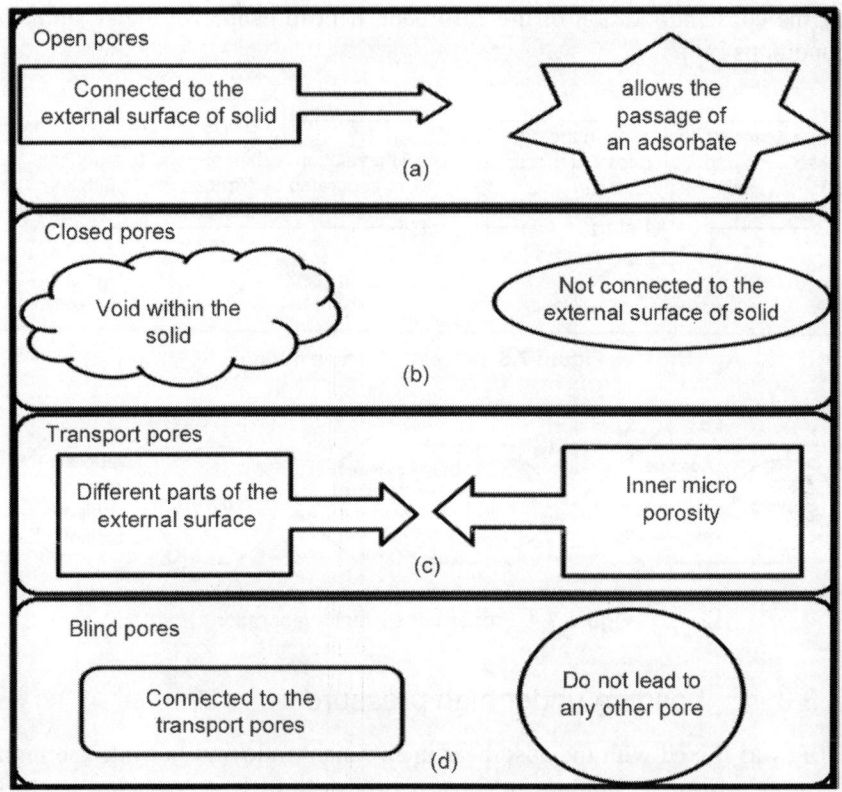

Figure 7.2 Characteristic feature of different pores: (a) open pores,
(b) closed pores, (c) transport pores, (d) blind pores

7.3   Methods of drug loading

7.3.1   Simple mixing

In this method, adsorbent is placed in the drug solution and stirred for suitable time using magnetic stirrer (Fig. 7.3). The solution is then allowed to stand for 1 h, separated and dried over 24 h at 60°C. This method is used for variety of drugs like ibuprofen, dexamethasone, griseofulvin, ranitidine and furosemide [13].

7.3.2   Solvent evaporation

Adsorbent is closely sieved in the range of 250–350 μm to nullify the effect due to variation in particle size (Fig. 7.4). Drug was loaded in solvent followed by the constant addition of the adsorbent, kept to evaporate under ambient conditions [11].

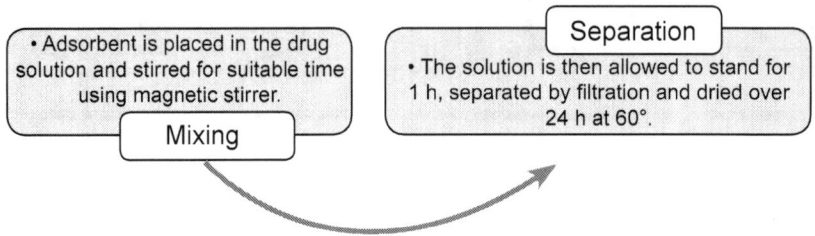

Figure 7.3 Process of simple mixing

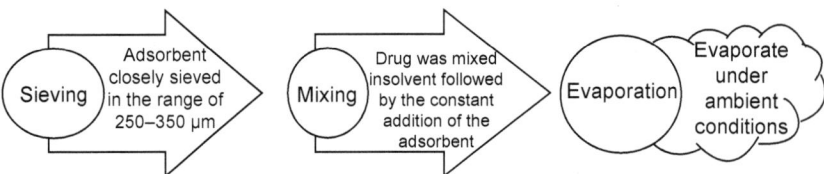

Figure 7.4 Process of solvent evaporation

7.3.3   Loading under high pressure

Drug was mixed with the adsorbent in sufficient ratio and put into the high pressure adsorption equipment for a period over 24 h. After being washed with a deionised water to get rid of unentrapped drug, the powder was dried in vacuum oven at 65°C for 5 h (Fig. 7.5). This method is used for the loading of Brilliant Blue [13].

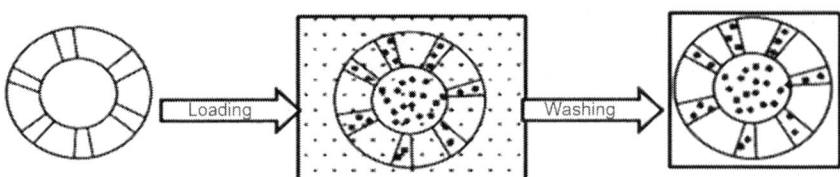

Figure 7.5 Diagrammatic representation of the drug loading in porous adsorbent under high pressure

7.3.4   Vacuum process

Adsorbent is placed in drug solution and the mixer evacuated for suitable time after which the vacuum was released. The adsorbent and drug solution were then allowed to stand for 1 h. Following this, solids separated using filter paper and dried for 24 h at 60°C (Fig. 7.6). Various drugs like diltiazem hydrochloride, benzoic acid, sodium benzoate are used for loading on the adsorbent [14].

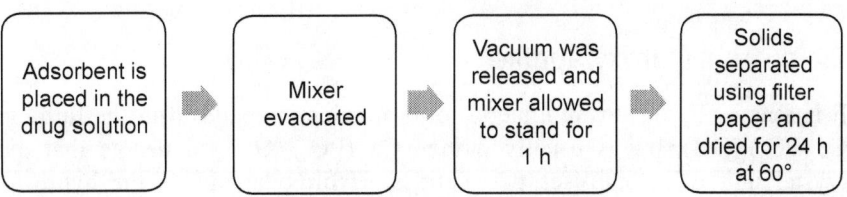

Figure 7.6 Vacuum process

7.3.5   Stirring in drug solution or suspension

Adsorbent is stirred in drug solution or suspension followed by drying in simple tray drier. Instead of using excess medium requisite minimum drug solution or suspension were loaded (Fig. 7.7). Therefore vacuum process is not essential and high yield can be achieved. This method can be applied to a variety of drugs like theophylline [15].

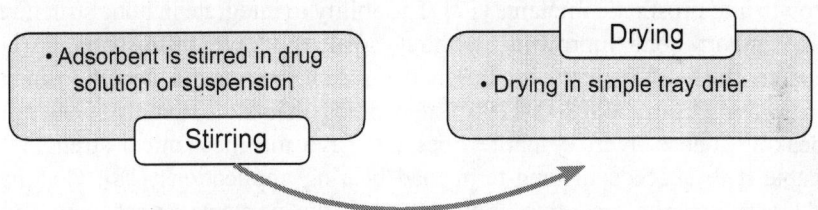

Figure 7.7 Process of stirring in drug solution or suspension

7.3.6   Layer-by-layer adsorption

The technique is usually performed in the aqueous solution at room temperature and thus it was suitable to encapsulate polypeptide and proteins drugs of poor stability. An LBL assembly of two oppositely charged polyelectrolytes at solid surfaces was developed as the alternating adsorption of these polyelectrolytes on a charged substrate due to their electrostatic attraction and the complex formation resulting in the defined macromolecular layer on the surfaces [16].

7.4   Novel applications of porous carrier

Even though porous carrier have got wide application in various areas like ceramics, paints, plastic industry, etc. These are some of carriers reported with many pharmaceutical applications.

7.4.1   Hydroxylapatite

It is a naturally occurring mineral form of calcium apatite with the formula $Ca_5(PO_4)_3(OH)$, but is usually written $Ca_{10}(PO_4)_6(OH)_2$ to denote that the crystal unit cell comprises two entities. Hydroxylapatite is the hydroxyl end member of the complex apatite group. The OH− ion can be replaced by fluoride, chloride or carbonate, producing fluorapatite or chlorapatite. Pure hydroxylapatite powder is white. Naturally occurring apatites can, however, also have brown, yellow, or green colorations, comparable to the discolorations of dental fluorosis. Up to 50 percent of bone is made up of a modified form of the inorganic mineral hydroxylapatite (known as bone mineral). Carbonated calcium-deficient hydroxylapatite is the main mineral of which dental enamel and dentin are comprised. Hydroxylapatite crystals are also found in the small calcifications (within the pineal gland and other structures) known as corpora arenacea or 'brain sand'. Thus, it is commonly used as a filler to replace amputated bone or as a coating to promote bone in growth into prosthetic implants [17]. The ability to integrate in bone structures and support bone ingrowths, without breaking down or dissolving (i.e. bioactive). Hydroxyapatite is a thermally unstable compound, decomposing at temperature from about 800–1200°C depending on its stoichiometry. Generally speaking, dense hydroxyapatite does not have the mechanical strength to enable it to succeed in long-term load bearing applications [18]. Coatings of hydroxyapatite are often applied to metallic implants (most commonly titanium/titanium alloys and stainless steels) to alter the surface properties. In this manner the body sees hydroxyapatite-type material acceptable. Without the coating, the body would see a foreign body and work in such a way as to isolate it from surrounding tissues. To date, the only commercially accepted method of applying hydroxyapatite coatings to metallic implants is plasma spraying. Hydroxyapatite may be employed in forms such as powders, porous blocks or beads to fill bone defects or voids. These may arise when large sections of bone have had to be removed (e.g. bone cancers) or when bone augmentations are required (e.g. maxillofacial reconstructions or dental applications). The bone filler will provide a scaffold and encourage the rapid filling of the void by naturally forming bone and provides an alternative to bone grafts. It will also become part of the bone structure and will reduce healing times compared to the situation, if no bone filler was used [18]. Chai et al. investigated on antibacterial activation of hydroxyapatite (HA) with controlled porosity by different antibiotics. A novel hydroxyapatite (HA) was elaborated with specific internal porosities for using as a bone-bioactive antibiotic (ATB) carrier material. The results also showed that for all three types of ATB (vancomycin, ciprofloxacin and gentamicin), adsorbed amount on the micro-porous HA were hugely higher than that on dense HA. The micro-porosity of test HA had also significantly prolonged the release time of antibiotics even under mimic physiological conditions. Furthermore, it also has primarily proved by a pilot test that the antibacterial efficiency of crude micro-porous HA could be further significantly improved by other methods of functionalization such as cold plasma technique [19]. Takahashi et al. studied the efficacy of recombinant human bone morphogenetic protein-2 (rhBMP-2) for enhancing anterior cervical spine interbody fusion when added to a porous hydroxyapatite (HA) graft was investigated. The results show that through osteogenesis at the fusion site, the addition of rhBMP-2 to a porous HA ceramic graft enhances the rate of anterior cervical fusion [20]. Cosijnsa et al. studied the porous hydroxyapatite tablets as carriers for low-dosed drugs. Tablets containing hydroxyapatite and a pore forming agent (50% (w/w) Avicel PH 200/20, 37.5% and 50% corn starch/37.5% sorbitol) were manufactured by direct compression followed by sintering. Sintering (1250°C) revealed tablets with an acceptable friability (<1%). Avicel PH 200 as pore forming agent resulted in tablets with highest porosity (50%) and the highest median pore diameter (5 μm). Aqueous drug solutions (metoprolol tartrate, riboflavin sodium phosphate) were spiked on the tablet surface. Drug release from the tablets was slow, independent of the drug. To accelerate drug release, tablets were manufactured using a modified gel casting technique yielding tablets with a median pore size of 60 μm and 80 μm. Release from these tablets was drastically increased indicating that the permeability of the tablets was influenced by the pore size, shape and connectivity of the porous network. Changing and controlling these parameters made it possible to obtain drug delivery systems providing different drug delivery behaviour [21].

7.4.2   Polystyrene beads

Polystyrene, poly(1-phenylethane-1,2-diyl)), abbreviated following ISO Standard PS is an aromatic polymer made from the aromatic monomer styrene, a liquid hydrocarbon that is commercially manufactured from petroleum by the chemical industry. Polystyrene is one of the most widely used kinds of plastic. Polystyrene is a strong plastic created from ethylene and benzene that can be injected, extruded, or blow molded, making it a very useful and versatile manufacturing material. Polystyrene is a "polymer of styrene." Polymers are large molecules consisting of adjoined identical molecules, and styrene is a colorless, oily liquid. When polystyrene is made, its structure is that of a rigid transparent thermoplastic, resembling stiff white foam. It is one of the most common types of plastic, and can be found in the home, in the office, at industrial sites, and just about any other place you would find plastics. Businesses rely on polystyrene for a number of uses, including manufacturing, packaging, and construction. It is found frequently in the food industry and used as a disposable transportation system to keep hot and cold foods at desired temperatures. Disposable and reusable items can be made from polystyrene as it is cheap but durable. Patil et al. formulated self-emulsifying system (SES) of lipophilic drug Loratadine by using solvent evaporation method as drug loading technique. His main objective is to predict the release pattern of drug from carrier if size of bead is varying [22]. Tiwari et al. reported that self-emulsifying drug delivery systems (SEDDSs) have gained exposure for their ability to increase solubility and bioavailability of poorly soluble drugs. They found that SEDDSs could efficiently improve oral absorption of the sparingly soluble drugs by rapid self-emulsification and, subsequently, dispersion in the absorption sites. SEDDSs possess unparalleled potential in improving oral bioavailability of poorly water-soluble drugs. Following their oral administration, these systems rapidly disperse in GI fluids, yielding micro- or nanoemulsions containing the solubilized drug. Owing to its miniscule globule size, the micro/nano-emulsifed drug can easily be absorbed through lymphatic pathways, bypassing the hepatic first-pass effect [23].

7.4.3   Mesoporous silica

Mesoporous silica is a form of silica and a recent development in nanotechnology. The most common types of mesoporous nanoparticles are MCM-41 and SBA-15. MSU-, KSW-, FSM-, HMM. Research continues on the particles that have applications in catalysis, drug delivery and imaging. The large surface area of the pores allows the particles to be filled with a drug or a cytotoxin. Ordered mesoporous silica (e.g. SBA-15, TUD-1, HMM-33, and FSM-16) also shows potential to boost the in vitro and in vivo dissolution of poorly water-soluble drugs. Many drug-candidates coming from drug discovery suffer from poor water solubility. An insufficient dissolution of these hydrophobic drugs in the gastrointestinal fluids strongly limits the oral bioavailability. One example is itraconazole which is an antimycoticum known for its poor aqueous solubility. Upon introduction of itraconazole-on-SBA-15 formulation in simulated gastrointestinal fluids, a supersaturated solution is obtained giving rise to enhanced trans-epithelial intestinal transport. Also the efficient uptake into the systemic circulation of SBA-15 formulated itraconazole has been demonstrated in vivo (rabbits and dogs). This approach based on SBA-15 yields stable formulations and can be used for a wide variety of poorly water-soluble compounds [18]. The structure of these particles allows them to be filled with a fluorescent dye that would normally be unable to pass through cell walls. The MSN (Mesoporous silica nanoparticles) material is then capped off with a molecule that is compatible with the target cells. When the MSNs are added to a cell culture, they carry the dye across the cell membrane. These particles are optically transparent, so the dye can be seen through the silica walls. The dye in the particles does not have the same problem with self-quenching that a dye in solution has. The types of molecules that are grafted to the outside of the MSNs will control what kinds of biomolecules are allowed inside the particles to interact with the dye [24].

Wei et al. investigated that mesoporous structure and strong basicity were successfully prepared by dispersing MgO on SBA-15 in three different methods, viz. impregnation, microwave irradiation and their combination, through magnesium acetate path. His objective is to study various attempts on preparing mesoporous basic material MgO/SBA-15. Yu et al. see the effect of Mesoporous SBA-15 molecular sieve as a carrier for controlled release of nimodipine [25].

Salonen et al. used mesoporous silicon that has many important properties advantageous to drug delivery applications. The small size of the pores confines the space of a drug and engages the effects of surface interactions of the drug molecules and the pore wall. Depending on the size and the surface chemistry of the pores, increased or sustained release of the loaded drug can be obtained. Drug loading from a solution at room temperature enables the use of porous silicon (PSi) also with sensitive therapeutic compounds susceptible to degradation, like peptides and proteins [26].

7.4.4   Alumino-silicate

Zeolites mean 'boiling stones' in Greek because of their ability to froth when heated to about 200°C. Zeolite is an alumino-silicate mineral with a porous crystalline structure that stays hard in water. They are tectosilicates exhibiting an open, three-dimensional structure containing cations needed to balance the electrostatic charge of the framework of silica and alumina tetrahedral and containing water. There are approximately 50 types of zeolite, each with its own chemical and physical properties. These properties make zeolite clay useful in many applications like soil amendments, odor control, water filtration media, etc. [27].

Barata-Rodrigues et al. developed a method where porous carbon was templated from inorganic materials such as zeolites (Tosoh H-Beta) and mesoporous molecular sieves. The main carbon precursor used was furfuryl alcohol, complemented in some materials by an additional treatment using propylene carbon vapor deposition. The structures of the templates and carbons were compared using electron microscopy and powder X-ray diffraction. Carbons were further characterized with elemental analysis and nitrogen adsorption. A careful study of the synthesized carbons' pore size distributions using DFT with various pore geometries was carried out and comparison with two commercial carbon adsorbents made, in order to assess the potential of such template carbons for gas separation and gas storage [28].

7.4.5   Polypropylene

Polypropylene (PP), also known as polypropene, is a thermoplastic polymer used in a wide variety of applications including packaging, stationery, plastic parts and reusable containers of various types, laboratory equipment, loudspeakers, automotive components, and polymer banknotes. An addition polymer made from the monomer propylene is rugged and unusually resistant to many chemical solvents, bases and acids. Most commercial polypropylene is isotactic and has an intermediate level of crystallinity between that of low-density polyethylene (LDPE) and high-density polyethylene (HDPE). Polypropylene is normally tough and flexible, especially when copolymerized with ethylene. The melt flow rate (MFR) or melt flow index (MFI) is a measure of molecular weight of polypropylene. There are three general types of polypropylene: homopolymer, random copolymer, and block copolymer. The co-monomer used is typically ethylene. Ethylene-propylene rubber or EPDM added to polypropylene homopolymer increases its low temperature impact strength. This material is often chosen for its resistance to corrosion and chemical leaching. Polypropylene, highly colorfast, is widely used in manufacturing carpets, rugs and mats to be used at home. Its most common medical use is in the synthetic, non-absorbable suture prolene, manufactured by Ethicon Inc. Polypropylene has been used in hernia and pelvic organ prolapse repair operations to protect the body from new hernias in the same location. A small patch of the material is placed over the spot of the hernia, below the skin, and is painless and is rarely, if ever, rejected by the body. However, a polypropylene mesh will erode over the uncertain period from days to years. Therefore, the FDA has issued several warnings on the use of polypropylene mesh medical kits for certain applications in pelvic organ prolapse, specifically when introduced in close proximity to the vaginal wall due to a continued increase in number of mesh erosions reported by patients over the past few years [29].

Streubel et al. developed a novel multiparticulate gastroretentive drug delivery system and to demonstrate its performance in vitro. Floating microparticles consisting of (i) polypropylene foam powder; (ii) verapamil HCl as model drug; and (iii) Eudragit RS, ethylcellulose (EC) or polymethyl methacrylate (PMMA) as polymers were prepared with an O/W solvent evaporation method. The release rate increased with increasing drug loading and with decreasing polymer amounts. The type of polymer significantly affected the drug release rate, which increased in the following rank order: PMMA < EC < Eudragit RS [30].

Sher et al. investigated by combining floating and pulsatile principles to develop drug delivery system, intended for chronotherapy in arthritis by using low-density porous carrier-based conceptual drug delivery system. This approach was achieved by using low-density microporous polypropylene, Accurel MP 1000®, as a multiparticulate carrier along with drug of choice ibuprofen. In solvent evaporation, methanol (M) and dichloromethane (DCM) were used [31].

7.4.6   Calcium silicate

Calcium silicate (often referred by its shortened trade name Cal-Sil or Calsil) is the chemical compound Ca_2SiO_4, also known as calcium orthosilicate and sometimes formulated $2CaO.SiO_2$. It is one of group of compounds obtained by reacting calcium oxide and silica in various ratios e.g. $3CaO.SiO_2$, Ca3SiO5, $2CaO.SiO_2$, Ca_2SiO_4, $3CaO.2SiO_2$, $Ca_3Si_2O_7$ and $CaO.SiO_2$, $CaSiO_3$ [32]. Calcium orthosilicate is a white powder with a low bulk density and high physical water absorption. In its pure form, calcium metasilicate is a white to off-white color capable of absorbing up to two and a half times its weight of water. In this form, the hydrated powder retains its ability to flow freely. Addition of a mineral acid, such as hydrochloric acid (HCl), results in the formation of a gel. It is used as an anti-caking agent and an antacid. A white free-flowing powder derived from limestone and diatomaceous earth, calcium silicate has no known adverse effects to health. It is used in roads, insulation, bricks, roof tiles, table salt and occurs in cements, where it is known as belite (or in cement chemist notation C_2S). Sharma et al. reported on the effect of porous calcium silicate as adsorption carrier for in the tablet containing poorly water-soluble drug (meloxicam). The present study was characterization of microparticles obtained by adsorption of meloxicam, on a porous silicate carrier Florite RE (FLR) and development of a tablet formulation using these microparticles, with improved drug dissolution properties. The prepared tablets showed acceptable mechanical properties. Drug dissolution was much faster in both acidic and basic medium from prepared tablets as compared with commercial tablet. The results suggest that FLR provides a large surface area for drug adsorption and also that a reduction in crystallinity of drug occurs. Increase in surface area and reduction in drug crystallinity result in improved drug dissolution from microparticles [33].

7.4.7   Silicon dioxide

SYLYSIA® is a wide range of synthetic, micronized amorphous silica gel different degrees of porosity and high purity. The chemical structure of SYLYSIA® is based on Silicon Dioxide SYLYSIA® has a sponge-like structure with different degrees of porosity that varies from low (0.4 ml/g) to high (1.8 ml/g) pore volumes. SYLYSIA® is characterized by high specific. It is dry, white micronized powder, amorphous, porous, tasteless and odorless, chemically inert; its true specific gravity is 2.15; Refractive Index: 1.46. All these properties are precisely controlled to meet a wide variety of applications available. SYLYSIA is characterized by large internal surface area with a large number of pores, forming a three-dimensional structure by the silicate.

SYLYSIA is also helpful in keeping capsules from sticking together. By impregnating the surface of a gelatin capsule with SYLYSIA silica, the contact area between capsules is reduced. The use of SYLYSIA also aids in the release of the tablet from a mold or press by minimizing contact area. The high porosity of SYLYSIA silica enables it to absorb nearly three times its weight in liquids. Liquid ingredients can be made into powders for integration into systems. SYLYSIA has been widely used as a powderizing agent or carrier for liquid medicines, fragrances, vitamin oil, and many other products. Additional cosmetic applications include bath salts and related base cosmetics. SYLYSIA shows easy handling characteristics while stabilizing fragrance, hue, separation, adhesion, thickening, and fluidity. SYLYSIA is also used in cream products in the pharmaceutical and cosmetic industries. A tailored, micronized silica, SYLOPURE, is used as a thickening and abrasive agent in toothpastes. SYLYSIA can act to increase adhesive strength and durability. SYLYSIA can be used as an adhesive thickener to stabilize viscosity for longer shelf life. In addition to its adhesion and thickening effects, SYLYSIA effectively prevents stickiness and separation while providing emulsifying stability in these applications. These effects are stable regardless of environmental changes. SYLYSIA 300 series silicas have very high moisture capacities at low relative humidity which makes them a particularly effective drying agent at relative high humidity.

Planinšek et al. impregnated a porous SiO_2 (Sylysia) with carvedilol from acetone solution to improve dissolution of this poorly water-soluble drug. Solvent evaporation in a vacuum evaporator and adsorption from acetone solution were the methods used to load various amounts of carvedilol into the Sylysia pores. The results showed that when the drug precipitated in a thin layer within the carrier the dispersion retained a high-specific surface area, micropore volume, and drug-release rate from the solid dispersion. Increasing the amount of drug in the solid dispersion caused particle precipitation within the pores that decreased the carrier's specific surface area and pore volume and decreased the release rate of the drug. The results also suggest that the amorphous form of carvedilol, the improved wettability and weak interactions between the drug and carrier in the solid dispersion also contribute to improved dissolution of the drug from the dispersion [34].

Kortesuo et al. evaluated the possibilities to control the release rate of dexmedetomidine (DMED) from different spray-dried silica gel microparticle formulations. Microparticles were prepared by spray drying a silica sol polymer solution containing the drug. Drug release was investigated both in vitro and in vivo. The influence of sol–gel synthesis parameters, like pH and the water/alkoxide ratio (r) of the sol, on the release behaviour of the drug was studied. Silica gel microparticles had a smooth surface. The release rate was the slowest from microparticles prepared with water/alkoxide ratio 3:5. The bioavailability of dexmedetomidine in dogs showed that the release was sustained from silica gel microparticles as compared with a subcutaneously administered reference dose of 0.1 mg [35].

7.4.8   Magnesium aluminometasilicate

Neusilin® [$Al_2O_3.MgO.1.7SiO_2.xH_2O$] is a fine white powder or granule of magnesium aluminometasilicate manufactured by Fuji Chemical Industry. Compared to other common excipients in the silicate family, Neusilin®'s superior physico-chemical properties can resolve formulation problems encountered with oily actives, improve the quality of tablets, powder flow, capsules and many more. The most suitable grade for converting oil to powder is Neusilin® US2.When the oil load is comparatively high, an addition of 0.5% to 2% Neusilin® UFL2 will improve flowability substantially. Neusilin® UFL2 alone at 0.5% can resolve sticking issues of oily formulations. Due to its large surface area and porous nature, Neusilin® US2 adsorbs high loads of oil and can be compacted into high quality tablets. Neusilin® US2 and UFL2 grades show higher oil adsorption capacity* when compared to

MCC or Colloidal silica. On addition to starch, the UFL2 particles stick to the surface and facilitate flow as in a 'roller blade' model. A 0.5% addition of UFL2 to potato starch vastly improves flowability. Amorphous form of Magnesium Alumino meta silicate with a neutral pH that can be used in both direct compression and wet granulation of solid dosage forms. Formulating poorly water-soluble drugs by solid dispersion leads to a remarkable improvement in dissolution and bioavailability. Neusilin® can potentially resolve problems associated with tabletting and improve efficiency of solid dispersion. Durgacharan et al. reported development of solid self micro emulsifying drug delivery system with Neusilin US2 for enhanced dissolution rate of Telmisartan. Aim of present study was to develop solid self micro emulsifying drug delivery system (S-SMEDDS) with Neusilin US2 for enhancement of dissolution rate of Telmisartan (TEL). SMEDDS was prepared using Oleic acid, Tween 80 and PEG 400 as oil, surfactant and cosurfactant respectively. For formulation of stable SMEDDS, micro emulsion region was identified by constructing pseudo ternary phase diagram containing different proportion of surfactant: co-surfactant (Km value 1:1, 2:1 and 3:1), oil and water. Prepared SMEDDS was evaluated for thermodynamic stability study, dispersibility tests, globule size and zeta potential. S-SMEDDS was prepared by adsorption technique using Neusilin US2 as solid carrier. Prepared S-SMEDDS was evaluated for flow properties, drug content, reconstitution properties, FT-IR, SEM, DSC and in-vitro dissolution study. Results showed that prepared liquid SMEDDS passed dispersibility test with good thermodynamic stability [36].

Kiran Kumar et al. reported on development of solid self emulsifying drug delivery systems containing efavirenz: in vitro and in vivo evaluation. Efavirenz is a drug with an absorption window. Its oral bioavailability is 40–50%. The aim of this work was to formulate solid self-emulsifying drug delivery system to improve the solubility and bioavailability of efavirenz. Seven S(M)EDDS formulations were prepared. The optimized liquid self-emulsifying drug delivery system formulation (F3) was converted into solid S(M)EDDS with free-flowing powder by adsorbing onto a solid carrier like Neusilin US2 for encapsulation. X-ray diffraction studies showed no physico-chemical interaction. The in-vitro dissolution characteristics and in-vivo bioavailability studies of optimized formulation and reference standard confirmed that better systemic absorption and bioavailability also found to be increased with optimized formulation [37]. A compilation of the work done on drug release modulation by use of porous carriers of pharmaceutical significance and a compilation of the patent on the basis of application of porous carriers are shown in Table 7.1 and Table 7.2.

Table 7.1 A compilation of the work done on drug release modulation by use of porous carriers of pharmaceutical significance

Adsorbent	Experimental work	Drugs incorporated
Hydroxyapatite	Enhancement of anterior cervical spine interbody fusion	Recombinant human bone morphogenetic protein-2 (rhBMP-2)
	Tablets containing hydroxyapatite and a pore forming agent (50% (w/w) Avicel PH 200/20, 37.5% and 50% corn starch/37.5% sorbitol) were manufactured by direct compression followed by sintering	Metoprolol tartrate, riboflavin sodium phosphate
Polystyrene beads	Self-emulsifying system (SES)	Loratadine
Mesoporous silica	Various attempts on preparing mesoporous basic material MgO/SBA-15 by using impregnation, microwave irradiation and their combination, through magnesium acetate path	–
	Effect of Mesoporous SBA-15 molecular sieve as a carrier for controlled release	Nimodipine
Clay and zeolites	Development of a method where porous carbon was templated from inorganic materials such as zeolites (Tosoh H-Beta) and mesoporous molecular sieves	–
Polypropylene	Development of a novel multiparticulate gastroretentive drug delivery system and to demonstrate its performance in vitro	Verapamil HCl
	Investigation by combining floating and pulsatile principles to develop drug delivery system, intended for chronotherapy in arthritis by using low-density porous carrier based conceptual drug delivery system	Ibuprofen
Calcium silicate	Effect porous calcium silicate as adsorption carrier for in the tablet containing poorly water soluble drug	–
Sylysia	To improve dissolution of this poorly water-soluble drug	Carvedilol
Neusilin	Preparation of liquid loadable tablets (LLT)	Cyclosporine

Table 7.2 A compilation of the patent on the basis of application of porous carriers

Rationality for invention of the patent	Title of the patent	Patent number	References
Objective to produce a controlled-release drug delivery composition comprising an effective amount of pharmaceutical medicament in a polymer substrate which can swell in a polar organic solvents	Crosslinked porous polymer for controlled drug delivery	US 4548990	[38]
The invention relate to an apparatus for applying by means of a slotted nozzle device formed between a stationary incoming nozzle device and a trailing nozzle lip	Coating porous carriers	US 6077351	[39]
Multi-phase calcium silicate hydrate having unique physical and chemical properties are prepared by hydrothermal reaction of specialized ratio of CaO and Sio2	Multi-phase calcium silicate hydrate, method for their preparation and improved paper and pigment products produced therewith	US 6726807 B1	[40]
A process through which hydroxyapatite can be engineered to function as an efficient vehicle for drug delivery from coating or self-supported microsphere	Biofunctional hydroxyapatite coating and microspheres for in situ drug encapsulation	US 6730324B2	[41]
A process for encapsulating an integrated circuit die using a porous carrier	Die encapsulation using a porous carrier	US 7015075B2	[42]
The invention provide a melt-blended film forming, thermosetting, powder coating composition for producing a high temperature resistant coating	Heat resistant powder coating	US 0224431 A1	[43]

7.5   Conclusions

Today the porous carriers have a major role to play in the pharmaceutical industry. Their inner structure consists of unidirectional channel like pores that forms a hexagonal pattern. The presence of porous structures like microporous, mesoporous and macroporous was found to be essential in providing the sustained drug delivery systems, floating drug delivery systems and improvement of poorly water-soluble drugs, etc. Therefore, in the years to come, there is going to be continued interest in the porous carriers to have better materials for drug delivery systems.

7.6   References

1. Ali J., Khar R.K., Ahuja A. A Textbook of Biopharmaceutics & Pharmacokinetics. 2002; 1: 225–27.

2. Li Z., Wen L., Shao L., Chen J. Fabrication of porous silica nanoparticles and their applications. *Control Release*. 2004; 98: 245–54.

3. Torchilin V.P. Recent advances with liposomes as pharmaceutical carriers. *Nat Rev Drug Discov*. 2005; 4: 145–60.

4. Allen T.M., Cullis P.R. Drug delivery systems: Entering the main stream. *Science*. 2004; 303: 1818–22.

5. Faruk C. Scale effect on porosity and permeability: Kinetics model and correlation. *AmInst Chem Engg J*. 2001; 47: 271–87.

6. Gurny R., Doelker E., Peppas N.A. Modelling of sustained release of water soluble drugs from porous, hydrophobic polymers. *Biomaterials*. 1981; 3: 27–32.

7. Charney C., Begu S., Tourne P., Nicole L., Lerner D., Devoisselle J.M. Inclusion of ibuprofen in mesoporous template silica: Drug loading and release property. *Eur J Pharm Biopharm*. 2003; 57: 533–40.

8. Sher P., Insavle G., Porathnam S., Pawar A.P. Low Density porous carrier drug adsorption and release study by response surface methodology using different solvents. *Int J Pharm*. 2007; 331: 72–83.

9. Sharma S., Sher P., Badve S., Atmaram P.P. Adsorption of Meloxicam on porous calcium silicate: Characterisation and tablet formulation. *AAPS Pharm Sci Tech*. 2005; 6: E618–25.

10. Salis A., Sanjust E., Solinas V., Monduzzi M. Characterization of Accurel MP 1004 polypropylene powder and its use as a support for lipase immobilization. *J Mol Cat B Enz*. 2003; 24–25: 75–82.

11. Streubel A., Sipemann J., Bodmeier R. Floating matrix tablets based on low-density foam powder: Effects of formulation and processing parameters on drug release. *Eur J Pharm Sci*. 2003; 18: 37–45.

12. Song S.W., Hidajat K., Kawi S. Functionalized SAB-15 material as carrier from controlled drug delivery: Influence of surface properties on matrix drug interactions. *Langmuir*. 2005; 21–95: 68–75.

13. Byrne R.S., Deasy P.B. Use of commercial porous ceramic particles for sustained drug delivery. *Int J Pharm*. 2002; 246: 61–73.

14. Li Z., Wen L., Shao L., Chen J. Fabrication of porous silica nanoparticles and their applications in release control. *J Control Release*. 2004; 98: 245–54.

15. Ohta K.M., Fuji M., Takeri T., Chikazawa M. Development of a simple method for the preparation of silica gel based controlled drug delivery system with high drug content. *Eur J Pharm Sci.* 2005; 26: 87–96.

16. Wang C., He C., Tong Z., Liu X., Ren B., Zeng F. Combination of adsorption by porous CaCO3 microparticles and encapsulation by polyelectrolyte multilayer films for sustained drug delivery. *Int J Pharm.* 2006; 308: 160–7.

17. Kundu B., Soundrapandian C., Nandi S.K., Mukherjee P., Dandapat N., Roy S., Datta B.K., Mandal T.K., Basu D., Bhattacharya R.N. Development of new localized drug delivery system based on ceftriaxone-sulbactam composite drug impregnated porous hydroxyapatite: a systematic approach for in vitro and in vivo animal trial. *Pharm. Res.* 2010; 27(8): 1659–76.

18. Kundu B., Lemos A., Soundrapandian C., Sen P.S., Datta S., Ferreira J.M.F., Basu D. Development of porous HAp and β-TCP scaffolds by starch consolidation with foaming method and drug-chitosan bilayered scaffold based drug delivery system. *J Mater. Sci.* 2010; 21(11): 2955–69.

19. Chai F., Hornez J.C., Blanchemain N., Neut C., Descamps M., Hildebrand H.F. Antibacterial activation of hydroxyapatite (HA) with controlled porosity by different antibiotics. Groupe de Recherchesur les Biomatériaux France. 2007; 24(5): 510–4.

20. Takahashi T., Tominaga E., Watabe N., Yokobori A.T., Sasada H., Yoshimoto T. Use of porous hydroxyl apatite graft containing recombinant human bone morphogenetic protein-2 for cervical fusion in a caprine model Tohoku University, Sendai. 1999; 90(2): 224.

21. Cosijns A., Vervaet C., Mullens S., Siepmann F., Hoorebeke L.V., Masschaele B., Cnudde V., Remon J.P. Porous hydroxyapatite tablets as carriers for low-dosed drugs. *European Journal of Pharmaceutics and Biopharmaceutics.* 2007; 67(2): 498–506.

22. Tiwari R., Tiwari G., Rai A.K. Self-emulsifying drug delivery system: An approach to enhance solubility. Department of Pharmaceutical Sciences. 2011.

23. Patil P., Paradkar A. Porous polystyrene beads as carriers for self-emulsifying system containing loratadine. *AAPS Pharm Sci Tech.* 2006; 7(1): E199–E205.

24. Brian G., Jennifer A., Lin Y., Victor S. Biocompatible mesoporous silica nanoparticles with different morphologies for animal cell membrane penetration. *Chemical Engineering Journal.* 2007; 137: 23–29.

25. Wei Y.L., Yand C., Zhu J.H. Attempts on preparing mesoporous basic material MgO/ SBA-15 Recent advances in the science and technology of zeolites and related materials. Proceedings of the 14th International Zeolite Conference Studies in Surface Science and Catalysis. 2004; 154(1): 878–885.

26. Salonen J., Kaukonen A.M., Hirvonen J., Lehto V.P. Mesoporous silicon in drug delivery applications Department of Physics, University of Turku, FI-20014 Turku, Finland.

27. Hemingway B.S., Robie R.A. Thermodynamic properties of zeolites: low-temperature heat capacities and thermodynamic functions for phillipsite and clinoptilolite. Estimates of the thermochemical properties of zeolitic water at low temperature. 1984; 69: 692–700.

28. Barata R.P.M, Mays T.J., Moggridge G.D. Structured carbon adsorbents from clay, zeolite and mesoporous aluminosilicate templates. *Carbon*. 2003; 41(12), 2003:2231–46.

29. FDA Public Health Notification: Serious Complications Associated with Transvaginal Placement of Surgical Mesh in Repair of Pelvic Organ Prolapse and Stress Urinary Incontinence. FDA. 2008.

30. Streubel A., Siepmann J., Bodmeier R. Floating microparticles based on low density foam powder. *International Journal of Pharmaceutics*. 2002; 241(2): 279–292.

31. Sher P., Ingavle G., Ponrathnam S., Pawar P.A. Low density porous carrier: Drug adsorption and release study by response surface methodology using different solvents. *Int. J Pharm*. 2007; 331(1): 72–83.

32. Taylor H.F.W. Cement Chemistry. Academic Press. 1990: 33–34.

33. Sharma S., Sher P., Badve S., Pawar A.P. Adsorption of Meloxicam on Porous Calcium Silicate: Characterization and Tablet Formulation. *AAPS Pharm Sci Tech*. 2005; 6(4): E618–E625.

34. Odon P., Kand B., Franc V. Carvedilol dissolution improvement by preparation of solid dispersions with porous silica. University of Ljubljana. 2011.

35. Pirjo K., Manja K., Minna K., Mika J., Tiina L., Lauri V., Sirpa L., Juha K. Effect of synthesis parameters of the sol–gel-processed spray-dried silica gel microparticles on the release rate of dexmedetomidine. *Biomaterials*. 2002; 23(13): 2795–2801.

36. Bhagwat D.A., D'Souza J.I. *Int. J Drug Dev Res*. 2012; 4: 398–407.

37. Kumar V.K., Devi M.A., Bhikshapathi D.V.R.N. Development of Solid Self Emulsifying Drug Delivery Systems Containing Efavirenz: In Vitro And In Vivo Evaluation; *Int J Pharm Bio Sci*. 2013; 4(1): 869–82.

38. Muller K.F., Heiber S.J. Crosslinked porous polymer for controlled drug delivery. Us 4548990.

39. Herzog P. Coating porous carriers. US 6077351.

40. Mathur V.K. Multi-phase calcium silicate hydrate, method for their preparation and improved paper and pigment products produced therewith. US 6726807 B1.

41. Trczynski T., Liu D.M., Yang Q. Biofunctional hydroxyapatite coating and microspheres for in situ drug encapsulation. US 6730324B2.

42. Fay O.R., Amrine C.S., Lish K.R. Die encapsulation using a porous carrier. Us 7015075B2.

43. Decker O.H., Zhou W.J. Heat resistant powder coating. US 0224431 A1.

8

Polymeric nanoparticles as carrier for controlled release

Muddana Eswara Bhanoji Rao[1*], Suryakanta Swain[2] and Sitty Manohar Babu[2]

[1]Department of Pharmaceutics, Roland Institute of Pharmaceutical Sciences,
Khodasingi, Berhampur (Ganjam), Odisha-7600 10, India
[2]Department of Pharmaceutics, Southern Institute of Medical Sciences,
College of Pharmacy, SIMS group of Institutions, Mangaldas Nagar, Vijyawada Road,
Guntur -522 001, Andhra Pradesh, India

8.1   Introduction

During the last two decades, considerable attention has been given to the development of novel drug delivery systems. The rationale for controlled drug delivery is to alter the pharmacokinetics and pharmacodynamics of drug substances in order to improve the therapeutic efficacy and safety through the use of novel drug delivery systems. Besides more traditional matrix or reservoir drug delivery systems, colloidal drug delivery systems have gained in popularity. The major colloidal drug delivery systems investigated include liposomes and polymeric nanoparticles [1]. These systems have been investigated primarily for site-specific drug delivery, for controlled drug delivery, and also for the enhancement of the dissolution rate / bioavailability of poorly water-soluble drugs. The primary routes of administration under investigation are parenteral routes; however, other routes such as the oral, ocular, or topical routes are also being investigated.

While numerous publications have appeared on liposomes from international research teams, the research on polymeric nanoparticles has been primarily performed by a few research groups in Europe (Speiser et al., Cauvreur at al., Davis et al., Gurny et al., Kreuter et al.). Nanoparticles are being investigated as an alternative colloidal drug delivery system that could potentially avoid some of the technical problems observed with liposomes. When compared to the fluid or semifluid liposome systems, the polymeric nanoparticle systems could be classified as solid particles.

Nanoparticles are colloidal particles with a size smaller than 1 mm. The active compound can be present in various physical states; it can be dissolved in the polymeric matrix, can be encapsulated, or can be adsorbed or attached to the surface of the colloidal carrier. The term nanoparticles encompass both nanocapsules and nanospheres (Fig. 8.1). Nanocapsules have a core-shell structure (a reservoir system), while nanospheres represent a matrix system. It is often difficult to distinguish between the different structures and morphologies of the particles; the term nanoparticles is thus used as a general term.

This chapter presents an overview of the different methods used to prepare polymeric nanoparticles and the physico-chemical methods used to characterize the nanoparticles. In vivo results with nanoparticles are not reviewed. Several excellent reviews exist on this subject [2, 3].

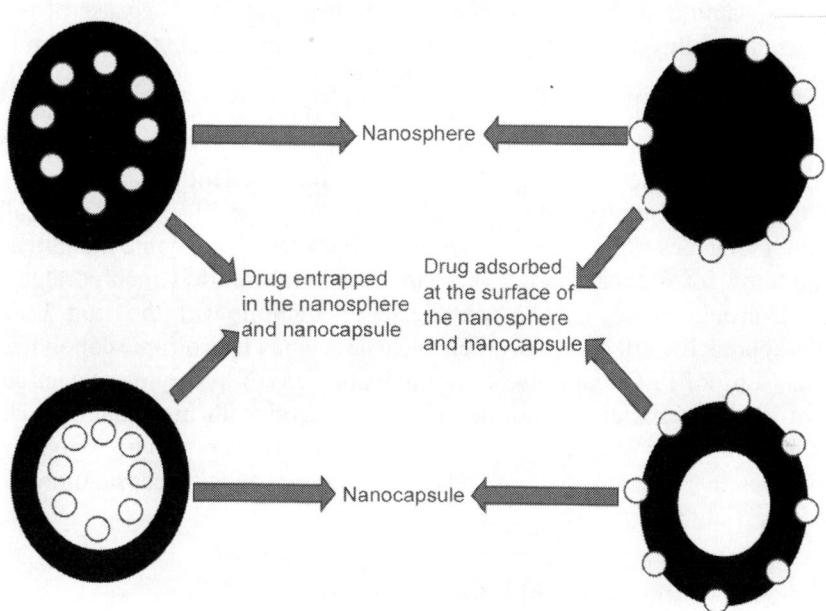

Figure 8.1 Types of drug-loaded nanoparticles

8.1.1   Structural compositions of polymeric micelles

Polymeric micelles have received considerable attention in the past two decades as a new multifunctional nanoplatform for the delivery of hydrophobic drugs. Polymeric micelles are nanosized (typically in the range of 20–100 nm) supramolecular constructs formed from the self-assembly of amphiphilic block copolymers in aqueous environments. In water, the hydrophobic segment of the block copolymer self-associates into a semisolid core, with the hydrophilic segment of the copolymer forming a coronal layer. The resulting core–shell architecture is important for drug delivery purposes; the hydrophobic core serves as a reservoir for water-insoluble drugs, and the outer shell protects the micelle from rapid clearance in circulation. Poly (ethylene glycol) (PEG) is most commonly used for the hydrophilic segment. PEG molecules are biologically inert. Moreover, they are shown to prevent nonspecific protein adsorption to the micelle surface prolonging the blood circulation time of the micelles. Other polymers, such as poly (N-vinyl pyrrolidone) and poly (N-isopropyl acrylamide), are also used as hydrophilic blocks but with much less frequency. Compared with the hydrophilic blocks, the chemistry of core-forming hydrophobic polymers is much more diverse. Polyesters and poly (L-amino acids) are the most widely used polymers because of their biocompatibility and biodegradability. Examples include, but are not limited to, poly(lactic acid) (both L-isomer, or PLA and D,L-isomer, or PDLLA); poly(e-caprolactone); poly(L-aspartic acid) (pAsp); and poly(L-glutamic acid). Recently, a new hydrotropic polymer design was reported to form polymeric micelles with high drug loading and excellent physical stability. This process involves the screening of hundreds of pharmaceutically safe molecules to identify candidate structures that enable heightened solubility for a chosen drug; then, the structural motif is incorporated as the hydrophobic segment to enhance its interactions with the drug. Using this approach, Park and co-workers have developed hydrotropic copolymers consisting of PEG and poly (4-(2-vinylbenzyloxy-N-picolylnicotinamide)) that provide efficient encapsulation of paclitaxel with high loading. This method might provide a universal strategy to produce tailor-made polymeric micelles that can achieve high drug loading and stable encapsulation of a wide variety of drugs [4].

8.1.2   Types of polymeric micelle

To date, the three major types of micelle delivery systems based on linear block copolymers are (1) common block copolymer micelle, (2) drug-conjugated block copolymer micelle, and (3) block ionomer complex micelle. PEG is most often used as a hydrophilic segment because of its flexibility, nontoxicity, and hydrophilicity. However, the options available for the hydrophobic block are much broader. For example, the AB-type block polymer PEGYb-polyester, such as PEGYb-poly (D,L-lactide) (PDLLA), PEGYb-poly(lactide-co-glycolide) (PLGA), and PEGYb-poly((-caprolactone) (PCL), is a popular family of block polymers used for drug delivery. PEOYb-poly (gbenzyl- L-glutamate) was developed later. Two long-chain fatty acyl groups in phospholipids are hydrophobic, and they have been used successfully as the hydrophobic co-reforming group. By changing the hydrophilic segments, a series of lipid derivatives have been reported for preparing drug-loaded micelles including PEGYphosphatidyl ethanolamine (PE), poly(2-alkyl-2-oxazoline)YPE, poly(acryloyl morpholine)YPE, PVPYPE, and polyglycerolYphosphatidylglycerol (Scheme 8) (58Y60). Recently, Chang et al. reported several kinds of amphiphilic diblock copolymers composed of PEG and hydrophobic poly[bis(ethyl glycinat- N-yl) phosphazene] or poly [bis(trifluoroethoxy) phosphazene] which were synthesized via controlled cation-induced polymerization of a phosphoranimine at an ambient temperature using a PEGYphosphoranimine macroinitiator. In a similar method, using methoxyethoxyethoxyl group (MEEP) as hydrophilic block, amphiphilic diblock polystyreneYb-poly-[bis(methoxyethoxyethoxy) phosphazene] (PSYb-PMEEP) and MEEP-phenyl/MEEP copolyphosphazene were synthesized and characterized [5].

8.1.3   Architecture properties relationship of polymeric micelle

From a chemistry standpoint, it is feasible to synthesize various block copolymers that form a micelle. When such a micelle is used as a drug carrier in the body, many performance-related issues must be addressed, including its static and dynamic stability, morphology, size and size distribution, biocompatibility, drug-loading capacity, release rate, circulation time, biodistribution, and endocytosis mechanism. To a large extent, these properties are determined by the architecture of the block copolymer [5].

8.2   Preparation of nanoparticles

The techniques used to prepare nanoparticles are generally classified in two groups. In the first group, nanoparticles are formed from preformed polymers. The polymers include both water-insoluble and soluble polymers of synthetic, semisynthetic, or natural origin. Alternatively, nanoparticles are not prepared from preformed polymers but through various polymerization reactions of lipophilic or hydrophilic monomers. The techniques used for the preparation of nanoparticles based on water-insoluble polymers are primarily derived from methods developed for the preparation of aqueous colloidal polymer dispersions, which are used in the coating of solid dosage forms [5]. Latexes are obtained by emulsion polymerization (e.g. acrylic latexes-Eudragits) and pseudolatexes through the emulsification of polymer solutions or melts (e.g., ethyl cellulose-Aquacoat, Surelease).

The pharmaceutical industry adopted the colloidal polymer dispersions from other industrial applications including paints, varnishes, adhesives, and paper coatings. The aqueous colloidal polymer dispersions have been developed to avoid problems associated with the use of organic polymer solutions during the coating of solid dosage forms. The polymer dispersions allow the formation of water-insoluble coatings through the coalescence of the colloidal polymer particles into a homogeneous film from a completely aqueous coating medium. Problems associated with organic solvents, such as industrial and environmental hazards, pollution, and increased costs could be avoided. Drug-containing colloidal polymer particles (nanoparticles) can be obtained through the incorporation of a drug substance during or after the preparation of the polymer dispersions. As with microencapsulation methods, there is not one universal technique to prepare nanoparticles. The choice of a particular preparation method and a suitable polymer depends on the physico-chemical properties of the drug substance, the desired release characteristics, the therapeutic goal, the route of administration, the biodegradability/biocompatibility of the carrier material, and regulatory considerations. From a technological point of view, the successful selection of a preparation method is determined by the ability to achieve high drug loadings, high encapsulation efficiencies, and high product yields, and the potential for easy scale-up. For example, methods with high encapsulation efficiencies but with only low drug loading capacity are limited to very potent drugs. Methods that result in nanoparticles with high drug loadings and high encapsulation efficiencies are preferred. The term nanoparticles often actually describe the suspended system of the nanoparticles in an aqueous phase, which is the colloidal polymer dispersion. Besides high drug loadings and encapsulation efficiencies, a preparation method should also be able to yield polymer dispersions with a high nanoparticles content, which also directly relates to the size of the dose that can be reasonably administered. The techniques for preparing nanoparticles from preformed polymers are presented first, followed by techniques to prepare nanoparticles by polymerization of various monomers.

8.3   Nanoparticles prepared from preformed polymers

Nanoparticles have been prepared from a variety of both water-soluble and water-insoluble polymers of synthetic, semisynthetic, and natural origin. The use of preformed polymers has several advantages when compared to use of nanoparticles prepared through the polymerization of monomers.

These include the use of polymers with well-characterized physic-chemical properties, established safety and approval standards, the absence of residual monomers or polymerization reactants (e.g., initiators or catalysts) and the lack of possible reactions between drugs and monomers. In addition, it is not possible to obtain nanoparticles by emulsion polymerization for polymers where suitable monomers are not available. These polymers include, in particular, polyesters such as polylactic acid (PLA) and polylactic / flycolic acid copolymers, cellulose derivatives such as ethyl cellulose acetate, and natural polymers such as gelatin or sodium alginate. Most techniques for preparing nanoparticles from preformed polymers are derived from conventional emulsification technologies, which have been used to prepare microparticles. The nanoparticles can be formed either by dispersing the polymer-drug phase into an external phase or by phase inversion techniques, whereby the external phase is first added to the internal phase containing the drug and the polymer. These techniques also allow the use of mixtures of drugs or polymers. The preparation techniques using polymers are classified based on the solubility of the polymer.

8.3.1   Nanoparticles prepared from water-insoluble polymers

Solvent evaporation method – water-immiscible organic solvents

The solvent evaporation method is a technique widely used for preparing biodegradable and non-degradable microspheres [6–8]. In this method, the drug and polymer are dispersed / dissolved in a volatile, water-immiscible organic solvent (e.g., methylene chloride, chloroform, ethyl acetate). This solution or dispersion is then emulsified in an external aqueous phase containing an emulsifying agent by using conventional emulsification equipment to form an oil-in-water (O/W) emulsion. Microspheres with a size range generally between 5 μm and 250 μm are obtained after solvent evaporation. To prepare nanoparticles and not microspheres, the O/W emulsion is homogenized under high shear with appropriate homogenization equipment (e.g., microfluidizer, sonication) prior to the precipitation of the polymer to further reduce the particle size of the internal organic phase into the colloidal size range (Fig. 8.2) [9–12]. Drug-polymer particles in the nanometer size range are formed after solvent diffusion into the aqueous phase and evaporation at the water/air-interface causing polymer and drug precipitation. The solvent removal process can be accelerated through heating and the application of a vacuum. The polymers used to prepare the nanoparticles include biodegradable polymers such as PLA and its copolymers and cellulose derivatives such as ethyl cellulose or cellulose acetate phthalate. In principle, any water-insoluble polymer being soluble in a water-immiscible organic solvent could be converted into colloidal polymer dispersion.

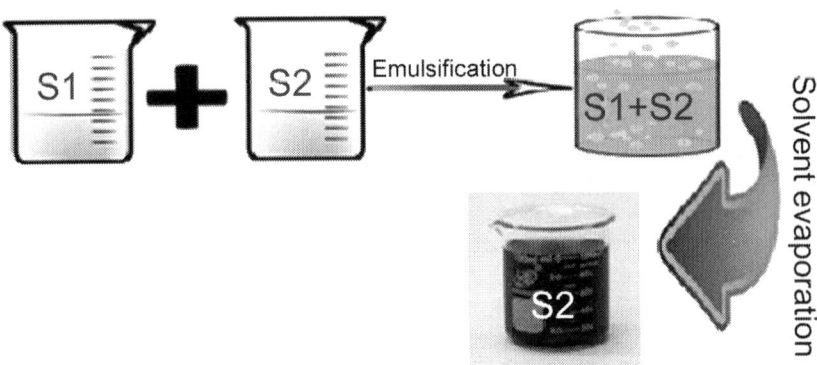

Figure 8.2 Preparation of nanoparticles by solvent evaporation (S1: organic solution of polymer; S2: nonsolvent of polymer)

The high-pressure emulsification-solvent evaporation method is limited to water-insoluble drugs. Unless specific binding exists, water-soluble drugs cannot be encapsulated because of complete portioning into the external aqueous phase during emulsification. One problem with this method, especially at higher drug loadings, is the appearance of drug crystals in the external aqueous phase. Drug crystallization in the external phase can be explained with the change in drug solubility in the aqueous phase during nanoparticle preparation [12]. The solubility of the drug in the external aqueous phase changes during the preparation of the nanoparticles. It first increases because of organic solvent diffusing from the internal organic phase into the external aqueous phase. After solvent evaporation, the drug solubility decreases and the amount of drug present in the aqueous phase exceeds the drug solubility, resulting in drug precipitation. The drug crystallization depends on the polymer, organic solvent, and type and concentration of surfactant used. The rate of solvent diffusion into the aqueous phase influences the rate of polymer precipitation. It is related to the water solubility and rate of evaporation of the organic solvent. A faster polymer precipitation results in a decrease in drug crystallization in the external phase. Another important factor for high drug entrapment (low crystallization in the external aqueous phase) is the drug-polymer compatibility. Drug drystallization in the aqueous phase occurs at higher drug loadings with polymers having a higher affinity for the drug. Drug used in excess of the solubility of the drug in the polymer precipitates in the external phase. In contrast to microparticles, where the drug can be either dissolved or dispersed in the polymeric matrix, the drug is only dissolved in the polymeric nanoparticle matrix. Increasing the concentration of sodium lauryl sulfate in the aqueous phase also enhances drug crystallization. The surfactant not only solubilizes the drug but also solubilizes the organic solvent during nanoparticle preparation. This results in higher amounts of drug crystallizing in the aqueous phase after solvent evaporation. Various processing and formulation variables influence the particle size of the resulting nanoparticles. The particle size of the solidified nanoparticles is determined by the size of the emulsified polymer-drug-solvent droplets, and hence depends on the homogenization equipment. Nanoparticles have been prepared with conventional laboratory homogenizers [9], by ultrasonication and by microfluidization [10–13]. Microfluidizer processing is a patented mixing technology available for the preparation of dispersed systems on both a laboratory and production-size scale. In this process, the liquid (e.g., O/W-emulsion) is pumped through microchannels to an impingement area at high operating pressures. Cavitation and the accompanying shear and impact are responsible for the particle size reduction within the "interaction chamber." Other pharmaceutical applications of microfluidization include the preparation of liposomes, microemulsions, and parenteral nutrition emulsions. With the microfluidizer, the particle size decreases with increasing operating pressure and increasing number of cycles [12]. The particle size of the nanoparticles is also affected by the type of surfactant and surfactant concentration, the viscosity of the organic polymer solution, and the phase ratio of the internal to the aqueous phase [12, 14]. As expected, the particle size decreases with decreasing viscosity of the organic phase (lower polymer concentration) and increasing poly (vinyl alcohol) concentration in the aqueous phase (up to a optimum concentration). Besides high-shear emulsification techniques, surfactants, which reduce the interfacial tension, are necessary in order to obtain particles in the submicron range. The surfactants or polymeric stabilizers used include, among others, various polysorbates, sodium lauryl sulfate, poly (vinyl alcohol), and gelatin. For nanoparticles to be administered by injection, purification of the nanoparticles from surfactants may be necessary. The emulsifiers can be removed by dialysis or by separating the nanoparticles from the aqueous phase by ultracentrifugation followed by washing steps with surfactant-free water. This, however, may lead to destabilization of the nanoparticle suspension. Nanoparticles of the PLA have been prepared using albumin as a colloidal stabilizer, resulting in a fully biodegradable system [15]. Nanoparticles can also be formed by phase inversion. In this method, an aqueous phase is dispersed into the drug-containing organic polymer solution in order to form water-in-oil (W/O) emulsion. Phase inversion occurs upon further addition of water. Various emulsifier combinations can be used. One popular method, which has been used to prepare commercial aqueous ethyl cellulose dispersion (Surelease) for coating purposes, is based on the in situ formation of the emulsifying agent [5]. Fatty acids, such as oleic or stearic acid, are added to the polymer melt. The aqueous phase containing an alkaline agent is added to the organic polymer phase, resulting in the ionization of the fatty acid at the interface.

Solvent evaporation method-water-miscible organic solvents

Replacing the organic solvents (e.g., methylene chloride) commonly used in the solvent evaporation method with less toxic solvents, such as ethanol or acetone, would be highly desirable from a toxicological point of view. However, the addition of solutions of polymers, such as PLA or ethyl cellulose, in these water-miscible solvents to an external aqueous phase would not result in nanoparticles under the same experimental conditions. Because of the complete miscibility of the organic solvent and water, large precipitates or agglomerates would form at the commonly used polymer concentrations. Various modified solvent evaporation methods, as described below, have been developed for water-miscible solvents. These methods are based either on the use of much lower polymer solution concentrations or on the use of polymers with stabilizing functional groups, or on a reduced miscibility of the solvent-water system based on a "salting-out mechanism." Liposomes, another colloidal delivery system, have been prepared with water-miscible solvents by a so-called solvent injection method [16]. The injection of a solution of the lipids in ethanol into an aqueous phase results in small unilamellar vesicles, whereas the injection of solutions of the lipid in water-immiscible solvents results in large unilamellar vesicles [17]. The preparation of polymeric nanoparticles with water-miscible solvents is derived from this technology (Fig. 8.3A). The methodology is similar to the solvent evaporation method described above, with only the water-immiscible organic solvent being replaced with a water-miscible solvent such as acetone or ethanol [18]. The drug-polymer solution is then poured into an external aqueous phase containing the emulsifying agent, to form droplets spontaneously. Nanospheres (a matrix-system) are formed after solvent evaporation. As mentioned above, only very dilute solutions of most polymers, in particular polymers without stabilizing functional groups can be used because of polymers; in particular polymers without stabilizing functional groups can be used because of polymer precipitation at higher concentrations.

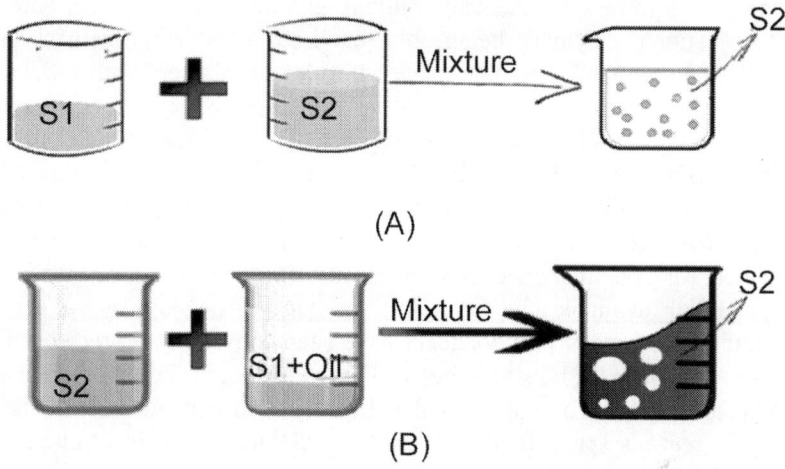

Figure 8.3 Preparative techniques of nanospheres (A) and nanocapsules (B)

Besides using the traditional solvent evaporation method, stealth nanoparticles have also been prepared with acetone by using poly(d,l-lactide)-methoxy (polyethylene glycol), a proprietary, more hydrophilic PLA-derivative [19]. In analogy to stealth liposomes, the circulation time of the nanoparticles could be increased because of the more hydrophilic character of the nanoparticles. Oligonucleotides have been entrapped within these stealth nanoparticles through the complexation with oligopeptides. The oligonucleotide/oligopeptide coprecipitates with the polymer after addition of an acetonic solution to water. However, it has yet to be proven that the complex is truly encapsulated within the nanoparticles or if it just coexists in submicron size beside the nanoparticles. Drug-containing acrylic nanoparticles can be formed spontaneously at high polymer solution concentrations (up to 40%) with pharmaceutically acceptable organic solvents, without the use of surfactants or sophisticated equipment, with polymers containing stabilizing functional groups [20]. The polymers used were acrylic copolymers with quaternary ammonium groups, Eudragit RS and RL 100. Aqueous latexes of these polymers are used extensively in the coating of solid dosage forms and have been prepared by emulsification of the solid polymer [21], or of solutions of the polymers in acetone [22], without the use of emulsifying agents. In contrast to the conventional solvent evaporation method, which uses primarily toxic, water-miscible solvents such as methylene chloride, water-miscible solvents such as acetone or ethanol are used with these acrylic polymers. The drug-polymer solution is simply added to an aqueous phase and the nanoparticles from spontaneously without high shear. Eudragit RS and RL 100 are cationic polymers, being poly (ethyl acrylate-methyl methacrylate-trimethyl ammonioethyl methacrylate chloride) copolymers with rations of 1:2:0:1 and 1:2:0.2. The nanoparticles are stabilized by the quaternary ammonium groups present in the polymer. The average particle size of nanoparticles prepared by microfluidization with the water-immiscible solvent, methylene chloride, is larger when compared to the size of spontaneously formed nanoparticles. As expected, the particle size increases with increasing polymer concentration because of increased solution viscosity. Commercially available aqueous colloidal polymer dispersions such as acrylic or cellulosic latexes or pseudolatexes have been converted into redispersible powders to reduce the bulk volume, to allow more flexible compounding, and to improve stability of polymers that hydrolyze in aqueous media during storage. Freeze or spray-dried Eudragit RS or RL nanoparticles can be easily resuspended because of the quaternary ammonium groups. The particle size closely matches the original particle size. Besides the parenteral route, potential application of colloidal polymer dispersions include the oral delivery of bad-tasting or irritating drugs, or topical administration in the form of brush-on nanosuspensions or free films for transdermal systems. Upon drying of nanosuspensions on the skin, films or individual nanoparticles could be formed depending on the minimum film formation temperature of the polymer-plasticizer-drug combination. Drug-containing films of water-insoluble polymers are generally prepared by casting and drying an organic drug-polymer solution or suspensions. To circumvent problems associated with the use of organic solvents, polymeric films have been prepared from aqueous polymer dispersions. Water-soluble drugs can be dissolved in the colloidal polymer dispersion prior to film casting. Several alternatives are available for the incorporation of water-insoluble drugs into colloidal polymer dispersions. Liquid lipophilic drugs have been emulsified directly into lattices prior to the preparation of films for transdermal application [23]. In a previous study, propranolol has been dispersed into an ethyl cellulose pseudolatex. Drug settling during the drying process results in a nonhomogenous drug distribution in the polymer film. A more homogenous distribution is obtained by dissolving the drug in a water-insoluble plasticizer, dibutyl sebacate, and emulsifying the drug-plasticizer solution into the latex prior to film casting [24]. Lidocaine has been incorporated into an ethyl cellulose pseudolates during the preparation of the colloidal dispersion by a sonication-solvent evaporation method. The drug is added to the organic polymer solution prior to the emulsification step. The pseudolatex forms a clear, flexible film on the skin with local anaesthetic activity [25]. Niwa et al. used a mixed-solvent system of methylene chloride and acetone to prepare PLGA-nanoparticles [26]. The addition of the water-miscible solvent, acetone, results in nanoparticles in the submicron range; this is not possible with only the water-immiscible organic solvent, under the experimental conditions. The addition of acetone decreases the interfacial tension between the organic and the aqueous phase and, in addition, results in the perturbation of the droplet interface because of the rapid diffusion of acetone in the aqueous phase. This technique is named "spontaneous emulsification solvent diffusion method." Nanocapsules of biodegradable polymers, such as PLA and PLA-copolymers or poly (e-caprolactone), have been prepared by an "interfacial polymer deposition mechanism" [27–33]. An additional component, water-immiscible oil, is added to the drug-polymer-solvent mixture (Fig. 8.3B). A solution of the polymer, the drug, and water-immiscible oil in a water-miscible solvent such as acetone is added to an external aqueous phase. The water-miscible organic solvent diffuses rapidly into the aqueous phase. The polymer precipitates at the oil/water interphase, surrounding the drug-containing oil core. It is hypothesized that the polymer acts as a surfactant stabilizing an O/W emulsion. This technique results in nanocapsules because of a core-shell structure. The drug must have a high solubility in the oil-solvent mixture in order to obtain nanoparticles with organic solvents, the disadvantage of this interfacial deposition method is the use of only very diluted polymer solutions. A polymer precipitate and not individual nanoparticles are formed at higher polymer concentrations because of the water-miscibility of the organic solvent. To obtain more concentrated dispersions, the polymer dispersion can be concentrated under reduced pressure or be converted into a redispersible powder by spray-or freeze-drying techniques. An interesting modification of the solvent evaporation method is the "salting-out procedure" (Fig. 8.4) [34–39]. The nanoparticles are prepared by adding an aqueous phase saturated with an electrolyte or nonelectrolyte to a solution of the polymer and drug in a water-miscible organic solvent under agitation until an O/W-emulsion forms. Poly(vinyl alcohol) has been added to the aqueous phase to act as a viscosity-enhancing and emulsifying agent. Saturating the aqueous phase reduces the miscibility of acetone and water by a salting-out process and allows the formation of an O/W emulsion from the water by a salting-out process and allows the formation of an O/W emulsion from the otherwise miscible phases. After the formation of the internal organic phase droplets, water is added to allow diffusion of the organic solvent into the external phase and precipitation of the polymer resulting in nanoparticle formation. The solubility/miscibility of acetone with the external aqueous phase is thereby gradually increased up to complete miscibility by diluting the initially saturated aqueous phase with water. The salt or nonelectrolyte can be removed by centrifugation and subsequent washing steps from the polymeric nanoparticles. Various polymers including cellulose acetate phthalate, methacrylic acid copolymers, ethyl cellulose and PLA have been used. Cellulose acetate phthalate nanoparticles have been investigated as an in-situ gelling ophthalmic drug delivery system [35]. The nanoparticles are insoluble in water; however, the enteric polymer dissolves/gels at the pH of the tear fluids. Cellulose acetate phthalate, an ester, is not stable and hydrolyzes in aqueous media. The polymer suspension could be converted in a redispersible nanoparticles powder, which, however, may not be a practical approach for ophthalmic drug delivery. Enteric polymers based on acrylates (e.g., Eudragit L) are stable in aqueous colloidal dispersion and could be used instead of cellulose acetate phthalate. The major advantages of the salting-out method are the avoidance of chlorinated solvents and surfactants commonly used with the conventional solvent evaporation method. Converting the polymer dispersion into a redispersible powder without surfactants could be a problem with nanoparticles prepared with polymers without surface-active functional groups.

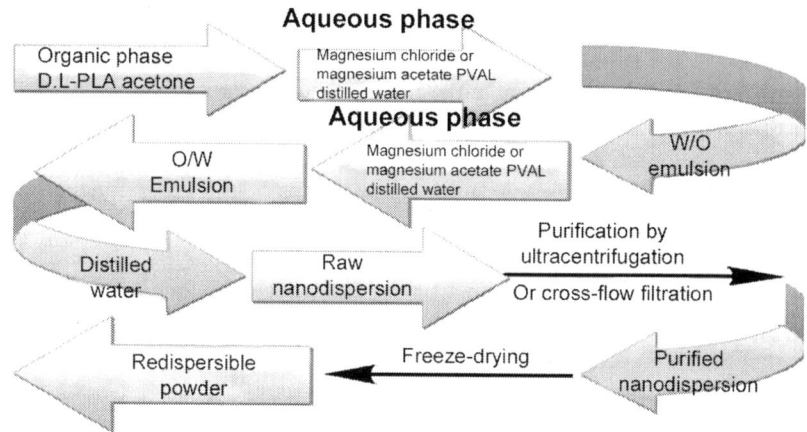

Figure 8.4 Preparative techniques of nanoparticles by salting out process

When compared to nanoparticles prepared by emulsion polymerization techniques, the particle size of nanoparticles obtained by solvent evaporation methods is generally larger and the particle size distribution is not monodisperse. The polymer dispersions prepared by emulsification techniques are, therefore, generally physically less stable than nanoparticles prepared by emulsion polymerization techniques. The use of high-shear homogenization or sonication techniques is energy consuming and could potentially result in the degradation of the drug or the polymeric carrier. With nanoparticles intended for parenteral use, either biocompatible/biodegradable surfactants have to be used or the surfactant has to be removed by techniques such as dialysis. This, however, may result in physical stability problems of the polymer dispersion. Besides physical stability, the chemical stability of not only the drug but also of the polymeric carrier in the aqueous medium has to be investigated. In particular a biodegradable polyester, such as PLA, is prone to hydrolysis [40]. With PLA nanoparticles, the conversion of the polymer dispersion to a redispersible powder by freeze-drying advised. In addition, with nanoparticles prepared by the solvent evaporation method, the issue or residual organic solvents must be addressed.

8.3.2   Nanoparticles prepared from hydrophilic polymers

W/O-emulsification methods

Nanoparticles of hydrophilic polymers (e.g., albumin, chitosan, gelatin, or carbohydrates) can be prepared by W/O-emulsification techniques (Fig. 8.5).

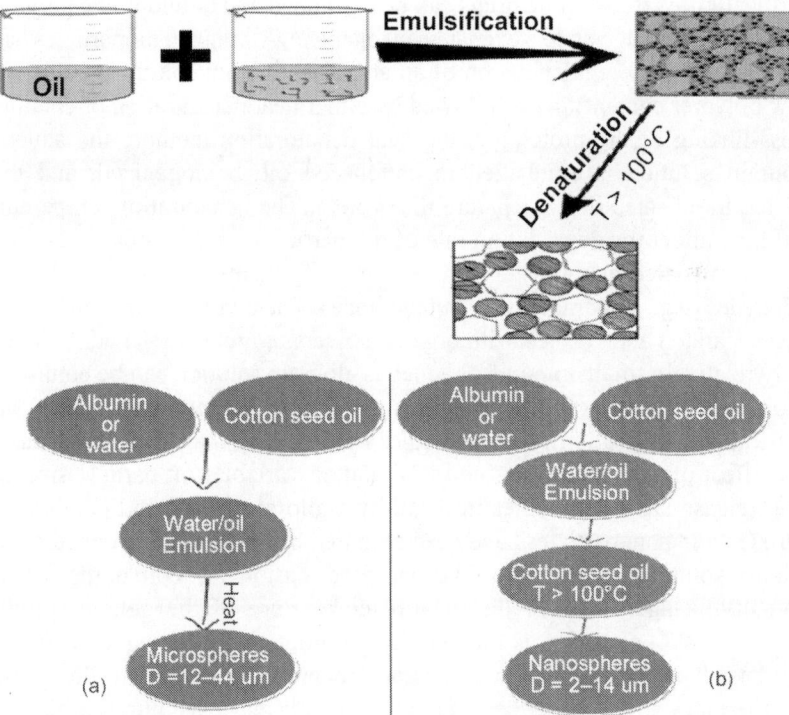

Figure 8.5 Preparative techniques of microspheres (a) and nanospheres (b) by thermal denaturation process.

This process has been developed for the preparation of albumin microspheres [41-44]; however, as with microspheres prepared by the solvent evaporation method, the use of high-shear homogenization equipment or ultrasonication allowed the formation of emulsions in the nanometer size range. Basically, an aqueous polymer solution is emulsified into an external, water-immiscible phase, such as an oil or organic solvent followed by homogenization. Upon water removal, the polymer droplets solidify. Upon contract with water, the resulting nanoparticles would completely dissolve because of their high water solubility. Insoluble nanoparticles can be obtained by further hardening/ insolubilizing the polymer through chemical cross-linking with aldehydes or other cross-linking agents or through denaturation at elevated temperature.

To obtain high encapsulation efficiencies, the drug must be insoluble in the external phase. Hydrophilic polymer nanoparticles prepared by W/O-emulsification techniques are, therefore, limited to water-soluble drugs. The drug can be added prior to the emulsification step into the aqueous polymer solution or it can be adsorbed into the nanoparticles after separation and purification. Water-soluble drugs can be entrapped and bound to the polymer, for example through an ion-exchange mechanism. Albumin nanoparticles have been prepared by emulsification of an aqueous albumin solution (up to 50% w/w polymer content) in oil followed by either heat denaturation or chemical cross-linking of the protein. In the heat denaturation method, the aqueous albumin solution is emulsified in cottonseed oil, homogenized, and then poured into heated oil to denature the protein. The denaturation temperature and time affect the degradation rate of the particles. The nanoparticles could also be prepared at room temperature through chemical cross-linking with aldehydes (e.g., glutaraldehyde, butanedione). The albumin nanoparticles are then separated and washed with organic solvents to remove the adhering oil. Only relatively small amounts of aqueous albumin solution can be emulsified into large amounts of oil, because significantly longer drying times and possible agglomeration result at higher ratios of internal to external phase. The effect of various process and formulation variables on particle size and drug release has been studies in detail by Gallo *et al.* [45] and Gupta *et al.* [46]. Gelatin nanoparticles have been prepared by emulsifying a concentrated gelatin solution (30%) into hydrogenated castor oil containing proper emulsifying agents above the gelatin temperature of the gelatin solution [47]. The W/O emulsion is then cooled in order to gel the aqueous gelatin droplets. It is then diluted with acetone to probably dehydrate the gelatin nanoparticles and to ease the removal of the oil phase by filtration through a membrane filter with a pore size of 50 nm. The nanoparticles are washed with acetone and insolubilized/hardened with a formaldehyde solution. The nanoparticles have an average diameter of less than 300 nm. The degradation of the nanoparticles and hence the drug release are dependent on the degree of cross-linking with glutaraldehyde. Besides emulsifying into an oil phase, the aqueous polymer solution can also be emulsified into organic solvents such as chloroform. Gelatin nanoparticles have been prepared with a mixture of chloroform or toluene as the external phase [48]. The use of organic solvents is, however, undesirable. Chitosan is one of the few naturally derived polysaccharides carrying basic functional groups. It is obtained through the deacetylation of chitin. Since it is only soluble in acidic media, chitosan micro- and nanoparticles are prepared by emulsifying aqueous solutions of chitosan in acetic acid into an external oil phase [49]. The cross-linking agent can be added to the aqueous chitosan solution prior to emulsification. The nanoparticles are obtained after solidification of the internal aqueous phase through the removal of water at elevated temperature and reduced pressure. Colloidal magnetite particles can be added to the aqueous phase prior to emulsification to produce externally guidable magnetic nanoparticles. Magnetic albumin nano- and microparticles have been prepared previously in a similar manner [50, 51]. The polysaccharides, chitosan and sodium alginate, form gels with counterions such as tripolyphosphate or calcium chloride. Nanoparticles can be prepared by emulsifying an aqueous polysaccharide solution into the oil phase followed by emulsification of an aqueous solution of the counterion solution. The drawbacks of the W/O-emulsification method include the use of large amounts of oils as the external phase, which must be removed by washing with organic solvents; heat stability problems of drugs; possible interactions of the cross-linking agent with the drug; and, as with all nanoparticles prepared by emulsification techniques, a fairly broad particle size distribution.

Aqueous phase separation techniques

Phase separation (e.g., precipitation or coacervation) of water-soluble macromolecules such as gelatin or albumin can be induced through pH-changes or the addition of desolvating agents such as salts or water-miscible organic solvents (e.g., ethanol, isopropanol). The preparation of gelatin microcapsules by simple or complex coacervation is well known. Gelatin and albumin nanoparticles have been prepared through desolvation of the dissolved micromolecules by either salts (e.g., sodium sulfate or ammonium sulfate) or ethanol [52–55]. This technique is similar to simple coacervation method. Bulk precipitation of the polymer must be avoided. The system generally has to be processed with a homogenizer to minimize particle-particle association. The particles can then be insolubilized through cross-linking with an optimum amount of aldehydes. Higher concentrations of cross-linking agent result in larger agglomerates. The coacervation and hardening process must be optimized to avoid the formation of larger agglomerates. Turbidity measurements are used to follow the phase separation/desolvation process and to establish three component phase diagrams. The hardening reaction can be terminated through the addition of sodium metabisulphite [55]. These phase separation methods avoid the use of oils as the external phase. However, the nanoparticle suspension must be purified from the desolvating and cross-linking agents. In addition, drug loadings may be low, unless specific binding of the drug to the carrier molecules or covalent linking occurs. To load the drug in the nanoparticles, the drug can either be dissolved in the aqueous phase before nanoparticle formation or it can be added to the cross-linked, but empty, nanoparticles. Gelatin nanocapsules containing triamcinolone acetonide have been prepared by an O/W-emulsification method, which, in principle, is similar to a simple coacervation technique [56]. A solution of sodium of sodium sulfate is added to induce dehydration of gelatin and coacervation around the chloroform droplets. The capsule walls are hardened with glutaraldehyde. The capsular structure of the nanoparticles is confirmed by transmission electron microscopy. Holes in the shell are the result of the removal of chloroform. They can be avoided by restricting the applied pressure during the freeze-drying process. Alginate nanoparticles have been prepared by the addition of an optimized amount of calcium chloride to a sodium alginate solution [57, 58]. This so-called microgel can be cross-linked with the oppositely charged polymer, poly-l-lysine, to form nanospheres. This technique uses the well-known gelcification phenomenon of sodium alginate with Ca-ions, however, at much lower concentrations than normally used for gel formation. Again, since only one phase is used, high drug loadings are difficult to obtain unless specific binding of the drug to the anionic polymeric carrier occurs. More than 50 mg doxorubicin, a cationic drug, can be loaded onto 100 mg sodium alginate. The drug can probably be released by an ion-exchange mechanism, which generally occurs quite rapidly.

Nanoparticles prepared through polymerization of monomers

The methods used to prepare nanoparticles by polymerization are derived from methods developed to prepare latexes. The monomers, which are primarily of acrylic origin, can be either dissolved or emulsified into the continuous phase, with the resulting polymer being insoluble in either case. The preparation of nanoparticles by polymerization reactions is generally classified according to the resulting polymer or the resulting particle structure. In this chapter, the preparation methods are divided into methods resulting in either nanospheres or nanocapsules.

8.3.3   Nanospheres

Poly (alkyl methacrylate) nanoparticles

These nanoparticles are prepared by polymerization of alkyl methacrylate monomers. The most frequently used monomer, methyl methacrylate, has a water solubility of approximately 1.5%. After dissolving it in the aqueous phase, the polymerization can be initiated either chemically by the addition of initiators such as ammonium or potassium peroxodisulfate at elevated temperature [59] or by high-energy radiation [60]. The last method has the advantage that no additional agents are needed. This simplifies the purification procedures. Nucleation of the polymerization is initiated directly in the monomer solution. Above a certain molecular weight, the oligomers precipitate to for aggregates that are stabilized by surfactant molecules. The nanoparticles are formed through the growth of the aggregates [61]. The molecular weight and the particle size of the nanoparticles increase with increasing monomer concentration and decrease slightly with increasing temperature and with increasing initiator concentration [59, 62]. Above a certain temperature or initiator concentration, the number of nucleating radicals remains constant, therefore resulting in a constant number of particles. Additional monomer, however, will increase the molecular weight of the resulting particles. Although poly(methyl methacrylate) (PMMA) nanoparticles are generally produced without surfactants, a more homogeneous particle size distribution can be obtained with the addition of hydrophilic macromolecules to the aqueous phase [2]. These molecules may also be bound to the surface of the nanoparticles. The active substance can be added to the aqueous phase either prior, during, or after the polymerization reaction or it can be chemically linked to the polymer particles. An overview of the incorporation modes and the entrapment efficiencies is described by Allemann et al. [63]. Hydrophilic drugs are predominantly added/adsorbed to the already formed nanoparticles, whereas lipophilic drugs can be dissolved in the monomer or in an organic solvent miscible with the monomer but immiscible with the aqueous phase. When adding drugs to the already formed nanoparticles, an important factor is the compatibility of the drug with the polymer dispersion. Potential interactions between the drug and polymer or surfactant can result in flocculation of the polymer suspension. Methyl methacrylate has been copolymerized with various other monomers, such as hydroxypropyl methacrylate or methacrylic acid, with the intention of preparing nanoparticles with different hydrophilicities in order to modify their circulation time and body distribution [64]. Thermal- and pH-sensitive poly(N-isopropylacrylamide-co-methacrylic acid) nanospheres have been prepared by a dispersion polymerization process [65]. The monomer mixture is dissolved in water. Potassium persulfate is used as an initiator, sodium lauryl sulfate as a stabilizer, and N,N'-methylenebisacrylamide as the cross-linking agent. The volume phase transition of the nanoparticles (swelling, shrinking), as measured by transmittance, is pH and temperature sensitive. It can be controlled by the ratio of the two monomers. The research on PMMA nanoparticles has declined in recent years because of their lack of acceptable biodegradability. More rapidly degrading polymers, such as polyesters or poly (cyano acrylates), are required for intravenous drug delivery. PMMA nanoparticles are primarily used for basic body distribution studies in animals in order to determine the fate of the nanoparticles within the body and the distribution in various organs as a function of time. This allows conclusions to be drawn with respect to site-specific drug delivery. In addition, they have also been used as adjuvants for vaccines [66]. The antigens can be added either prior to the polymerization reaction, or, in order to avoid potential destruction of the antigen, to the already-polymerized nanoparticles. The adjuvant effect increases with decreasing particle size and increasing hydrophobicity of the nanoparticles.

Poly (alkyl cyanoacrylate) nanoparticles

In recent years, the interest in polymeric nanoparticles has shifted from poly(alkyl methacrylate) to poly(alkyl cyanoacrylate) nanoparticles because of their biodegradability and a simpler polymerization procedure [67]. They are prepared by emulsion polymerization. Emulsion polymerization is a frequently used method for the preparation of nanoparticles from monomers [2]. In this method, monomer is emulsified in an immiscible external aqueous phase containing a surfactant. Above the critical micelle concentration, micelles form and are able to solubilize the monomer molecules. In addition to micelle formation, the surfactants also adsorb on the monomer emulsion droplets and stabilize the emulsion and then the polymeric nanoparticles. The polymerization reaction can be initiated within the micelles, or, with more soluble monomers, also in the continuous phase. In the first case, the monomer molecules diffuse from the emulsion droplets through the aqueous phase to the micelles. The solubilized monomer molecules within the micelles are then polymerized to form the polymer dispersion. Polymerization and chain growth are maintained by further monomer molecules diffusing to the growing polymer. The emulsified monomer droplets act as a reservoir for the monomer. After the polymerization reaction, the emulsifier molecules stabilize the colloidal polymer dispersion against physical instability (coagulation, flocculation, coalescence). Later it was found that emulsion polymerization could be carried out without any emulsifier molecules being present. It was concluded that the initiation of the polymerization occurred in the solvent phase. The polymerization, therefore, initiated with dissolved monomer molecules. Initially, the growing polymer molecules are still dissolved in the external phase. After reaching a critical molecular weight, the molecules become insoluble, and phase separation and nanoparticle formation occurs. A nanoparticle, therefore, consists of a large number of individual polymer molecules. In general, alkylcyanoacrylate monomers have a lower water solubility than alkyl methacrylate monomers. The nanoparticles are, therefore, prepared by emulsion polymerization, whereby the water-insoluble monomers are emulsified into the external phase under agitation. The monomer droplets are usually stabilized through the addition of surfactants or steric stabilizers to the aqueous phase (e.g., 1% w/w dextran 70, poloxamers). The nanoparticles are formed through an anionic polymerization reaction, initiated by the OH-ions present in water [2, 3]. This reaction proceeds very rapidly. To decrease the polymerization rate, the pH of the aqueous phase is adjusted to acidic values (pH 3.5) with diluted hydrochloric or citric acid. The pH-value of the aqueous phase influences the particle size and the molecular weight of the nanoparticles; the lower the pH, the smaller the particles and the lower the molecular weight [68]. Bubbling sulfur dioxide through the monomer shortly before the polymerization reaction produces even smaller particles [69]. The concentration of sulfur dioxide is varied through the bubbling time. Increasing the concentration of sulfur dioxide from 0% to 3% based on the monomers results in a decrease in nanoparticle size from 160 mm to 18 mm. Zeta potential measurements show a significant increase in negative charge in the presence of sulfur dioxide. The higher net charge also explains the better physical stability of the polymer dispersion. The higher net charge also explains the better physical stability of the polymer dispersion. In addition, the nanoparticles can be prepared without the stabilizer, dextran, which can potentially cause anaphylactoid reactions. To enable transendothelial passage, nanoparticles smaller than 50 nm have been prepared by optimizing the surfactant concentration (Pluronic F-68, investigate range, 0.2–10%) and the pH of the aqueous phase [70]. The particle size decreases with an increase from 0.2% to 3% Pluronic F-68, and then remains fairly unchanged at higher surfactant concentrations. Smaller nanoparticles are obtained at lower HCl concentrations, which is consistent with the anionic polymerization mechanism, and lower monomer concentrations. The addition of cosolvents, such as acetone or methanol, to the aqueous phase in order to solubilize poorly soluble drugs increases the nanoparticle size dramatically, for example to more than 300 nm at 30% methanol. Both hydrophilic (ampicillin) and lipophilic (dexamethasone) drugs can be adsorbed efficiently. Stabilizers, which are added to stabilize the monomer emulsion and also the nanoparticle suspension, can significantly influence the particle size and the molecular weight of the nanoparticles [71, 72]. The molecular weight varies between 15,300 for Pluronic F-68 and 4,177 for Triton X100 [68], and increases with increasing surfactant concentration. A higher molecular weight can also be the result of the incorporation of the stabilizers (dextrans or poloxamers) into the polymer chain [72]. The particle size of the nanoparticles can also be affected by the monomer concentration and the stirring speed. Interestingly, the particle size grows with increasing stirring rate [71]. This is attributed to the higher kinetic energy overcoming the interfacial energy barrier of the particles, thus resulting in coalescence. Colloidal polymer dispersions may be converted into a redispersible powder by freeze-drying. Freeze-drying and redispersion of the nanoparticles in water result in similar particle size distribution and drug loading when compared to the original polymer dispersions [73]. Among colloidal polymeric drug carriers, alkyl cyanoacrylate nanoparticles have been most extensively investigated, both in vitro and in vivo. A summary of the drugs used and the incorporation method and encapsulation efficiency is given in Allemann et al [63]. As with the other polymerization methods, the drugs can be incorporated prior or after the polymerization reaction. The encapsulation efficiency has been shown to be higher with monomers having longer side chains and by adding the drug prior to the polymerization reaction [74–76]. Lipophilic drugs can be incorporated by dissolving them in the monomer or in an organic solvent to be added to the monomer. The encapsulation efficiency depends on the partitioning behavior of the drug between the monomer/polymer and the aqueous phase. It generally decreases with increasing theoretical drug loading (the amount of drug added to the polymerization medium). Ionizable hydrophilic compounds can be entrapped within cyanoacrylate nanoparticles through the formation of ion pairs with oppositely charged counterions. Vidarabine has been incorporated in the presence of the anionic surfactant, dioctylsulfosuccinate [77] and antisense oligonucleotides have been complexed with hydrophobic cations such as quantisense oligonucleotides have been complexed with hydrophobic cations such as quaternary ammonium salts. The oligonucleotides are added to the already polymerized nanoparticle suspension [78]. Only 1–2% of oligonucleotides are adsorbed in the absence of the counterions, whereas complete adsorption can be achieved in the presence of the counterions. It is suggested that the formed ion pairs are adsorbed on the nanoparticles by hydrophobic interactions between the hydrophobic moiety of the cation and the alkyl chains of the poly (cyanoacrylate) polymer. The extent of the adsorption increases with increasing oligonucleotide chain length and depends on the hydrophobic character of the counterions. Harmia et al. have investigated the sorption of pilocarpine onto nanoparticles as a function of polymer type, counterion, added electrolyte, and surfactants [79]. The adsorption follows Langmuir and Freundlich isotherms and is enhanced by using the less soluble drug salt. Nanoparticles with a high specific surface area are prepared from less hydrophobic polymers and an increasing electrolyte concentration. The time at which the peptide, a growth-hormone-releasing factor, is added after the initiation of the polymerization reaction is critical [80]. If it is added shortly after the polymerization is started, it is highly associated with the polymer; however, it also becomes covalently bound to the polymer. With regard to the drug release mechanism, the peptide is released by erosion of the polymeric matrix rather than by passive diffusion through the matrix. In the absence of esterases, no drug is released. High encapsulation efficiencies at high drug loadings, as is frequently obtained with microparticles, are more difficult to achieve with colloidal drug carriers. The solids content of the polymer dispersion is often quite dilute (<1% w/w). The biodegradability of cyanoacrylate nanoparticles can be controlled by using different monomers. The rate of biodegradation is inversely proportional to the length of the side chain length [81]. For this reason, butyl, isobutyl, or isohexyl cyanoacrylate monomers are preferably used today rather than methyl and ethyl cyanoacrylate monomers. The degradation of the nanoparticles has been described to be controlled by surface erosion [82]. The side chains are enzymatically hydrolyzed. The backbone is not degraded, but gradually becomes more hydrophilic and ultimately water soluble. It has been shown that the drug release from nanospheres is controlled by the erosion of the polymer [80, 83]. In addition, the toxicity of the monomers decreases with increasing chain length. The degradation of the nanoparticles will also be affected by the molecular weight of the polymer and its particle size. The pH of the degradation medium also affects the degradation. Poly (ethyl cyanoacrylate) nanoparticles do not show significant degradation at pH 1.2, but degrade more rapidly with an increase the pH of the medium [84]. This may have implications, in particular for the oral application of nanoparticles due to the pH gradient in the gastrointestinal tract. Nanoparticles from copolymers can be prepared by merely mixing different monomers [85]. The polymerization reaction is an anionic process; basic drugs can, therefore, act as polymerization starters [77]. They can be incorporated into the polymer chain. If the drug–monomer interaction unwanted, the drug can be added to the finished nanoparticles after polymerization. To alter the natural body distribution of the nanoparticles (passive targeting) and their rapid clearance through the reticuloendothelial system, colloidal magnetites have been incorporated into polycyanoacrylate [86] or albumin nanoparticles [50, 51]. After intravenous administration, the nanoparticles can then be trapped in specific locations with an external magnetic field. The body distribution can also be altered or prolonged through the coating of the nanoparticles with, for example, hydrophilic surfactants (active tarteting). In analogy to stealth liposomes or PLA nanoparticles, PEG has been adsorbed to poly (alkyl cyanoacrylate) nanoparticles during the nanoparticle formation [87].

Polyacrylamide nanoparticles

Nanoparticles can be prepared from water-soluble monomers emulsified in an external organic phase [88–90]. For example, polyacrylamide nanoparticles have been prepared by solubilizing acrylamide and the cross-linking agent, N,N' bisacrylamide with surfactants in an organic phase (W/O microemulsion). The polymerization reaction can be initiated within the inverse micelles chemically by using N,N,N',N'-tetramethyl-ethylene-diamine and potassium peroxodisulfate as initiators or through gamma-, ultraviolet-, or light-irradiation. The drugs are generally added prior to the polymerization reaction. This method is primarily suitable for water-soluble drugs being insoluble in the external phase. The encapsulation efficiency depends on the distribution of the drug between the external phase and the internal monomer/nanoparticle phase. The drawback of the method of preparing polyacrylamide nanoparticles is the use of toxic monomers, organic solvents (e.g., hexane, chloroform), and surfactants necessary to stabilize the microemulsion. Purification techniques are required to remove these toxic components and, if used, the initiators. Because of these factors and the nondegradability of the particles, interest in this technique is decreasing.

8.4   References

1. J. Kreuter, (ed.), *Colloidal Drug Delivery Systems*, Marcel Dekker, New York, 1994.

2. J. Kreuter, Nanoparticles, In *Colloidal Drug Delivery Systems*, J. Kreuter, (ed.) Marcel Dekker, New York, 1994, pp 219–342.

3. F. Puisieux, G. Barratt, G. Couarraze, P. Couvreur, J.P. Devissaguet, C. Dubernet, E. Fattal, H. Fessi, and C. Vauthier, Polymeric micro- and nanoparticles as drug carriers. In: *Polymeric Biomaterials*, (S. Dumitriu. Ed.), Marcel Dekker, New York, 1994, pp. 749–794.

4. Huabing Chen, Chalermchai Khemtong, Xiangliang Yang, Xueling Chang and Jinming Gao. Nanonization strategies for poorly water-soluble drugs. Drug discovery today. March 1–2, 2010, doi:10.1016/j.drudis.2010.02.009.

5. J.W. McGinity (ed.), Aqueous *Polymeric Coating for Pharmaceutical Applications*, Marcel Dekker, New York, 1989.

6. L.R. Beck, D.R. Cowsar, D.H. Lewis, R.J. Cosgrove, C.T. Riddle, S.L. Lowry, and T.A. Epperly, A new long-acting injectable microcapsule system for the administration of progesterone, *Fertility and Sterility*, 31:545–551 (1979).

7. S. Benita, J.P. Benoit, F. Puisieux, and C. Thies, Characterization of drug-loaded poly(d,1-lactide) microspheres, *J. Pharm. Sci.*, 73:1721–1724 (1984).

8. R. Bodmeier, H. Chen, P. Tyle, and P. Jarosz, Pseudoephedrine HCl microspheres formulated into an oral suspension dosage form, J. Control. Rel., 15:65–77 (1991).

9. R. Vurny, N.A. Peppas, D.D. Harrington, and G.S. Banker. Development of biodegradable and injectable lattices for controlled release of potent drugs, *Drug. Dev. Ind. Pharm.*, 7:1–25 (1981).

10. H.J. Krause, A. Schwarz, and P. Rohdewald, Polylactic acid nanoparticles, a colloidal drug delivery system for lipophilic drugs, *Int. J. Pharm.*, 27:145–155 (1985).

11. F. Koosha, R.H. Muller, S.S. Davis, and M.C. Davies, The surface chemical structure of poly(β-hydroxybutyrate) microparticles produced by solvent evaporation process, *J. Control. Rel.*, 9:149 (1989).

12. R. Bodmeir, and H. Chen, Indomethacin polymeric nanosuspensions prepared by microfluidization, *J. Control. Rel.*, 12:223–233 (1990).

13. F. Koosha, R.H. Muller, and C. Washington, Production of polyhydroxybutyrate (PHB) nanoparticles for drug targeting, *J. Pharm. Pharmacol.*, 39:136P (1987).

14. P.D. Scholes, A.G.A. Coombers, L. Illum, S.S. Davis, M. Vert, and M.C. Davies, The preparation of sub-200 nm poly(lactide-co-glycolide) microspheres for site-specific drug delivery, *J. Control. Rel.*, 25:145–153 (1993).

15. T. Verrecchia, G. Spenlehauer, D.V. Bazile, A. Murry-Brelier, Y. Archimbaud, and M. Veillard. Non-stealth (poly(lactic acid/albumin)) ajd stealth (poly(lactic acid-polyethyleneglycol)) nanoparticles as injectable drug carriers, *J. Control. Rel.*, 36: 49–61 (1995).

16. S. Batzri, and E.D. Korn. Single bilayer liposomes prepared without sonication, *Biochimica et Biophysica Acta*, 443: 629–634 (1973).

17. D. Deamer, and A.D. Bangham. Large volume liposomes by an ether vaporization method, *Biochimica et Biophysica Acta* 298:1015–1019 (1976).

18. H. Fessi, J.P. Devissaguet, F. Puisieux, and C. Thies. Procede de preparation de systems colloidaux dispersible d'une substance, sous forme de nanoparticles, French Patent 2,608,988, (1986).

19. C. Emile, D. Bazile, F. Herman, C. Helene, and M. Veillard. Encapsulation of oligonucleotides in stealth Me.PEG-PLA50 nanoparticles by complexation with structured oligopeptides. In: *Proc. 7th Int. Pharmaceutical Technol.* Conf., Budapest, Hungary, pp 461–462 (1985).

20. R. Bodmeier, H. Chen, P. Tyle, and P. Jarosz. Spontaneous formation of drug-containing acrylic nanoparticles, *J. Microencapsulation*, 8(2): 161–170 (1991).

21. K.O.R. Lehmann. Chemistry and application properties of polymethacrylate coating systems, In: Aqueous Polymeric Coating for Pharmaceutical Applications, (J.W. McGinity, ed.) Marcel Dekker, New York, pp. 153–245.

22. R.K. Chang, J.C. Price, and C. Hsiao. Preparation and preliminary evaluation of Eudragit RL and RS pseudolatices for controlled drug release, *Drug Dev. Ind. Pharm.*, 15:361–372 (1989).

23. R. Lichtenberger, K. Wendel, and H.P. Merkle. Polymer films from aqueous lates dispersions as carriers for transdermal delivery of lipophilic drugs. In: *Proc. 15th Int. Symp. On Controlled Release Bioactive Materials*, Basel, Switzerland, pp. 147–148 (1988).

24. R. Bodmeir, and O. Paeratakul. Drug release from polymeric films and laminates prepared from aqueous latexes, *Proc. 5th Int. Pharmaceutical Technol. Conf.*, Paris, France, pp. 61–68 (1989).

25. S. Buyukyaylaci, Y.M. Joshi, G.E. Peck, and G.S. Banker. Polymeric dispersions as a new topical drug delivery system. In: *Recent Advances in Drug Delivery Systems*, Plenum Press, New York, pp. 291–306 (1984).

26. T. Niwa, H. Takeuchi, T. Hino, N. Kunou, and Y. Kawashima. Preparations of biodegradable nanospheres of water-soluble and insoluble drugs with d,l-lactide/glycolide co-polymer by a novel spontaneous emulsification solvent diffusion method, and the drug release behavior, *J. Control.* Rel., 25:89–98 (1993).

27. H. Fessi, F. Puisieux, and J.P. Devissaguet. Procede de preparation de systems colloidaux dispersible d'une substance, sous forme de nanocapsules, European patent 274–961 (1987).

28. N. Ammoury, H. Fessi, J.P. Devissaguet, F. Puisieux, and S. Benita. Physicochemical characterization of polymeric nanocapsules and in vitro release evaluation, *S.T.P. Pharma Sci.*, 5:647–651 (1989).

29. N. Ammoury, H. Fessi, J.P. Devissaguet, M. Allix, M. Plotkine, and R.G. Boulu. Effect on cerebral blood flow of orally administered indomethacin-loaded poly(isobutyl-cyanoacrylate) and poly(d,l-lactide) nanocapsules, *J. Pharm. Pharmacol.*, 42:558–561 (1990).

30. N. Ammoury, H. Fessi, J.P. Devissaguet, M. Dubrasquet, and S. Benita. Jejunal absorption, pharmacological activity, and pharmacokinetic evaluation of indomethacin-loaded poly(d-i-lactide) and poly(Isobutyl-cyanoacrylate) nanocapsules in rats, *Pharm. Res.*, 8:101–105 (1991).

31. H. Fessi, F. Puisieux, J.P. Devissaguet, N. Ammoury, and S. Benita. Nanocapsule formation by interfacial polymer deposition following solvent displacement, *Int. J. Pharm.*, 55:R1-R-4 (1989).

32. N. Ammoury, H. Fessi, J.P. Devissaguet, F. Puisieux, and S. Benita. In vitro release kinetic pattern of indomethacin from poly (d,l-lactide) nanocapsules, *J. Pharm. Sci.*, 79(9): 763–767 (1990).

33. L. Marchal-Heussler, H. Fessi, J.P. Devissaguet, M. Hoffman, and P. Maincent. Colloidal drug delivery systems for the eye. A comparison of the efficacy of three different polymers : Polyisobutylcyanoacrylate, polylactic-co-glycolic acid, poly-epsiloncaprolacton, *S.T.P. Pharma Sci.*, 2:98–104 (1992).

34. C. Bindschaedler, R. Furny, and E. Doelker. Process for preparing a powder of water-insoluble polymer, which can be redispersed in a liquid phase, the resulting powder and utilization thereof, Swiss Patent 1947/88 (1988).

35. H. Ibrahim, C. Bindschaedler, E. Doelker, P. Buri, and R. Gurny. Concept and development of ophthalmic pseudo-latexes triggered by pH, *Int. J. Pharm.*, 77:211–219 (1991).

36. H. Ibrahim, C. Bindschaedler, E. Doelker, P. Buri, and R. Gurny. Aqueous nanodispersions prepared by a salting-out process, *Int. J. Pharm.*, 87:239–246 (1992).

37. E. Allemann, R. Gurny and E. Doelker. Preparation of aqueous polymeric nanodispersions by a reversible salting-out process, influence of process parameters on particle size, *Int. J. Pharm.*, 87:247–253 (1992).

38. E. Allemann, J.-C. Leroux, R. Gurny, and E. Doelker. In vitro extended-release properties of drug loaded poly (dl-lactic acid) nanoparticles produced by a salting-out procedure, *Pharm. Res.*, 10(12):1732–1737 (1993).

39. E. Allemann, E. Doelker, and R. Gurny. Drug loaded polylactic acid) nanoparticles produced by a reversible salting-out process: purification of an injectable dosage form, *Eur. J. Pharm. Biopharm.*, 39:13–18 (1992).

40. D. Lemoine, C. Francois, V. Berlage, F. Kedzierewicz, P. Maincent, M. Hoffman, and V. Preat. Stability study of poly (d,1-lactide), poly(d,1-lactice-co-glycolide) and poly($\sum$-caprolactone) nanoparticles. *Proc. 7th Int. Pharmaceutical Technol. Conf.*, Budapest, Hungary, pp. 481–482 (1995).

41. B. Ekman, and I. Sjoholm. Incorporation of macromolecules in microparticles: Preparation and characteristics, *Biochem.* 15: 5115 (1976).

42. A.F. Yapel, Jr. Albumin microspheres: Heat and chemical stabilization In: *Methods in Enzymology: Part A, Drug and Enzyme Tarteting*, (K.J. Widder and R. Green, eds.), Academic Press, Orlando, 1985, pp 3–18.

43. W.E. Longo, and E.P. Goldberg. Hydrophilic albumin microspheres. In: *Methods in Enzymology: Part A, Drug and Enzyme Targeting*, (K.J. Widder and R. Green, eds.), Academic Press, Orlando, 1985, pp. 18–26.

44. E. Tomlinson, and J.J. Burger. Incorporation of water-soluble drugs in albumin microspheres, In: *Methods in Enzymology: Part A, Drug and Enzyme Targeting*, (K.J. Widder and R. Green, eds.), Academic Press, Orlando, 27–43.

45. J.M. Gallo, C.T. Hung, and D.G. Perrier. Analysis of albumin microsphere preparation, *Int. J. Pharm.*, 22:63–74 (1984).

46. P.K. Gupta, C.T. Hung, and D.G. Perrier. Albumin microspheres II. Effect of stabilization temperature on the release of adriamycin, *Int. J. Pharm.*, 33:147 (1986).

47. T. Yoshioka, M. Hashida, S. Muranishi, and H. Sezaki. Specific delivery of mitomycin C to the liver, spleen, and lung: Nano-and microspherical carriers of gelatic, *Int. J. Pharm.*, 8:131–141 (1981).

48. Y. Tabata and Y. Ikada. Synthesis of gelatin microspheres containing interferon, *Pharm. Res.*, 6; 422 (1989).

49. E.E. Hassan, R.C. Parish, and J.M. Gallo. Optimized formulation of magnetic microspheres containing the anticancer agent, oxantrazole, *Pharm. Res.*, 9:390 (1992).

50. K. Widder, G. Flouret, and A. Senyei. Magnetic microspheres: Synthesis of a novel parenteral drug carrier, *J. Pharm. Sci.*, 68:79–82 (1979).

51. K. Widder, R.M. Morris, R.G. Poore, P.H. Howards, and A. Senyei. Selective targeting of magnetic albumin microspheres containing low-dose doxorubicin: Total remissioning Yoshida Sarcoma-bearing rats, *Eur. J. Cancer and Clinical Oncology*, 19:141–147 (1983).

52. J.J. Marty, R.C, Oppenheim, and P. Speiser. Nanoparticles – a new colloidal drug delivery system, *Pharm. Acta Helv.*, 53:17–22 (1978).

53. M. El-Samaligy and P. Rohdewald. Triamcinolone diacetate nanoparticles, a sustained release drug delivery system suitable for parenteral administration, *Pharm. Acta Helv.*, 57:201 (1982).

54. M. El-Samaligy, and P. Rohdewald. Reconstituted collagen nanoparticles, a novel drug carrier delivery system, *J. Pharm. Pharmacol.*, 35:537–539 (1983).

55. R.C. Oppenheim. Solid colloidal drug delivery systems: Nanoparticles, *Int. J. Pharm.*, 8:217–234.

56. H.J. Krause, and P. Rohdewald. Preparation of gelatin nanocapsules and their pharmaceutical characterization, *Pharm. Res.*, 2:239–243 (1985).

57. M. Rajanorivony, C. Vauthier, G. Couarraze, F. Puisieux, and P. Couvreur. Development of a new drug carrier made from alginate, *J. Pharm. Sci.*, 82(9): 912–917 (1993).

58. J.E. Diederichs, C. Vauthier, H. Alphandary, P. Couvreur, and R.H. Muller. Formation process and determination of microviscosity of alginate nanoparticles. In: *Proc. 21st Int. Symp. On Controlled Release of Bioactive Materials*, Nice, France, pp. 511–512 (1994).

59. U.E. Berg, J. Kreuter, P.P. Speiser, and M. Soliva. Herstellung and in vitro-Prufundvon polymeren Adjuvantien fur Impfstoffe, *Pharm. Ind.*, 48:75 (1986).

60. J. Kreuter, and H.J. Zehnder. The use of 60Co-g-irrdiation for the production of vaccines, *Radiat. Effects.*, 35:161 (1978).

61. C. Vauthier-Holzscherer, S. Benabbou, G. Spenlehauer, M. Veillard, and P. Couvreur. Methodology for the preparation of ultra-dispersed polymer systems, *S.T.P. Pharma Sci.*, 1(2):109–116 (1991).

62. J. Kreuter. Evaluation of nanoparticles as drug-delivery systems. I. Preparation methods, *Pharm. Acta Helv.*, 58:196 (1983).

63. E. Allemann, R. Gurny, and E. Doelker. Drug-loaded nanoparticles-preparation methods and drug targeting issues, *Eur. J. Pharm. Biopharm.*, 39(5):173–191.

64. A. Rolland, D. Gibassier, P. Sado, and R. Le Verge. Methodologie de preparation de vecteurs nanoparticulaires a base de polymers acryliques, *J. Pharm. Belg.*, 41:83–93 (1986).

65. X.Y. Wu, and P.L. Lee. Preparation and characterization of thermal- and pH-sensitive nanospheres, *Pharm. Res.*, 10(10): 1544–1547 (1993).

66. J. Kreuter. Possibilities of using nanoparticles as carriers for drugs and vaccines, *J. Microencapsulation*, 5(2):115–127 (1988).

67. P. Couvreur, and C. Vauthier. Polyalkylcyanoacrylate nanoparticles as drug carrier : present state and perspectives, *J. Control. Rel.*, 17:187–198 (1991).

68. G. Puglisi, G. Giammona, M. Fresta, B. Carlisi, N. Micali, and A. Villari. Evaluation of polyalkylcyanoacrylate nanoparticles as a potential drug carrier: preparation, morphological characterization and loading capacity, J. Microencapsulation. 10(3):353–366 (1993).

69. V. Lenaerts, P. Raymond, J. Juhasz, M.A. Simard, and C. Jolicoeur. New method for the preparation of cyanoacrylic nanoparticles with improved colloidal properties, *J. Pharm. Sci.*, 78(12):1051–1052 (1989).

70. B. Seijo, E. Fattal, L. Roblot-Treupel, and P. Couvreur. Design of nanoparticles of less than 50 nm diameter : preparation, characterization and drug loading, *Int. J. Pharm.* 62:1–7 (1990).

71. S.J. Douglas, L. Illum, and S.S. Davis. Particle size and size distribution of poly(butyl-2-cyanoacrylate) nanoparticles. I. Influence of physicochemical factors, *J. Colloid Interface Sci.*, 101:149 (1985).

72. S.J. Douglas, L. Illum, and S.S. Davis. Particle size and size distribution of poly(butyl-2-cyanoacrylate) nanoparticles. II. Influence of stabilizers, *J. Colloid Interface Sci.*, 103:154 (1985).

73. C. Verdun, P. Couvreur, H. Vranckx, V. Lenaerts, and M. Roland. Development of a nanoparticle controlled-release formulation for human use, *J. Control, Rel.*, 3:205–210 (1986).

74. N. Bapat, and M. Boroujerdi. Uptake capacity and adsorption isotherms of doxorubicin on polymeric nanoparticles: effect of methods of preparation, *Drug Dev. Ind. Pharm.*, 18:65–77 (1992).

75. M.J. Alonso, C. Losa, P. Calvo, and J.-L. Vila Jato. Approaches to improve the association of amikacin sulphate to poly(alkylcyanoacrylate) nanoparticles, *Int. J. Pharm.*, 68:69–76 (1991).

76. P. Couvreur, B. Kante, M. Roland, and P. Speiser. Adsorption of antineoplastic drugs to polyalkylcyanoacrylate nanoparticles and their release in calf serum, *J. Pharm. Sci.*, 68:1521–1524 (1979).

77. V. Guise, J.Y. Drouin, J. Benoit, J. Mahuteau, P. Dumont, and P. Couvreur. Vidarabine-loaded nanoparticles: A physicochemical study, *Pharm. Res.*, 7:736–741 (1990).

78. C. Chavany, T.L. Doan, P. Couvreur, F. Puisieux, and C. Helene. Polyalkylcyanoacrylate nanoparticles as polymeric carriers for antisense oligonucleotides, *Pharm. Res.*, 9(4): 441–449 (1992).

79. T. Harmia, P. Speiser, and J. Kreuter. Optimization of pilocarpine loading onto nanoparticles by sorption procedures, *Int. J. Pharm.*, 33:45–54 (1986).

80. J.L. Grangier, M. Puygrenier, J.C. Gautier, and P. Couvreur. Nanoparticles as carriers for growth hormone releasing factor, *J. Control. Rel.*, 15:3–13 (1991).

81. F. Leonard, R.K. Kulkarni, G. Brandes, J. Nelson, and J.J. Mameron. Synthesis and degradation of poly(alkylcyanoacrylates), *J. Appl. Polym. Sci.*, 10:259–272 (1966).

82. R. Nuller, C. Lherm, J. Herbort, and P. Couvreur. In vitro model for the degradation of alkylycyanoacrylate nanoparticles, *Biomaterials*, 11:590–595 (1990).

83. V. Nenaerts, P. Couvreur, D. Christiaens-Leyh, E. Joiris, M. Roland, B. Rollman, and P. Speiser. Degradation of polyisobutyl-cyanoacrylate nanoparticles, *Biomaterials*, 5:65–68 (1984).

84. A. Piskin, A. Tuncel, A. Denizli, E.B. Denkbas, H. Ayhan, H. Cicek, and K.T. Xu. Nondegradable and biodegradable polymeric particles, in Diagnostic Biosensor

Polymers, (A.M. Usmani, and N. Akmal, eds.), American Chemical Society, Washington, DC, pp. 222–237 (1994).

85. B. Kante, P. Couvreur, G. Dubois-Crack, C. De Meester, P. Guiot, M. Roland, M. Mercier, and P. Speiser. Toxicity of polyalkylcyanoacrylate nanoparticles I: Free nanoparticles, *J. Pharm. Sci.*, 7:786–790 (1982).

86. A. Ibrahim, P. Couvreur, M. Roland, and P. Speiser. New magnetic drug carrier, *J. Pharm. Pharmacol.*, 35:59–61 (1982).

87. C. Vauthier, M.I. Popa, F. Puisieux, and P. Couvreur. Evaluation of potentiality to graft PEG to poly(alkylcyanoacrylate) nanoparticles in the course of the formation of the nanoparticles in absence of surfactants. Proc. *22nd Int. Symp. On Controlled Release of Bioactive Materials*, Seattle, WA 592–593 (1995).

88. G. Birrenbach, and P.P. Speiser. Polymerized micelles and their use as adjuvants in immunology, *J. Pharm. Sci.*, 65:1763–1766 (1976).

89. B. Ekman, and I. Sjoholm. Improved stability of proteins immobilized in microparticles prepared by a modified emulsion polymerization technique, *J. Pharm. Sci.*, 67:693 (1978).

90. P. Edman, B. Ekman, and I. Sjoholm. Immobilization of proteins in microspheres of biodegradable polyacryldextran, *J. Pharm. Sci.*, 69:838 (1980).

Liposomes as novel drug delivery vehicle

**Satya Prakash Singh*[1], Chinam Niranjan Patra[2], Suryakanta Swain[2]
and Vaseem Ahamad Ansari[1]**

*[1]Faculty of Pharmacy, Integral University, Lucknow – 226026, India.
[2]Department of Pharmaceutics, Roland Institute of Pharmaceutical Sciences,
Berhampur, Odisha – 760010, India.*

9.1 Introduction

The name liposome is derived from two Greek words: 'Lipos' meaning fat and 'Soma' meaning body. A liposome can be formed into variety of sizes as unilamellar or multi-lamellar construction, and its name relates to its structural building blocks, phospholipids, and not to its size. Liposomes were first described by British haematologist Dr Alec D Bangham in 1961 (published 1964), at the Babraham Institute, in Cambridge. They were discovered when Bangham and R. W. Horne were testing the institute's new electron microscope by adding negative stain to dry phospholipids. The resemblance to the plasmalemma was obvious, and the microscope pictures served as the first real evidence for the cell membrane being a bilayer lipid structure. Liposomes are defined as "simple microscopic vesicles in which an aqueous volume is entirely enclosed by a membrane composed of lipid molecule." Various amphipathic molecules have been used to form liposome. The drug molecules can either be encapsulated in aqueous space or intercalated into the lipid bilayer (Fig. 9.1).

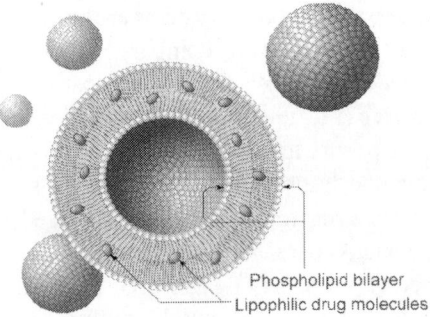

Phospholipid bilayer
Lipophilic drug molecules

Figure 9.1 Liposomes with lipophilic drugs incorporated in the phospholipid bilayer
[*Source:* The University of Tromsø website]

Over the past few decades, liposomes have received widespread attention as a carrier system for therapeutically active compounds, due to their unique characteristics such as capability to incorporate hydrophilic and hydrophobic drugs, good biocompatibility, low toxicity, lack of immune system activation, and targeted delivery of bioactive compounds to the site of action. Additionally, some achievements since the discovery of liposomes are controlled size from micro scale to nanoscale and surface-engineered polymer conjugates functionalized with peptide, protein, and antibody.

Although liposomes have been extensively studied as promising carriers for therapeutically active compounds, some of the major drawback for liposomes used in pharmaceutics is the rapid degradation due to the reticuloendothelial system (RES) and inability to achieve sustained drug delivery over a prolonged period of time. New approaches are needed to overcome these challenges. Two polymeric approaches have been suggested thus far. The first approach involves modification of the surface of liposomes with hydrophilic polymers such as polyethylene glycol (PEG), while the second one is to integrate the pre-encapsulated drug-loaded liposomes within depot polymer-based system. A study conducted by Stenekes and coworkers reported the success of using temporary depot of polymeric materials to control the release of the loaded liposomes for pharmaceutical applications. This achievement leads to new applications, which requires collaborative research among pharmaceuticals, biomaterials, chemistry, molecular, and cell biology. Numerous studies in this context have been reported with temporary depot delivery system to control the release of pre-encapsulated drug-loaded liposomes. This system was developed to integrate the advantages while avoid the disadvantages of both liposome-based and polymeric-based systems. The liposome-based systems are known to possess limitations such as instability, short half-life, and rapid clearance. However, they are more biocompatible than the polymer-based systems.

On other hand, the polymer-based systems are known to be more stable and provide improved sustained delivery compared to liposome-based systems. However, one of the major setbacks is poor biocompatibility which is associated with loss of the bioactive (i.e., the drug) during fabricating conditions such as heat of sonication or exposure to organic solvents. The benefits of a composite system, however, include improvement of liposome stability, the ability of the liposome to control drug release over a prolonged period of time, and preservation of the bioactiveness of the drugs in polymeric-based technology. In addition, increased efficacy may be achieved from this integrated delivery system when compared to that of purely polymeric-based or liposome-based systems. The chapter focuses on different types of liposome-based technology and depot polymeric scaffold technologies, various methods for embedding drug-loaded liposomes within a depot, and various approaches reported to control the rate of sustained drug release within depot systems over a prolonged period of time. The common problem for liposome therapy is that liposomes are taken up readily by the cells of the reticuloendothelial system. Even though reversible endothelial blockade can be imposed with relative ease, it is not accompanied by any significant increase in liposome uptake by tissues outside of the mononuclear phagocyte system. The retention and uptake of circulating liposomes by reticuloendothelial cells and blood monocytes is a major obstacle to experimental efforts to 'target' liposomes to other cell types.

SUV (small unilamellar vesicle) liposomes are cleared from the circulation more slowly than larger MLV (multilamellar vesicle) liposomes. However, there is no evidence that SUV liposomes or liposomes with a prolonged half-life in the circulation accumulate to any significant degree in sites other than the liver, spleen and bone marrow. Moreover, reported intravenous injection of liposomes at regular intervals is accompanied by long-term paralysis of reticula-endothelial function. Since many of the proposed clinical applications involve multiple-dose treatment protocols, this problem represents a potentially serious drawback. Besides, there are no appreciable methods either for targeting liposomes to specific sites in vivo or for ensuring the delivery of liposomal contents to target cells. Although considerable amount of work has been reported on passive targeting to the reticulo-endothelial system, not much is known about active targeting to specific subsets of circulating blood cells or to vascular endothelium. Thus, efficient delivery systems suitable for site-specific targeting are being sought.

One of the most prolific areas of liposome applications is in biochemical investigations of conformation and function of membrane proteins. These are the so-called reconstitution studies and purified membrane proteins, such as ion pumps (sodium potassium- or calcium-ATPases), or glucose transport proteins are reconstituted in their active form into liposomes and then studied.

Furthermore, cell communication largely depends on the traffic of vesicles. Nerve impulses travel between synapses and neurons in synaptic vesicles carrying neurotransmitters. Various proteins regulate directions, addresses, docking, internalization or fusion of these vesicles with a great efficacy. Cells secrete and ingest macromolecules via exo- and endocytosis, respectively as well as transport molecules to and within Golgi apparatus by the use of vesicles. Obviously, our increasing knowledge of these processes will shed more light on the function of living cells as well as offer some solutions in case of its disfunction. The benefits and limitations of liposome drug carriers critically depend on the interaction of liposomes with cells and their fate in vivo after administration. In vitro and in vivo studies of the interactions with cells have shown that the predominant interaction of liposomes with cells is either simple adsorption or subsequent endocytosis. Fusion with cell membranes is much rarer. The fourth possible interaction is exchange of bilayer constituents, such as lipids, cholesterol, and membrane bound molecules with components of cell membranes. The body protects itself with a complex defense system. Upon entering into the body, larger objects cause thrombus formation and eventually their surface is passivated by coating with biomacromolecules while smaller particles, including microbes, bacteria, and colloids are eaten up by the cells of the immune system. This response of the immune system has triggered substantial efforts in the development of biocompatible and non-recognizable surfaces and has also, on the other hand, narrowed the spectrum of applications of microparticulate drug carriers only to targeting of the very same cells of the immune system.

Advantages of liposome

- Liposomes have increased efficacy as well as therapeutic index of drug.
- Liposome is increased stability via encapsulation.
- Liposomes are biocompatible, biodegradable, non-toxic, flexible and non-immunogenic for systemic and non-systemic administrations.
- Liposome helps in reduction of toxicity of the encapsulated drugs like Amphotericin B, Taxol.
- Liposomes are useful for reduction in exposure of sensitive tissues to toxic drugs.
- Liposomes can be used for the delivery of both hydrophilic and hydrophobic drugs.
- Protects the encapsulated drug from the external environment.

Disadvantages

- Production cost is high.
- Leakage and fusion of encapsulated drug / molecules may occur.
- Phospholipid may undergo oxidation and hydrolysis like reaction during storage.
- Short half-life and low solubility.

9.2   Structural component of liposomes

9.2.1   Lipids in Liposome

Phospholipids

Glycerol-containing phospholipids are most commonly used component of liposomal formulation. It represents more than 50% of weight of lipid in biological membranes. These are derived from Phosphatidic acid. The backbone of the molecule is glycerol moiety. At C_3 position –OH group is esterified to phosphoric acid. –OH at C_1 and C_2 are esterified with long chain fatty acid giving rise to the lipidic nature and one of the remaining –OH groups of phosphoric acid may be further esterified to a wide range of organic alcohols including glycerol, choline, ethanolamine, serine and inositol. Thus the parent compound of the series is the phosphoric ester of glycerol. Examples of phospholipids are – Phosphatidyl choline (Lecithin) – PC, Phosphatidyl ethanolamine (cephalin)-PE, Phosphatidyl serine (PS), Phosphatidyl inositol (PI), Phosphatidyl Glycerol (PG). For stable liposomes, saturated fatty acids are used (Fig. 9.2).

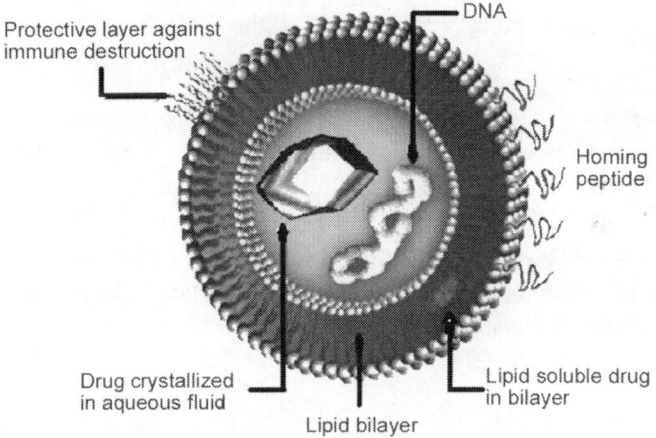

Figure 9.2 Liposome for drug delivery

Sphingolipids

Backbone of liposomes is sphingosine or a related base. These are important constituents of plant and animal cells. Most common Sphingolipids are Sphingomyelin, Glycosphingo lipids. Gangliosides – found on grey matter, used as a minor component for liposome production. This molecule contain complex saccharides with one or more sialic acid residues in their polar head group and thus have one or more negative charge at neutral pH. These are included in liposomes to provide a layer of surface charged group.

Cholesterols

Cholesterol and its derivatives are often included in liposomes for
- decreasing the fluidity or microviscocity of the bilayer
- reducing the permeability of the membrane to water-soluble molecules
- stabilizing the membrane in the presence of biological fluids such as plasma. (this effect used in formulation of i.v. liposomes)

Liposomes without cholesterol are known to interact rapidly with plasma protein such as albumin, transferrin, and macroglobulin. These proteins tend to extract bulk phospholipids from liposomes, thereby depleting the outer monolayer of the vesicles leading to physical instability. Cholesterol appears to substantially reduce this type of interaction. Cholesterol has been called the mortar of bilayers, because by virtue of its molecular shape and solubility properties, it fills in empty spaces among the Phospholipid molecules, anchoring them more strongly into the structure. The –OH group at 3rd position provides small Polar head group and the hydrocarbon chain at C17 becomes non-polar end by these molecules, the cholesterol intercalates in the bilayers.

Synthetic phospholipids

Examples of saturated phospholipids are – Dipalmitoyl phosphatidyl choline (DPPC), Distearoyl phosphatidyl choline (DSPC), Dipalmitoyl phosphatidyl ethanolamine (DPPE), Dipalmitoyl phosphatidyl serine (DPPS), Dipalmitoyl phosphatidic acid (DPPA), Dipalmitoyl phosphatidyl glycerol (DPPG)

Examples of unsaturated phospholipids are – Dioleoyl phosphatidyl choline (DOPC), Dioleoyl phosphatidyl glycerol (DOPG)

Polymeric materials

Synthetic phospholipids with diactylenic group in the hydrocarbon chain polymerizes when exposed to U.V., leading to formation of polymerized liposomes having significantly higher permeability barriers to entrapped aqueous drugs. For e.g., for other polymerisable lipids are – lipids containing conjugated diene, Methacrylate, etc. Also several polymerisable surfactants are also synthesized.

Polymer-bearing lipids

Stability of repulsive interactions with macromolecules is governed mostly by repulsive electrostatic forces. This repulsion can be induced by coating liposome surfaces with charged polymers. Non-ionic and water compatible polymers like polyethylene oxide, polyvinyl alcohol, and polyoxazolidines confers higher solubility. But adsorption of such copolymers containing hydrophilic segments with hydrophobic part leads to liposome leakage, so best results can be achieved by covalently attaching polymers to phospholipids; for e.g., Diacyl Phosphatidyl Ethanolamine with PEG polymer linked via a carbon at succinate bond.

Cationic lipids

E.g. DODAB/C – Dioctadecyl dimethyl ammonium bromide or chloride
      DOTAP – Dioleoyl trimethyl ammonium propane (chloride salt).

Other substances

- Variety of other lipids of surfactants are used to form liposomes
- Many single chain surfactants can form liposomes on mixing with cholesterol
- Non-ionic lipids
- A variety of Polyglycerol and Polyethoxylated mono and dialkyl amphiphiles used mainly in cosmetic preparations
- Single and double chain lipids having fluoro carbon chains can form very stable liposomes
- Sterylamine and Dicetyl phosphate incorporated into liposomes so as to impart either a negative or positive surface charge to these structures.

9.3   Classification of liposomes

Liposomes are classified on the basis of structure, method of liposome preparation and composition and application.

9.3.1   Based on structure

Unilamellar vesicles

Small unilamellar vesicles (SUV): size ranges from 20 nm to 40 nm
Medium unilamellar vesicles (MUV): size ranges from 40 nm to 80 nm
Large unilamellar vesicles (LUV): size ranges from 100 nm to 1000 nm

Oligolamellar vesicles (OLV)

These are made up of 2–5 bilayers of lipids surrounding a large internal volume

Multilamellar vesicles (MLV)

They have several bilayers (more than 5). They can compartmentalize the aqueous volume in an infinite numbers of ways. They differ according to way by which they are prepared. The arrangements of bilayers may be like onion, i.e. arrangements of concentric spherical bilayers of LUV/MLV enclosing a large number of SUV etc.

9.3.2   Based on liposome preparation method

REV: Single or oligolamellar vesicles made by Reverse-Phase Evaporation Method

MLV-REV: Multilamellar vesicles made by Reverse-Phase Evaporation Method

SPLV: Stable Plurilamellar Vesicles

FATMLV: Frozen and Thawed Multilamellar vesicles

VET: Vesicles prepared by extrusion technique

DRV: Dehydration-rehydration method

9.3.3   Based upon composition and application

Conventional Liposomes (CL): Neutral or negatively charged phospholipids and Cholesterol.

Fusogenic Liposomes (RSVE): Reconstituted Sendai virus envelopes

pH sensitive Liposomes: Phospholipids such as PE (Phosphatidyl ethanolamine) or DOPE with either CHEMS (Cholesteryl hemisuccinate) or OA

Cationic Liposomes: Cationic lipids with DOPE

Long Circulatory (Stealth) Liposomes (LCL): They have polyethylene glycol (PEG) derivatives attached to their surface to decrease their detection by phagocyte system (reticuloendothelial system; RES). The attachment of PEG to liposomes decreases the clearance from blood stream and extends circulation time of liposomes in the body. The attachment of PEG is also known as pegylation.

Immuno-Liposomes: CL or LCL with attached monoclonal antibody or recognition sequence.

9.4   Methods of liposome preparation

General methods of preparation. All the methods of preparing the liposomes involve four basic stages:
- Drying down lipids from organic solvent
- Dispersing the lipid in aqueous media
- Purifying the resultant liposome
- Analyzing the final product

9.5   Method of liposome preparation and drug loading

The following methods are used for the preparation of liposome:
1. Passive loading techniques
2. Active loading techniques

9.5.1   Passive loading techniques include three different methods

- Mechanical dispersion method
- Solvent dispersion method
- Detergent removal method (removal of non-encapsulated material)

Mechanical dispersion method

The following are types of mechanical dispersion methods:
- Sonication
- French pressure cell: extrusion
- Freeze-thawed liposomes
- Lipid film hydration by hand shaking, non-hand shaking or freeze drying
- Micro-emulsification
- Membrane extrusion
- Dried reconstituted vesicles

Sonication

Sonication is perhaps the most extensively used method for the preparation of SUV. Here, MLVs are sonicated either with a bath-type sonicator or a probe sonicator under a passive atmosphere. The main disadvantages of this method are very low internal volume/encapsulation efficacy, possible degradation of phospholipids and compounds to be encapsulated, elimination of large molecules, metal pollution from probe tip, and presence of MLV along with SUV.

There are two sonication techniques:

Probe sonication – The tip of a sonicator is directly engrossed into the liposome dispersion. The energy input into lipid dispersion is very high in this method. The coupling of energy at the tip results in local hotness; therefore, the vessel must be engrossed into a water/ice bath. Throughout the sonication process upto 1 h, more than 5% of the lipids can be de-esterified. Also, with the probe sonicator, titanium will slough off and pollute the solution.

Bath sonication – The liposome dispersion in a cylinder is placed into a bath sonicator. Controlling the temperature of the lipid dispersion is usually easier in this method, in contrast to sonication by dispersal directly using the tip. The material being sonicated can be protected in a sterile vessel, dissimilar the probe units, or under an inert atmosphere.

French pressure cell: extrusion – French pressure cell involves the extrusion of MLV through a small orifice. An important feature of the French press vesicle method is that the proteins do not seem to be significantly deceptive during the procedure as they are in sonication. An interesting comment is that French press vesicle appears to recall entrapped solutes significantly longer than SUVs, produced by sonication or detergent removal. The method involves gentle handling of unstable materials. The method has several advantages over sonication method. The resulting liposomes are rather larger than sonicated SUVs. The drawbacks of the method are that the high temperature is difficult to attain, and the working volumes are comparatively small (about 50 mL as the maximum).

Freeze-thawed liposomes – SUVs are rapidly frozen and thawed slowly. The short-lived sonication disperses aggregated materials to LUV. The creation of unilamellar vesicles is as a result of the fusion of SUV throughout the processes of freezing and thawing. This type of synthesis is strongly inhibited by increasing the phospholipid concentration and by increasing the ionic strength of the medium. The encapsulation efficacies from 20% to 30% were obtained.

Solvent dispersion method

    (a)   Ether injection (solvent vaporization) – A solution of lipids dissolved in diethyl ether or ether-methanol mixture is gradually injected to an aqueous solution of the material to be encapsulated at 55–65°C or under reduced pressure. The consequent removal of ether under vacuum leads to the creation of liposomes. The main disadvantages of the technique are that the population is heterogeneous (70–200 nm) and the exposure of compounds to be encapsulated to organic solvents at high temperature.

(b) Ethanol injection – A lipid solution of ethanol is rapidly injected to a huge excess of buffer. The MLVs are at once formed. The disadvantages of the method are that the population is heterogeneous (30–110 nm), liposomes are very dilute, the removal of all ethanol is difficult because it forms into azeotrope with water, and the probability of the various biologically active macromolecules to inactivate in the presence of even low amounts of ethanol is high.

9.5.2   Reverse phase evaporation method

This method provided a progress in liposome technology, since it allowed for the first time the preparation of liposomes with a high aqueous space-to-lipid ratio and a capability to entrap a large percentage of the aqueous material presented. Reverse-phase evaporation is based on the creation of inverted micelles. These inverted micelles are shaped upon sonication with a mixture of a buffered aqueous phase, which contains the water-soluble molecules to be encapsulated into the liposomes and an organic phase in which the amphiphilic molecules are solubilized. The slow elimination of the organic solvent leads to the conversion of these inverted micelles into viscous state and gel form. At a critical point in this process, the gel state collapses, and some of the inverted micelles were disturbed. The excess of phospholipids in the environment donates to the formation of a complete bilayer around the residual micelles, which results in the creation of liposomes. Liposomes made by reverse phase evaporation method can be made from numerous lipid formulations and have aqueous volume-to-lipid ratios that are four times higher than hand-shaken liposomes or multilamellar liposomes. Briefly, first, the water-in-oil emulsion is shaped by brief sonication of a two-phase system, containing phospholipids in organic solvent such as isopropyl ether or diethyl ether or a mixture of isopropyl ether and chloroform with aqueous buffer. The organic solvents are detached under reduced pressure, resulting in the creation of a viscous gel. The liposomes are shaped when residual solvent is detached during continued rotary evaporation under reduced pressure. With this method, high encapsulation efficiency up to 65% can be obtained in a medium of low ionic strength for example 0.01 M NaCl. The method has been used to encapsulate small, large, and macromolecules. The main drawback of the technique is the contact of the materials to be encapsulated to organic solvents and to brief periods of sonication. These conditions may possibly result in the breakage of DNA strands or the denaturation of some proteins. Modified reverse phase evaporation method was presented by Handa et al., and the main benefit of the method is that the liposomes had high encapsulation efficiency (about 80%).

Detergent removal method (removal of non-encapsulated material)

(a) Dialysis

The detergents at their critical micelle concentrations (CMC) have been used to solubilize lipids. As the detergent is detached, the micelles become increasingly better-off in phospholipid and lastly combine to form LUVs. The detergents were removed by dialysis. A commercial device called LipoPrep (Diachema AG, Switzerland), which is a version of dialysis system, is obtainable for the elimination of detergents. The dialysis can be performed in dialysis bags engrossed in large detergent free buffers (equilibrium dialysis).

(b) Detergent (cholate, alkyl glycoside, Triton X-100) removal of mixed micelles (absorption)

Detergent absorption is attained by shaking mixed micelle solution with beaded organic polystyrene adsorbers such as XAD-2 beads (SERVA Electrophoresis GmbH, Heidelberg, Germany) and Bio-beads SM2 (Bio-Rad Laboratories, Inc., Hercules, USA). The great benefit of using detergent adsorbers is that they can eliminate detergents with a very low CMC, which are not entirely depleted.

Gel-permeation chromatography

In this method, the detergent is depleted by size special chromatography. Sephadex G-50, Sephadex G-l00 (Sigma-Aldrich, MO, USA), Sepharose 2B-6B, and Sephacryl S200-S1000 (General Electric Company, Tehran, Iran) can be used for gel filtration. The liposomes do not penetrate into the pores of the beads packed in a column. They percolate through the inter-bead spaces. At slow flow rates, the separation of liposomes from detergent monomers is very good. The swollen polysaccharide beads adsorb substantial amounts of amphiphilic lipids; therefore, pretreatment is necessary. The pre-treatment is done by pre-saturation of the gel filtration column by lipids using empty liposome suspensions.

Coating liposomes

Coating liposomes with PEG reduces the percentage of uptake by macrophages and leads to a prolonged presence of liposomes in the circulation and, therefore, make available abundant time for these liposomes to leak from the circulation through leaky endothelium. A stealth liposome is a sphere-shaped vesicle with a membrane composed of phospholipid bilayer used to deliver drugs or genetic material into a cell. A liposome can be composed of naturally derived phospholipids with mixed lipid chains coated or steadied by polymers of PEG and colloidal in nature. Stealth liposomes are attained and grown in new drug delivery and in controlled release. This stealth principle has been used to develop the successful doxorubicin-loaded liposome product that is presently marketed as Doxil (Janssen Biotech, Inc., Horsham, USA) or Caelyx (Schering-Plough Corporation, Kenilworth, USA) for the treatment of solid tumors. The concerning on the application of stealth liposomes has been on their potential to escape from the blood circulation.

Drug loading in liposomes

Drug loading can be attained either passively (i.e., the drug is encapsulated during liposome formation) or actively (i.e., after liposome formation). Hydrophobic drugs, for example amphotericin B taxol or annamycin, can be directly combined into liposomes during vesicle formation, and the amount of uptake and retention is governed by drug-lipid interactions. Trapping effectiveness of 100% is often achievable, but this is dependent on the solubility of the drug in the liposome membrane. Passive encapsulation of water-soluble drugs depends on the ability of liposomes to trap aqueous buffer containing a dissolved drug during vesicle formation. Trapping effectiveness (generally <30%) is limited by the trapped volume delimited in the liposomes and drug solubility. On the other hand, water-soluble drugs that have protonizable amine functions can be actively entrapped by employing pH gradients, which can result in trapping effectiveness approaching 100%.

9.6   Characterization of liposomes

After preparation and before use in immunoassay the liposome must be characterized. Evaluation could be classified into three broad categories which are physical, chemical and biological methods. The physical methods include various parameters, which are size, shape, surface features, lamellarity, phase behaviors and drug release profile. Ma et al. evaluated structural integrity of liposomal phospholipids membrane by a new technique of gamma-ray perturb angular correlation (PAC) spectroscopy. In this In-label diethyene triamine penta acetic acid (DTPA) derivative dipalmitoyl phosphatidyl ethanolamine (DPPE) lipid were incorporated in the SUVs. This helps in the continuous non- invasive monitoring of the microenvironment of the lipid bilayer. Chemical characterization includes those studies which established the purity and potency of various liposomal constituents. Biological characterization is helpful in establishing the safety and suitability of formulation for the in vivo use for therapeutic application. The characteristics of the carrier through appropriate choice of membrane components, size and charge determines the final behavior of liposomes both in vitro and *in vivo* as well (Table 9.1).

Table 9.1 Characterization of liposomes with their quality control assays

A. Biological characterization

S. no.	Characterization parameters	Instrument for analysis
1	Sterility	Aerobic/anaerobic culture
2	Pyrogenicity	Rabbit fever response
3	Animal toxicity	Monitoring survival rats

B. Chemical characterization

1	Phospholipids concentration	HPLC/Barrlet assay
2	Drug concentration	Assay method
3	Cholesterol concentration	HPLC/cholesterol oxide assay
4	Phospholipids per oxidation	UV observance
5	Cholesterol auto-oxidation	HPLC/TLC
6	Phospholipids hydrolysis	HPLC/TLC
7	pH	pH meter
8	Anti-oxidant degradation	HPLC/TLC
9	Osmolarity	Osmometer

C. Physical characterization

1	Vesicle shape, and surface morphology	TEM and SEM
2	Vesicle size and size distribution	Dynamic light scattering ,TEM
3	Surface charge	Free-flow electrophoresis
4	Electrical surface potential and surface pH	Zeta potential measurement and pH sensitive probes
5	Phase behavior	DSC, freeze fracture electron microscopy
6	Lamellarity	$P^{31}NMR$
7	Percent capture	Mini column centrifugation, gel exclusion
8	Drug release	Diffuse cell/ dialysis

9.7   Application of liposome

The field of liposome research has expanded considerably over last 30 years. It is now possible to engineer a wide range of liposome of varying size, phospholipid composition, cholesterol composition, surface morphology suitable for wide range of applications. Liposomes interact with cells in many ways to cause liposomal components to be associated with target cells. The liposome carrier can be targeted to liver and spleen and distinction can be made between normal and tumors tissue using tomography. In case of transdermal drug delivery system, liposome has a great application. Liposomal drug delivery system when used to target the tumor cells leads to reduction in the toxic effect and enhances the effectiveness of drugs. The targeting of the liposome to the site of action takes place by the attachment of amino acid fragment, such as antibody or protein or appropriate fragments that target specific receptors cell. Liposomal DNA delivery vectors and further enhancement in the form of LPD-I and LPD-II are some of the safest and potential most versatile transfer vectors which are used to date.

Several modes of drug delivery application have been purposed for the liposomal drug delivery system, few of them are as follows:

- Enhance drug solublisation (Amphotericin-B, Minoxidil, Paclitaxels, and Cyclosporins)
- Protection of sensitive drug molecules (Cytosine arabinosa, DNA, RNA, Anti-sense oligo-nucleotides, Ribozymes)
- Enhance intracellular uptake (Anticancer, anti viral and antimicrobial drugs)
- Altered pharmacokinetics and bio-distribution (prolonged or sustained released drugs with short circulatory half life)
- Several recent applications of liposomal drug delivery system are as follows:

9.7.1   Liposomes used for respiratory drug delivery system

Liposome is widely used in several types of respiratory disorders. Liposomal aerosol has several advantages over ordinary aerosol which are summarized as follows:

- Sustained release
- Prevention of local irritation
- Reduced toxicity and
- Improved stability in the large aqueous core.
- Several injectable liposome based product are now in the market including ambisome, Fungisome and Myocet. To be effective, liposomal drug delivery system for the lung is dependent on the

following parameters:
- Lipid composition
- Size
- Charge
- Drug and Lipid ratio and
- Method of delivery

The recent use of liposome for the delivery of DNA to the lung means that a greater understanding of their use in macromolecular delivery via inhalational route is now emerging. Much of this new knowledge, including new lipids and analytical techniques, can be used in the development of liposome based protein formulations. For inhalation of liposome the liquid or dry form is taken and the drug release occurs during nebulization. Drug powder liposome has been produced by milling or by spray drying.

9.7.2   Delivery of nucleic acids and DNA

Soon after the first animal experiments began to show improved therapeutic outcomes for small molecule therapeutics, came the realization that liposomes could also be effective delivery systems for DNA, and for nucleic acid-based therapeutics such as antisense oligonucleotides (as ODN) and siRNA. In vivo delivery of polynucleic acids using lipid-based systems began with an early report that a liposomally encapsulated plasmid for rat insulin could result in gene expression following intravenous injection, and an early Phase I clinical trial for liposomal c-raf-1 as ODN (oligodeoxy ribonucleotide). This was followed by the demonstration by Felgner and others that fusogenic cationic lipids could be complexed with plasmid, and facilitate efficient transfection of cells in vitro. An explosion of studies then followed, to exploit the potential of gene therapy both in vitro and in vivo. Despite intensive effort, however, and the synthesis of hundreds of different cationic lipids, gene expression could only be observed following local, as opposed to systemic injection, and the toxic side effects of cationic lipids became increasingly evident. Other issues were the large size of the cationic lipid-DNA complexes and the high surface charge of these systems, which combine to result in rapid clearance from the circulation.

9.7.3   Liposomes used as ocular drug delivery system

Generally eye is protected by three highly efficient mechanisms (a) an epithelial layer that is a formidable barrier to penetration (b) tear flow

(c) the blinking reflex. All three mechanisms are responsible for poor drug penetration into the deeper layers of the cornea and the aqueous humor and for the rapid wash out of drugs from the corneal surface. Enhanced efficacy of liposomes encapsulated idoxuridine in herpes simplex infected corneal lesions in rabbits was first reported in 1981. Lee in 1985 concluded that ocular delivery of drugs could be either promoted or impeded by the use of liposome carriers, depending on the physiochemical properties of the drugs and lipid mixture employed. Ganglioside-containing liposomes and wheat germ agglutinin, a lectin that has a high binding affinity for both cornea and ganglioside, were tested for corneal adhesion. Corneal binding as well as accumulation and transcorneal flux of carbachol was enhanced 2.5–3.0-fold over 90 min exposure times. Davies et al. proposed the use of mucoadhesive polymers, carbopol 934P and carbopol 1342 to retain liposomes at the cornea. While precorneal retention times were indeed significantly enhanced under appropriate conditions liposomes even in the presence of the mucoadhesive had migrated toward the conjuctival sac with very little activity remaining at the corneal surface.

9.7.4   Liposomes for brain targeting

The biocompatible and biodegradable behaviour of liposomes have recently led to their exploration as drug delivery system to brain. Liposomes with a small diameter (100 nm) as well as large diameter undergo free diffusion through the Blood Brain Barrier (BBB). However, it is possible that a small unilamellar vesicles (SUVS) coupled to brain drug transport vectors may be transported through the BBB by receptor mediated or absorptive-mediated transcytosis. Similarly, cationic liposomes which were developed recently showed these structures to undergo absorptive-mediated endocytosis into cells. Whether cationic liposomes successfully undergo absorptive-mediated transcytosis through the BBB has not yet been determined. The transport of substances through BBB by liposomes was extensively studied. The important finding issues from their studies are that the addition of the sulphatide (a sulphur ester of galactocerebroside) to liposome composition increases their several recent applications ability to cross BBB. The neutropeptides, leu-enkephaline and mefenkephalin kyoforphin normally do not cross BBB when given systemically. The anti-depressant amitriptylline normally penetrate the BBB, due to versatility of this method. Nanoparticles (NP) were fabricated with different stabilizers. It was found that amitriptyline level was significantly enhanced in brain when the substance was adsorbed onto the NP and coated or particle stabilized with polysorbate 85.

9.7.5   Liposome as anti-infective agents

Intracellular pathogen like protozoal, bacterial, and fungal reside in the liver and spleen and thus to remove these pathogen the therapeutic agent may be targeted to these organ using liposome as vehicle system. The disease like leishmaniasis, candidiasis, aspergelosis, histoplasmosis, erythrococosis, gerardiasis, malaria and tuberculosis are targeted by the respective therapeutic agent using liposome as carrier shown in Table 9.2.

Table 9.2 Liposomal preparations for infective disease

Active constituents application	**Active targeting approach**
Pentamidin	Leishmaniasis
Antisense oligo-nucleotides	Leishmaniasis
Anamycin	Leishmaniasis
Asiaticoside	Tuberculosis and leprosy
Rifampicin	Tuberculosis
Passive targeting approach	
Amphotericin B	Meningitis, Leishmaniasis, Candidiasis
Praziquantal	Macrophage activation
Sparfloxacin	M. avium, M. Intracellularie complex
Gentamycin	Staphylococcal pneumonias

Use of Amphotericin B, a polyene antibiotic, in the treatment of systemic fungal infection is associated with extensive renal toxicity. Amphotericin B act by the mechanism, in which it binds to sterol in the membrane of sensitive fungi, thus increasing the membrane permeability. The toxicity of this compound is due to non-specificity and binding to the mammalian cell cholesterol. Recently, the first preparation of Amphotericin B (ambisome) in the form of liposome had passed all clinical trials, and now it is used for the treatment of fungal infections. Liposomal Amphotericin B, by passively targeting the liver and spleen, reduces the renal and general toxicity at normal dose but renal toxicity appears when the drug is given at elevated dose due to the saturation of liver and spleen macrophages. Liposome can also be targeted to lungs by coating vesicle with ostearoyl amylopectin, polyoxylethylene or mono-sialogangliocyte. The encapsulation of antitubercular agent like isoniazid and rifampicin in lung-targeted liposome modulates the toxicity and improve the efficiency of these drugs. Various formulations of the Amphotericin (Table 9.3) had been approved by several clinical trials and are now marketed at different European countries.

Table 9.3 Various liposomal preparation of Amphotericin B

Preparation	Drug	Targeted site
Liposome (Am Bisome)	Amphotericin B	Systemic fungal infection, visceral lieshmaniaisis
Liposome (Amphocil)	Amphotericin B	Systemic fungal infection
Liposome (ABLC)	Amphotericin B	Systemic fungal infection

9.7.6   Liposome in tumour therapy

The long-term therapy of anticancer drug leads to several toxic side effect. The liposomal therapy for targeting the tumour cell have been revolutionized the world of cancer therapy with least side effect. It has been said that the small and stable liposome are passively targeted to different tumour because they can circulate for longer time and they can extravasate in tissue with enhanced vascular permeability. Liposome macrophage uptake by liver and spleen hampered the development of liposome as drug delivery for over 20 years. Several formulations of liposomal anticancer drug which are in clinical use are given in Table 9.4.

Table 9.4 Various intravenous liposomal antibiotics/anti-neoplastics

Preparation	Drug	Targeted site
Liposome (Doxil)	Doxorubicin	Kaposi' sarcoma
Liposome (EVACTTM)	Doxorubicin	Refractory tumour, Metastatic breast cancer
Liposome (DaunoXome)	Daunosome	Advanced Kaposi' sarcoma, breast, small cell lung cancer, leukaemia and solid tumour
Liposome	Nystatin	Systemic fungal infection
Liposome	Anamycin	Kaposi' sarcoma, refractory breast cancer
Liposome (VincaXome)	Vincristine	Solid tumour
Liposome (Mikasome)	Amikacin	Serious bacterial infection

Doxil is the liposomal formulation of doxorubicin, intravenous, chemotherapeutic agent. Doxil is prepared by the new technology called stealth technology, stealth liposome. These are the long circulatory liposome which is prepared by several means. Caelyx and myocet are the liposomal formulations of doxorubicin. Caelyx is used for treatment of metastatic ovarian cancer but now in advanced breast cancer. Myocet's approved for metastatic breast cancer.

9.8   Conclusion

Almost from the time of their discovery in 1960s and the demonstration of their entrapment potential, liposome vesicle have drawn attention of researchers as potential carriers of various bioactive molecules that could be used for therapeutic applications in human and animals. Many factors contribute to their success as drug delivery vehicles. Liposomes solubilise lipophillic drug candidates that would otherwise be difficult to administer intravenously. The encapsulated drug is inaccessible to metabolizing enzyme; conversely, body component such as erythrocyte and tissue injection site are not directly exposed to full dose of the drug. Liposomes can cross the BBB because of the lipophillic nature of the phospholipids, so even the hydrophilic drugs (cannot easily cross the BBB) might be formulated as liposomes. Liposome can prolong the drug action by slowly releasing the drug in the body. Targeting option changes the distribution of the drug in the body. Liposomes are prepared by various methods in which the most common method applied for research purpose are film method and dehydration rehydration method and many more. The new developments in the liposome are the specific binding properties of a drug-carrying liposome to a target cell (tumor cell and specific molecules), stealth liposomes for targeting hydrophilic (water soluble) anticancer drugs like doxorubicin, mitoxantrone which leads to decrease in side effects because the drug is mostly concentrated at the site of action. There is even greater promise in future for marketing of more sophisticated and highly stabilized liposomal formulations. In future the liposomal drug delivery system will revolutionize the vesicular systems with wide application especially in the treatment of cancer.

9.9   Summary

Almost from the time of their discovery i.e. in early 1960s by Bangham, the demonstration of entrapment potential of liposomal vesicles had attracted researchers as potential carriers of various bioactive molecules that could be used for therapeutic applications. Liposomes are spherical-shaped nanovesicles which are biodegradable and essentially non-toxic in nature. Due to its versatile nature, it can encapsulate both hydrophilic and hydrophobic materials, and can be utilized as drug carriers in drug delivery systems. As a result, numerous improvements have been made, thus making this technology potentially useful for the treatment of certain diseases in the clinics. It is clear from numerous pre-clinical and clinical studies that drugs, such as antitumor drugs, packaged in liposomes exhibit reduced toxicities, while retaining, or gaining enhanced, efficacy. This results, in part, from altered pharmacokinetics, which leads to drug accumulation at disease sites, such as tumors, and reduced distribution to sensitive tissues. The success of liposomes as drug carriers has been reflected in a number of liposome-based formulations, which are commercially available or are currently undergoing clinical trials.

9.10   References

[1] Kamble R., Pokharkar V.B., Badde S., Kumar A. Development and characterization of liposomal drug delivery system for nimesulide. *Int. J. Pharm. Sci.* 2(4): 87–89 (2010).

[2] Bangham A.D. Liposomes. Marcel Dekker, 1, 1–26 (1983).

[3] Jain N.K. Controlled and novel drug delivery. CBS Publishers, 1(1), 304–326 (2007).

[4] Cornelius F. Functional reconstitution of the sodium pump. Kinetics and exchange reactions performed by reconstituted Na/K ATPase. *Biochim Biophys Acta.* 1071, 19–66 (1991).

[5] Villalobo A. Reconstitution of ion-motive transport ATPases in artificial lipid membranes, *Biochim. Biophys. Acta* 1071, 1–48 (1991).

[6] Himanshu A., Sitasharan P., Singhai A.K. Liposomes as drug carriers. IJPLS 2(7): 945–951 (2011).

[7] Felgner J.H., Kumar R., Sridhar C.N., Wheeler C.J., Tsai Y.J., Border R., Ramsey P., Martin M., Feigner P.L. Enhanced gene delivery and mechanism of studies with a novel series of cationic lipid formulations. *J. Biol. Chem.* 269, 2550–2561 (1994).

[8] Rongen HAH, Bult A, Bennekom WP Van. J. Immuno. Methods. 204:105-133 (1997).

[9] Kersten G.F.A., Crommelin D.J.A. Liposomes and ISCOMS as vaccine formulations. Biochim. Biophys. Acta. 1241, 117–138 (1995).

[10] Horton K. Disertation for degree of Advanced Studies in Chemical Engineering, University Rovira I Virgili, 2003.

[11] Remington. The Science and Practice of Pharmacy. Volume I, 21st Edition, B.I Publishers Pvt Ltd, pp. 314–316.

[12] Sharma A., Straubinger R.M. Novel taxol formulations: preparation and characterization of taxol-containing liposomes. *Pharm. Res.* 11, 889–896 (1994).

[13] Vyas S.P., Khar R.K. Targeted and Controlled Drug Delivery, Novel carrier system. CBS Publisher. 1(1): 173–206.

[14] Emanuel N., Kedar E., Bolotin E.M., Smorodinsky N.I., Barenholz Y. Preparation and characterization ofdoxorubicin-loaded sterically stabilized immunoliposomes. *Pharm. Res.* 13, 352–359 (1996).

[15] Riaz M. Liposome preparation method. *Pak J Pharm Sci* 9(1): 65–77 (1996).

[16] Kataria S., Sandhu P., Bilandi A., Akanksha M., Kapoor B., Seth G.L., Bihani S.D. Stealth liposomes: a review. IJRAP 2(5): 1534–1538 (2011).

[17] Mayer L.D., Bally M.B., Hope M.J., Cullis P.R. Techniques for encapsulating bioactive agents in to liposomes. *Chem Phys Lipids*, 40: 333–345 (1985).

[18] Song H., Geng H.Q., Ruan J., Wang K., Bao C.C., Wang J., Peng X., Zhang X.Q., Cui D.X. Development of polysorbate 80/phospholipid mixed micellar formation for docetaxel and assessment of its in vivo distribution in animal models. *Nanoscale Res Lett,* 6: 354 (2011).

[19] Mozafari M.R. Liposomes: An overview of manufacturing techniques. *Cell Mol Biol Lett,* 10(4): 711–719 (2005).

[20] Ohsawa T., Miura H., Harada K. Improvement of encapsulation efficiency of water-soluble drugs in liposomes formed by the freeze-thawing method. *Chem Pharm Bull* 33(9): 3945–3952 (1985).

[21] Samad A., Sultana Y., Aqil M. Liposomal Drug Delivery Systems: An Update Review. *Current Drug Delivery,* 4: 297–305 (2007).

[22] Shaeffer H.E., Brietfeller J.M., Krohn D.L. Lectin mediated attachment of liposomes to cornea. Influence of transcorneal drug flux. *Invest Opthalmol Vis Sci,* 23, 530–533 (1982).

10

Magnetically modulated drug delivery systems

Suryakanta Swain[1*], Chinam Niranjan Patra[2],
Muddana Eswara Rao[2] and Prafulla Kumar Sahu[3]

[1]*Southern Institute of Medical Sciences, College of Pharmacy, Department of Pharmaceutics, SIMS Group of Institutions, Mangaldas Nagar, Vijyawada Road, Guntur-522 001, Andhra Pradesh, India*
[2]*Roland Institute of Pharmaceutical Sciences, Department of Pharmaceutics, Khodasinghi, Berhampur-760 010, Ganjam, Odisha, India*
[3]*Department of Pharmaceutical Analysis, Raghu College of Pharmacy, Dakamarri, Bheemunipatnam, Visakhapatnam, Andhra Pradesh, 531162, India.*

10.1   Introduction

Pharmaceutical drug targeting is considered as the approach of delivering the drugs molecules to specific receptors, targeting tissues or any other specific parts of the body to one need to target the drug molecule exclusively. Various nonmagnetic micro/nano carriers (nanoparticles, microspheres and micro particles) have been successfully utilized for drug targeting but such as they show poor site specificity and rapidly clear off by the reticuloendothelial system (RES) under normal circumstances. In this case, magnetism plays an important role. Magnetic particles composed of magnetite and are well tolerated by the body. Magnetic fields are believed to be harmless to biological system and adaptable to any part of the body. Up to 60% of an injected dose can be deposited and released in a controlled manner in selected "non-reticuloendothelial" organs. Thus magnetic micro carriers are developed to overcome two major problems encountered in drug targeting namely RES clearance and site specificity [1]. These are various approaches to targeted drug delivery, which are broadly classified into three categories: the physical or mechanical approaches, the biological approach or the chemical approach. Magnetism plays an important role in different application in health care. Magnetic particles composed of magnetite are well tolerated by the body. Magnetic nanoparticles usually exist or can be prepared in the form of a single domain or a super paramagnetic magnetite (Fe_3O_4), greigite (Fe_3S_4), maghemite (r-Fe_2O_3), iron, nickel etc. Synthetic magnetite materials have many applications in optics, electronic and energy storage. Magnetism has applications in numerous fields like diagnostics, drug targeting, molecular biology, cell isolation, cell purification, hyperthermia, and radioimmunoassay [2]. The major principle of magnetic drug targeting is magnetically targeted drug delivery by particulate carriers is an efficient method of delivering drugs to localized disease sites. High concentration of chemotherapeutic or radiological agents can be achieved near the targeted site without any toxic effects to normal surrounding tissue.

10.2   Magnetic drug delivery systems: Basics

Magnetic drug delivery by particulate carriers is a very efficient method of delivering a drug to localized disease site. Magnetic drug transport technique is based on the fact that the drug can be either encapsulated into a magnetic microsphere or nanosphere or conjugated on the surface of the micro or nanosphere. When the magnetic carrier is intravenously administered, the accumulation takes place within area to which the magnetic field is applied and often augmented by magnetic agglomeration. The accumulation of the carrier at the target site allows them to deliver the drug locally. Efficiency of accumulation of magnetic carrier on physiological carrier depends on physiological parameters, e.g. particle size, surface characteristic, field strength and blood flow rate, etc. The magnetic field helps to extravasate the magnetic carrier into the targeted area. Very high concentration of chemotherapeutic agents can be achieved near the target site without any toxic effect to normal surrounding tissue or to whole body. It is thus possible to replace large amounts of drug targeted magnetically to localized disease site, reaching effective and up to several fold increased drug levels (Fig. 10.1) [3].

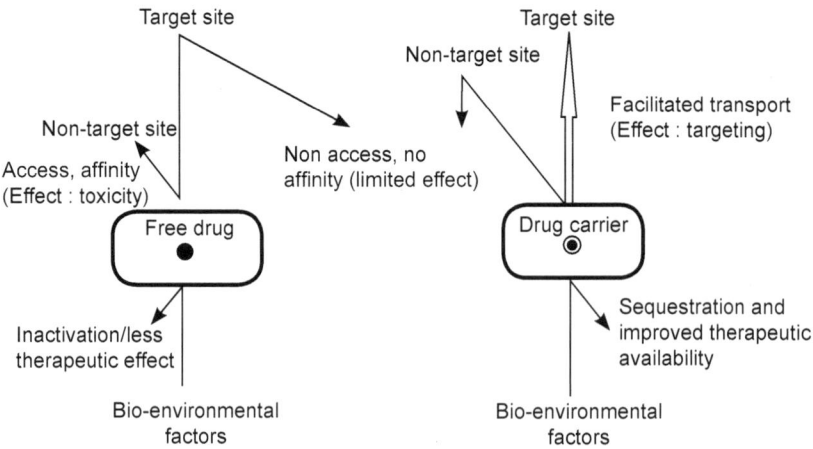

Figure 10.1 Targeting strategy and mechanism of free drug and drug carrier

Magnetic micro carriers are site specific and by localization of these micro carriers in the target area, the problem of their rapid clearance by RES is also surmounted. Linear blood velocity in capillaries is 300 times less i.e. 0.05 cm/s as compared to arteries, so much smaller magnetic field, 6–8 Koe, is sufficient to retain them in the capillary network of the target area. Other benefits includes avoidance of acute toxicity directed against endothelium and normal parenchyma cell, controlled release within target tissue for intervals of 30 minutes to 30 hours as desired, adaptable to any part of body. This drug delivery system reduces circulating concentration of free drug by a factor of 100 or more. Magnetic carrier technology appears to be a significant alternative for the bimolecular malformation (composition, inactivation or deformation). In case of tumor targeting, microsphere can internalize by tumor cells due to its much increased phagocytic activity as compared to normal cells. So, the problem of drug resistance due to inability of drugs to be transported across the cell membrane can be surmounted and adaptable to any part of the body. Apart from these benefits, this novel approach suffers from certain drawbacks, i.e. drug cannot be targeted to deep-seated organs in the body, so this approach is confined to the targeting of drugs in superficial tissue only like skin, superficial tumor or to joints only. Magnetic targeting is an expensive, technical approach and requires specialized manufacturer and quality controlled system. The magnet must have relatively constant gradients in order to avoid local overdosing with toxic drugs. It needs specialized magnet for targeting, advanced technique for monitoring, and trained personnel to perform the procedure. A large fraction (40–60%) of the magnetite, which is entrapped in carriers, is deposited permanently in target tissues. Due to this limitation magnetic drug targeting is likely to be approved only for very severe diseases that are refractory to other approaches [4].

10.3   Magnetically modulated micro or nano carriers

Magnetic micro carriers are the supramolecular particles that are small enough to circulate through capillaries without producing embolic occlusion (less than 4 micrometers), but are sufficiently susceptible to become captured in micro vessels and dragged into the adjacent tissues by magnetic field of 0.5–0.8 tesla (T). These micro carriers include microspheres, liposome, cells, nanoparticles, etc. Biological response modifiers (BRMs) alter host, tumor as well as microbial responses in four ways, such as augmentation of host effectors mechanisms directed against tumor cells or microorganisms, decrease in host responses that interferes that tumor resistance by a quantitative increase in endogenous effectors resistance by an increase in endogenous effectors molecules or redirecting their sites and duration of action, augmentation of tumor sensitivity to host cells by re-differentiating tumor cells and increase in host tolerance of conventional cancer treatment [5].

10.3.1   Magnetic microspheres

The use of magnetic force for the site-specific delivery by using albumin microspheres containing magnetic appears be a promising strategy. Significant improvements delivery systems compare with conventional and nonmagnetic microspheres drug regimens. In the presence of a suitable magnetic field, the microspheres are internalized by the endothelial cells of the target tissue in healthy as well as tumour bearing animals [4, 6]. A labeled marker contained a magnetic microsphere was significantly fixed at a part of the rat tail where a magnetic field was applied after injection into the tail artery. The localization of magnetic microspheres in the head and hind leg was successfully achieved by arterial infusion of the samarium-cobalt permanent magnet. The magnetically carried microspheres and liposomes were employed to selectively transport the curare like drugs (pyrocurine, diadonium) to the muscles of one of the limbs of the cat. The use of magnetic microspheres produced no changes in systemic arterial blood pressure, local blood flow, EEG, or ECG. The method also diminished the respiratory depression produced by the curare like substances when they are routinely used for body muscle relaxation [7–9].

10.4   Drugs selection for magnetic microparticles

In the selection of a drug for formulation of magnetic microspheres, following points are taken into consideration, the drug is so dangerous or labile that cannot allow it to circulate freely in the blood stream, the agent is so expensive, that cannot afford to waste 99.9% of it, requires a selective, regional effect to meet localized therapeutic objective, requires an alternative formulation essential to continue treatment in patient whose systemic therapy must be temporarily discontinued due to life threatening toxicity directed at selective organs [8].

10.5   Techniques for preparation of magnetic microparticle

There are mainly two techniques, which are commonly employed for preparation magnetic microspheres such as, phase separation emulsion polymerization (PSEP) and continuous solvent evaporation [1]. The various steps performed during phase separation by emulsification polymerization and continuous solvent evaporation is explained in Fig. 10.2 and 10.3, respectively. Formulations are made by preparing an aqueous mixture of water-soluble drug or (lipophilic drug water-soluble adducting agents), matrix material (typically albumin, but alternatively carbohydrate) emulsifying and 10 nm Fe_3O_4 particles (as magnetite) emulsifying the mixture in biodegradable oil, sonicating or shearing to produce sub micrometer sized spheres, stabilizing the matrix by heating or chemical cross linking, depending upon the polymer system used extracting the oil with a volatile organic solvent (typically hexane or ether), and lyophilizing the preparation of dryness. Non-magnetic microspheres and unincorporated magnetic microparticle are separated from the mixture by the following techniques. The magnetic microspheres are separated from non-magnetic ones by exposing the mixture to magnetic

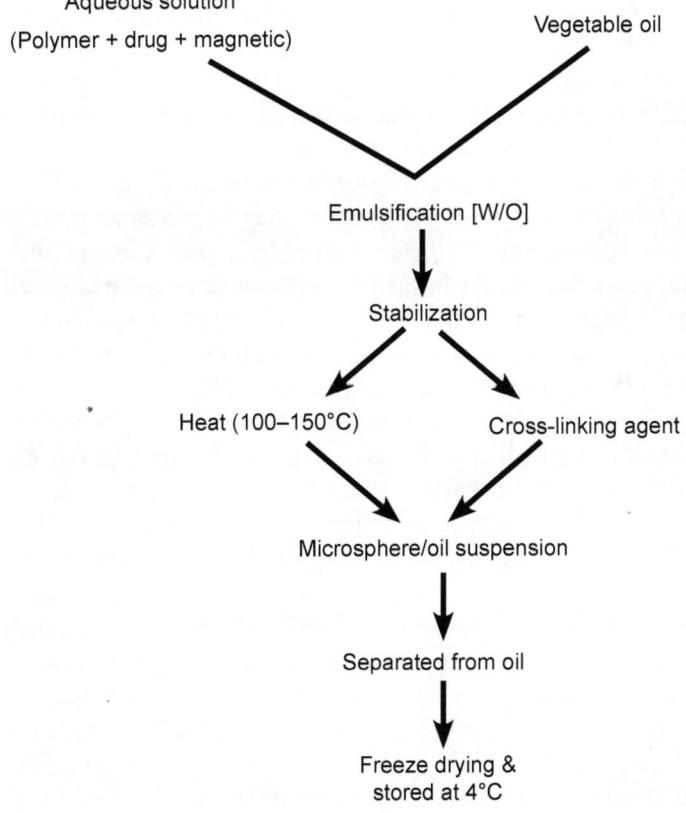

Figure 10.2 Preparative steps of magnetically responsive microspheres by phase separation emulsion polymerization

field and collecting the particles retained near the magnet. In one method is based on a repeated gentle stirring of the preparation with a bar magnet (permanent magnetic recovery tools). The magnet is withdrawn from the solution and the microspheres and unincorporated magnetite attached to it are washed off with a jet of 0.15 M sodium chloride. In another method, a suspension in a suitable medium is transferred in a funnel, the end of which is connected through a rubber tube, to a straight glass tube kept at an angle of 45 degree to the vertical axis. The suspension flowing from a funnel at a controlled rate is exposed to a magnet placed around the mid length of the glass tube. Microspheres are accumulated after the removal of magnetic field, by the flow of dispersion medium. Unincorporated magnetite can be removed from the magnetically responsive mixture by repeated washing with 0.5 M HCl or by transferring the microspheres in ether suspension in the presence of 300 G bar magnet placed adjacent to the decanting vessel. The free magnetite can be preferentially retained in the vessel due to its high magnetic responsiveness relative to the magnetic microspheres are then air dried [1]. Figure 10.4 explains the assembly used for separation of magnetic microspheres from the non-magnetic materials. The properties of magnetic particles for bio separation consist of one or more magnetic cores with a coating matrix of polymers, silica or hydroxyl apatite with terminal functionalized groups. The magnetic core generally consists either of magnetite (Fe_3O_4) or magnetite (γFe_2O_3) with super paramagnetic or ferromagnetic properties. Some magnetic cores can also be made with magnetic ferrites, such as cobalt ferrite or manganese ferrite. Super para magnetism is when the dipole moment of a single-domain particle fluctuates rapidly in the core due to the thermal excitation so that there is no magnetic moment for macroscopic time scales. Thus, these particles are non-magnetic when an external magnetic field is applied but do develop a mean magnetic moment in an external magnetic field. In contrast, ferromagnetism means that the particles have a permanent mean magnetic moment. Here, the larger effective magnetic anisotropy suppresses the thermally activated motion of the core moments. Advantages of the super paramagnetic particles are easy resuspension, large surface area, slow sedimentation and uniform distribution of the particles in the suspension media. Once magnetized, the particles behave like small permanent magnets, so that they form aggregates or lattice due to magnetic interaction. Advantages of ferromagnetic particles are very strong magnetic properties and therefore the fast separation with an external magnetic field even in viscous media. Ferromagnetic particles are generally recommended for the separation of DNA or RNA (SiMAG or MP-DNA), whereas super paramagnetic particles are more suitable for all other applications [11].

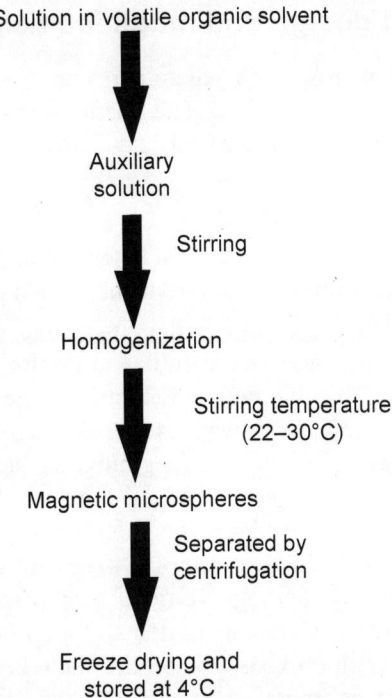

Figure 10.3 Preparative steps of magnetically responsive microspheres by continuous solvent evaporation

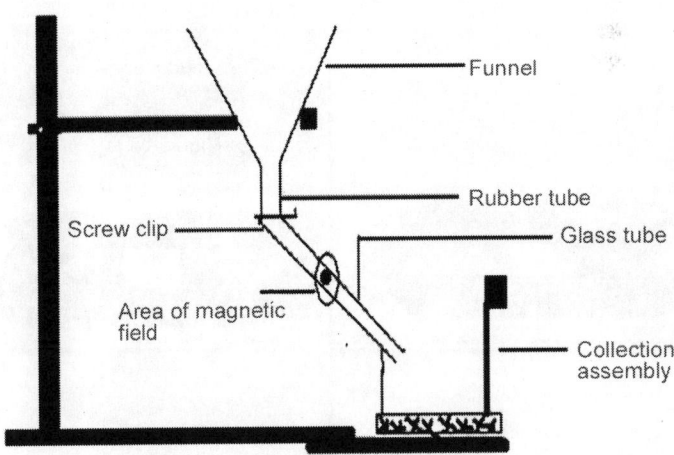

Figure 10.4 Assembly used for separation of magnetic microsphere from non-magnetic materials

10.6   Magnet design and magnetic field

Targeting of magnetic microspheres generally occurs from the forces exerted by the applied gradient magnetic field. The relationship of magnetic force to field gradient and magnetic moment of particles is expressed by the following general equation:

$$F = M\Delta H \qquad (10.1)$$

Where, F = Force on particles, M = Magnetic moment of particles after saturation of the magnetization and ΔH = Magnetic field gradient.

It is apparent from the equation that increased magnetic moments offer forces sufficient for the extra vascular migration at proportionately lower field gradients. The magnetic moments of the microspheres can be increased in following three ways such as magnetizing the sphere to saturation level prior to vascular targeting, clustering magnetite at the centre of each sphere to producer larger macro domains and lastly by substituting one of the newer ferromagnetic materials that has higher susceptibility that Fe_3O_4 [1]. Quadripolar magnetic field for improved magnetic field improved homogenicity of microsphere localization. Uniform gradient region is shown as dotted and high field region as cross hatched. A zero field region exists at the mid plane between pole pieces. To avoid microsphere aggregation in the infusion syringe, injection is performed near the zero field regions. The schematic representation of the quadripolar magnetic field and their basis of magnetic drug targeting are shown in Figs. 10.5 and 10.6, respectively [11].

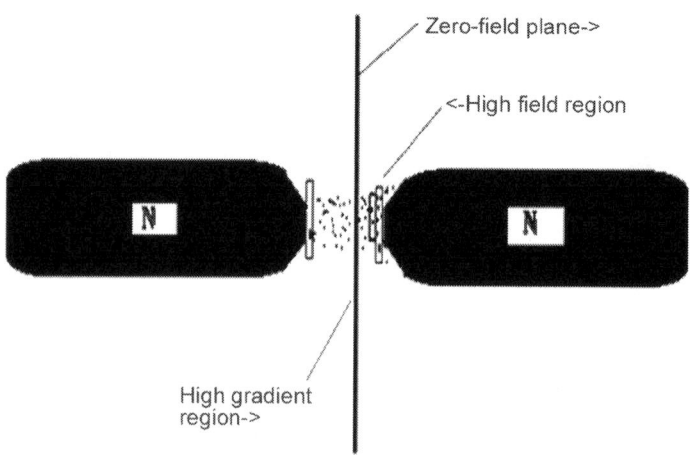

Figure 10.5 Schematic diagram of quadripolar magnetic field

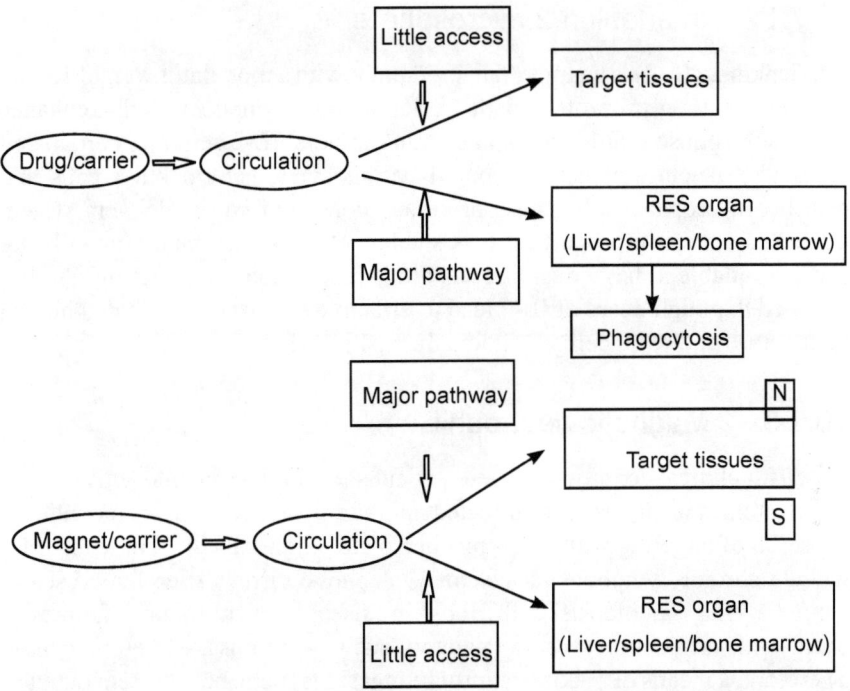

Figure 10.6 Basics of magnetic drug targeting

10.7   Variants of magnetic microsphere

10.7.1   Chemo attractant microsphere

Neutrophil chemo attractant, f-met-leu-phe, was the first selected biomodulator for magnetic targeting due to its lethal property when administered freely in the circulation at a concentration exceeding 2×10^{-7} M. It requires a site specific and local delivery to modulate inflammation and is a small, bacterially derived peptide, stable during microspheres preparation. It can be entrapped successfully in microparticles. F-met-leu-phe microspheres have been used either in disease modeling or clinical therapy. In pulmonary medicine these delivery systems are used as an experimental method to test new agents that prevent neutrophil degradation of lung elastin in smokers, means of studying the contribute of acute alveolar damage to acute type respiratory distress syndrome and as an adjuvant method in treating invasive pulmonary aspergillus's suffering patients [12].

10.7.2   Interleukin-2 microsphere

Interleukin-2 (IL-2) is an important modulator with a modulator weight 15000. It is chemically glycoprotein made by activating T lymphocytes that enhance immune response in infection and certain tumours. IL-2 activates certain cell types like T-helper cells, cytotoxic T-lymphocytes, natural killer cells and possibly macrophages. It is an appropriate molecule for drug delivery system due to its attractive properties. It is stable, active in mice and human being and is available in large quantities as product of genetic engineering. Studies revealed that high doses of IL-2 lead to effective against disseminated human melanomas and renal cell carcinoma therapy [13–16].

10.7.3   Magnetic neutrophil

In certain clinical conditions, where patient sera contains chemo active factor inactivators and neutrophils directed inhibitors of chemotaxis, an indirect approach of targeting white cells by chemo attraction falls [4]. These disorder include chronic lymphocytic leukemia, alcoholic cirrhosis, crohn's disease, hemodialysis, sarcoidosis and Hodgkin disease, even though failure of chemotaxis is not observed in all patients, such conditions are life threatening. Therefore, a means of making neutrophilingest magnetic base system ought to be developed, so that the sites of severe infection can be selectively approaches for therapy [11].

10.7.4   Biodistribution and tissue distribution of magnetic microsphere

By injecting microspheres in physiological solution containing 0.1% (w/v) tween 80 or a viscosity-enhancing agent, such as 50% (w/v) dextran, aggregation in target vessel can be minimized. For spheres smaller than 3 micrometer, initial (5–30 min) biodistribution is a function of the dose relative to the capacity of target capillaries, degree to which the magnetic field overlaps microvessels supplied by the injection vessel, extent of venous shunting before microspheres reach the field and the flow rates in target vessels (Table 10.1). Ranny et al. (1985) performed a study that is, large doses of trace-labelled spheres (6.6 mg/kg, sufficient to produce effective levels of fmet-leu-phe-peptide) were infused intravenously in to rates over a 3 minutes interval, allowed to travel at native flow rates into the central venous system through the right heart in to bath lungs, and captured by an 0.55 T magnetic field (gradient of 0.01 T/mm), which enveloped the right thorax and the middle half of the left thorax [12]. The splenic and renal concentrations of the spheres in the later study were slightly higher because there was greater vascular shunting due to incomplete magnetic envelopment of the left lung. The list of various drugs which can be used for formulating magnetic microsphere is shown in (Table 10.2).

Table 10.1 Name of organs and blood component with injected dose in tissues and their weight of magnetic microspheres

Name of organs and blood component	Injected dose in tissue (%)	Weight of magnetic microspheres (µg/gram)
Right Lung	15.9	195
Left Lung	11.2	135
Heart	3.3	67
Liver	45	37
Spleen	7.5–7.6	26–27
Kidney	11.3	24
Blood or plasma	3.3–3.4	2.9–3
Circulating white blood cells	<0.1	<0.004

Table 10.2 Therapeutic indication and evaluation techniques of magnetic based drug delivery systems

Magnetic-based drug delivery	Drug or excipeints used	Therapeutic indication	Magnetic materials	Evaluation techniques	References
Magnetic microspheres	Indomethacin, Eudragit-L100 & Magnetite	NSAIDS	$FeSO_4$ and $FeCl_3$ solution, NaOH solution	SEM, VSM, X-RD, In-vitro release study	[17]
	Oxantrazole, Chitosan, extra heavy mineral oil, Ferro fluid, glutral aldehyde	Anti-cancer agent	Fe_2O_3	Drug loading efficiency, % yield, microsphere morphology and particle size	[18]
	Doxorubicin, Human serum albumin, magnetite, cotton seed oil, diethyl ether	Anti-cancer agent	$FeSO_4$, $FeCl_3$ and NaOH solution	In-vivo tumour targeting	[19]
	Diclofenac sodium, Human serum albumin, magnetite, Neodymium magnet, 8000 G field, cotton seed oil	Anti-arthritic agent	$FeSO_4$, $FeCl_3$ and NaOH solution	FT-IR, SEM, DSC, In-vitro drug release study	[20]

Contd...

Contd...

Magnetic-based drug delivery	Drug or excipeints used	Therapeutic indication	Magnetic materials	Evaluation techniques	References
Magnetic nanoparticles	Probucol, Polyvinylpyrrolidon (PVP) / sodium dodecyl sulfate (SDS)	Hypolipidemic	$FeSO_4$	Dynamic light scattering and cryogenic-scanning electron microscopy and (13)C in situ solid-state NMR	[21]
	Paclitaxel, Poly-acrylic acid (PAA)	Anti-cancer agent	Fe_3O_4	Dynamic light scattering	[22]
	Insulin, alginate/chitosan	Antidiabetics	Iron oxide	*In-vitro* insulin release studies, scanning electron microscopy (SEM), transmission electron microscopy (TEM)	[23]
Magnetic liposomes	Paclitaxel, ophenanthroline and phospholipid	Anti-cancer agent	Ferrous and ferric salts in	UV-Visible spectrophotometer, dynamic light scattering (DLS), Zeta Potential, polydispersivity index	[24]
	Doxorubicin, PEG	Anticancer agent	Magnetite	MRI	[25]
	Eggphosphatidyl choline (egg-PC) and cholesterol, Lauric acid	Hyperthermia treatment of cancers	Manganese ferrite ($MnFe_2O_4$) magnetic fluid	Size, poly-dispersity, zetapotential, TEM	[26]
	Dextran, dipalmitoylphosphatidyl-choline (DPPC).	Antithrombotic (anti-platelet), to reduce blood viscosity, and as a volume expander in anemia	Fe3O4	Fluorescence detector	[27]
Magnetic niosomes	Methotrexate	Anti-neoplastic	Fe_3O_4	Transmission electron microscope	[28]
	Paclitaxel	Anti-neoplastic	Fe_3O_4	FT-IR, XRD and VSM, TEM and DLS.	[29]

Contd...

Contd...

Magnetic-based drug delivery	Drug or excipeints used	Therapeutic indication	Magnetic materials	Evaluation techniques	References
Magnetic microcapsule	Poly(N-isopropylacrylamide	Tissue engineering and controlled drug delivery and hydrogel	Fe_3O_4	X-ray diffraction	[30]
	Porous polysulfone (PSF), tributyl phosphate (TBP)	Thermoplastic polymer and used for superior replacement for polycarbonates	Fe_3O_4	Fourier transform infrared (FT-IR), scanning electron microscope (SEM), vibrating sample magnetometer (VSM) and thermogravimetric analysis (TGA)	[31]

10.7.5   Magnetic nanoparticles

Magnetic nanoparticles are particles in nano size range containing polymers, drug along with ferromagnetic particles (magnetite). Applications of magnetic particles include immunoassays, drug targeting, drug transporting, and biosensing [32]. Ferromagnetic iron dextran nanoparticles were prepared by reacting a mixture ferrous chloride and ferric chloride with dextran polymers under alkaline conditions. The particles of average size range (30–40 nm) showed a little non-specific binding to cells and had a magnetic moment. Magnetically responsive nanoparticles were prepared from enzymatically hydrolyzed starch and magnetite. Using this magnetic nanoparticles (average size range 100–300 nm) coupled monoclonal mouse anti-rat Ig kappa light chain antibody, a very high depletion of surface Ig positive cells (mostly B-cells) from one million peripheral blood mononuclear cells should be achieved. The separation efficiency of this technique was evaluated by flow cytofluorometric analysis and the technique has been reported to permit the detection of a small number of surface Ig positive among 10000 negative cells. Indomethacin bearing magnetic nanoparticle of polymethylemethacrylate, were prepared by the polymerization technique. The controlled growth of ferric hydroxide particles in the presence of non-ionic surfactant was affected to obtained nano-sizes particles, and these were subsequently heated to obtained magnetite. The effect of various particles, i.e. monomer concentration and magnetite concentration as well as the stirring rate was studied to characterize the particle size and its distribution. Then *in vivo* magnet responsiveness and kinetics of distribution of these magnetic and plain nanoparticles were characterized and reported up to 60 min post injection time, 60-fold higher concentration in target tail segment was recorded which resulted in considerably reduced drug concentration in other organs as evinces by data from control rats [11]. Figure 10.7 explains the bounding of nanoparticles to tumor cells.

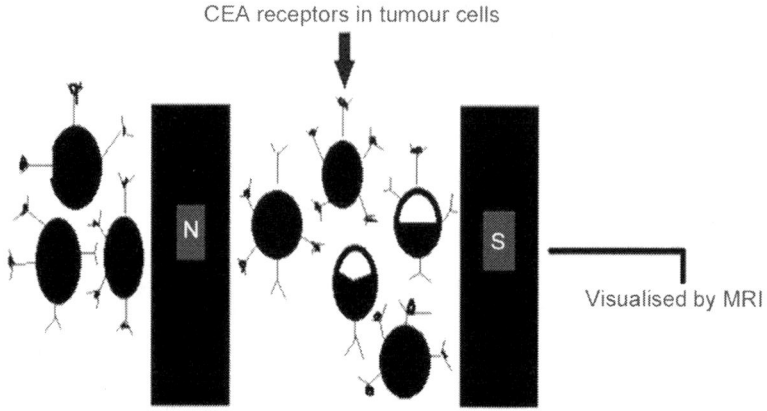

CEA receptors in tumour cells

Visualised by MRI

Figure 10.7 Targeting approach of nanoparticles to the tumor cells (antigen-antibody reaction)

10.7.6   Magnetic liposomes

Liposomes are simple microscopic vesicles in which lipid bilayer structures are present with an aqueous volume entirely enclosed by a membrane, composed of lipid molecule. There are a number of components present in liposomes, with phospholipids and cholesterol being the main ingredients but in case of magneto liposomes magnetite is one of the component of the liposomes [33]. Generally these are magnetic carrier which can be prepared by entrapment of Ferro fluid within core of liposomes [34, 35]. Magnetoliposome can also be produced by covalent attachment of ligands to the surface of the vehicles or by incorporation of target lipids in the matrix of structural phospholipids [36]. Alternatively magnetoliposomes are prepared using the phospholipids vesicle as a nanoreactor for the in situ precipitation of magnetic nanoparticles [37]. Vesicles are also prepared containing didodecyl methyl ammonium bromide; contain an ionic magnetic fluid [38]. These magnetoliposomes were effectively used for site-specific targeting, cell sorting and as magnetic resonance contrast enhancing agent. Thermo sensitive magnetoliposomes can release the entrapped drug after selective heating caused by the electromagnetic fields as shown in Fig. 10.8 [33, 39–43]. Antibody coated magnetoliposomes for hyperthermia treatment of cancer were prepared by coating phospholipid on to magnetic particles [44] and magnetically responsive polymerized liposomes as potential oral delivery vehicles for complex molecules such as protein and peptide to protect them from gastrointestinal environment and targeting them to the payer's patches [45] are depicted in Fig. 10.9.

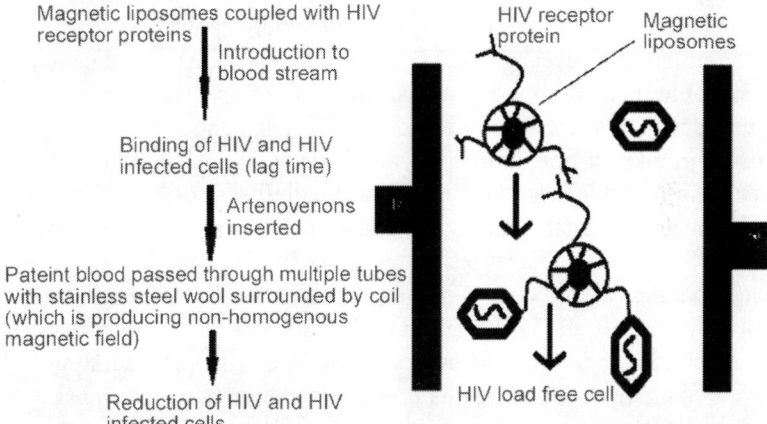

Figure 10.8 Schematic presentation of HIV treatment by magneto liposome

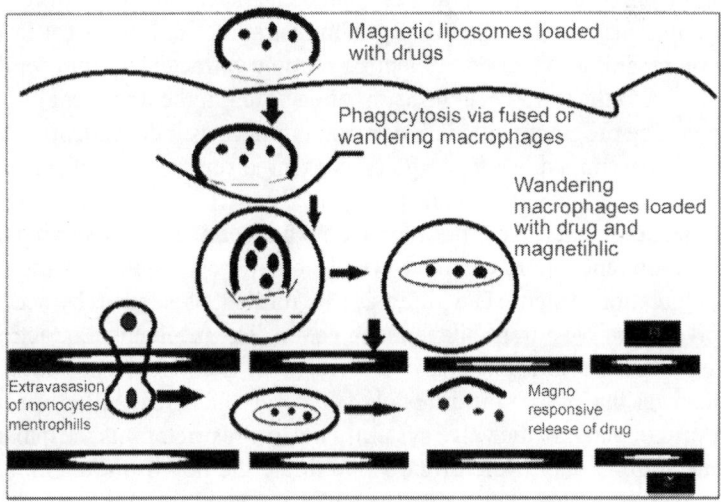

Figure 10.9 Carrier cellular co-ordinate targeting strategy

10.7.7   Magnetically programmable infusion pump

Magnetic technology is widely used for external programming of cardiac pacemakers. The same principle was adapted to an implantable infusion pump. The development of such pumps to a prototype stage, the newer method of radio frequency (rf) signaling could improve the magnetic approach because of greater programming flexibility and bi-directional transmission capacity in rf-programmable pump, the receiver is initially switched to a programmable made magnet located in the extracorporeal programming head. When the magnetic field and recognizable rf pulse sequence are applied simultaneously, the reprogramming occurs. Some of the experimental applications of programmable pumps are: continuous, time pulsed or circadian infusion of antitumor agents into the systemic and portal veins, control of pain in cancer patients by intrathecal or epidural infusion of morphine and treatment of motor specificity in multiple sclerosis with intraspinalinfusion of baclofen (GABA binder). The programmable infusion pump is an implantable, battery-powered device that stores and dispenses drugs according to instructions received from an externally applied programmer/control unit. The programmer or control unit was not assessed relative to the utilization of MRI because it is not intended for use in the MRI environment. The primary material used for this pump is titanium. Notably, this programmable infusion pump has suture loops that are used to anchor it as it is implanted in a subcutaneous pocket, typically located in the lower abdomen. The catheter is part of the infusion system used with the pump to deliver drugs to the intrathecal space. The programmable infusion pump and associated catheter are indicated for long-term continuous or intermittent drug administration via intrathecal infusion of morphine sulfate in the treatment of chronic intractable pain (benign or malignant) and the intrathecal infusion of baclofen in the treatment of severe spasticity. The programmable infusion pump functions independently, once it has been programmed. This pump contains three separate chambers such as an outer chamber, which contains a chemically inert liquid gas mixture used as a propellant to exert a constant force onto a second chamber serving as the drug reservoir, and a third chamber, which contains electronics, a battery, and a flow regulation system. The pump can be refilled, as needed, by accessing the refill septum via a transcutaneous injection. The mechanism of action for the programmable infusion pump involves a constant force provided by a propellant gas that pushes the drug out of the reservoir, across a filter, through a flow restrictor and into the valve system. The flow restrictor sets the maximum flow rate and the duty cycle of the valve titrate the flow from the minimum to maximum flow rate. The valve system is activated using a ceramic (non-magnetic) actuator to minimize the influence of external magnetic fields on the actual flow-control mechanism. The drug passes through the flow regulation system to the outlet. The flow restrictor limits the maximum flow rate that will be delivered by the programmable infusion pump. The remaining drug volume within the reservoir is detected by a fill level sensor, which determines the volume of the reservoir chamber and converts the calculation to determine the remaining drug volume [46]. The major advantages of this system are drug output can be increased to compensate for biological tolerance to pain medications. Catheter tips can be inserted into very small spaces or vessels were polymer slabs and even injectable microspheres will not fit, pump reservoirs require infrequent filling, only ones every 2–6 weeks depending on the drug stability and the pumps are designed to for up to 2 years on their original batteries [47]. This device require for drug stability in solution at 30°C, relative high expense, bulkiness of central pump unit, failure rate of up to 12%, occasional plugging of the outflow catheter, minor difficulty in accessing the injection port [48].

10.7.8   Magnetic emulsion

The chemical composition of the used magnetic emulsion includes 23% w/w of charged surfactant and 77% w/w of iron oxide (Fe_2O_3). The amount of organic phase in the ferrofluid droplets is lower than 5%. The preparative steps involve formation of O/W magnetic emulsion droplets as shown in Fig. 10.10. The major preparative steps of magnetic emulsion are briefly described as follows:

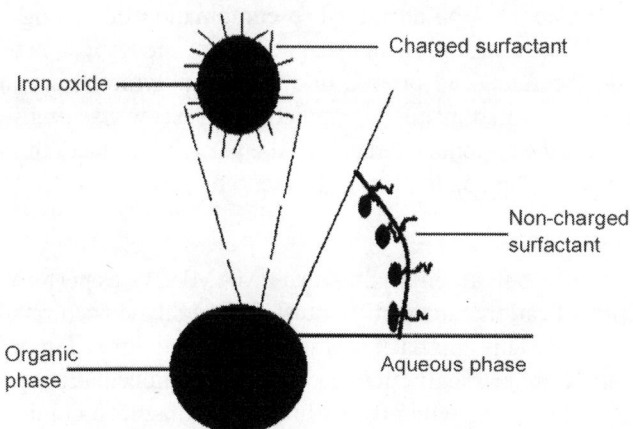

Figure 10.10 Schematic illustration for the formation of o/w magnetic emulsion droplets

Encapsulation

The encapsulation was performed via layer-by-layer polyelectrolyte adsorption process. The first cationic polyelectrolyte was adsorbed onto magnetic emulsion using well appropriate polycation concentration. After washing step using Milli-Q water, the obtained cationic magnetic particles were then introduced in high concentrated polyanion solution. The final magnetic particles were washed using Milli-Q water before any colloidal characterization. The preparation details are below reported. The amino-containing magnetic emulsion can further be converted into a cationic charge bearing emulsion system by adsorbing polyethyleneimine (PEI) onto the negatively charged magnetic nanodroplets. Five ml of the magnetic dispersion (5% w/v) were added to 10 ml of PEI (50 g/L) and stirred 15 min. To remove the excess of PEI, the dispersion was washed using Milli-Q water [48].

Polycation adsorption

Various experiments were performed in order to point the optimal conditions leading to stable cationic magnetic emulsion. For low PEI concentration (below 2–5 mg of PEI/mg of solid magnetic emulsion), the magnetic emulsion flocculates rapidly, whereas for high PEI concentration (above 5 mg of PEI/ mg of solid magnetic emulsion), stable magnetic emulsion was obtained. The observed aggregation can be attributed to the bridging flocculation process induced by the PEI. The colloidal stabilization observed for high PEI concentration is due to sterical stabilization process. The adsorption here was investigated at a fixed pH by the PEI solution (i.e. pH 10). The adsorption process of such polymer (low or no charged at pH 10) onto negatively charged magnetic emulsion can be attributed to combination of hydrogen bounding between carboxylic group on the magnetic emulsion and amine groups of the PEI and the forced adsorption due to high polymer concentration. The zeta potential of coated magnetic emulsion after a few washing with Milli-Q water, revealed the cationic character of the particles surface. In fact, the zeta potential measured at acidic pH was found to be positive [48].

Polyanion adsorption

The adsorption of polyanion (hydrolyzed PMAMVE) was performed via drop wise addition of cationic magnetic emulsion in highly concentrated polyanion polymer solution. This approach was selected and optimized in order to avoid aggregation of magnetic particles via bridging flocculation process. The adsorption of polyanion onto positively charged magnetic emulsion is due to the attractive electrostatic interaction. The adsorption of the polyanion onto elaborated cationic magnetic emulsion was evidenced by investigating the zeta potential. The zeta potential was found to be inversed as expected. The observed zigzag profile of zeta potential versus the adsorbed polyelectrolyte layer revealed the surface modification. The thickness layer of the adsorbed polycation is too thin to be evidenced by TEM analysis. Due to the slight polydispersity of the seed magnetic emulsion, the hydrodynamic thickness cannot be easily deduced from the hydrodynamic diameters measured by quasi-elastic light scattering [48]. The emulsion showed high retention by a magnetic field in-vitro. After i.v injection in the rat, the magnetic emulsion was mainly localized in the lungs by application of an electromagnet over the chest. Therefore, magnetic emulsions appear to have potential in conferring site specificity to certain chemotherapeutic agents [13].

Anionic magnetic emulsification

The adsorption of anionic polyelectrolyte onto amino-containing magnetic nanodroplets was performed as follows (Fig. 10.2). 15 ml of the hydrolyzed PMAMVE (11 g/L) solution and 20 ml of Milli-Q water was first mixed and 45 ml of cationic magnetic emulsion (~0.5%) was introduced drop wise along with mechanical stirring. The obtained magnetic particles were washed via magnetic separation and redispersed in Milli-Q water before any characterization and utilization [48].

Magnetic resealed erythrocyte

Local thrombosis in animal arteries was prevented by means of magnetic targeting of aspirin loaded red cells. Thrombosis was investigated in 18 dogs and 16 rabbit's arteries by surgically inverting a vascular wall flap into its lumen. A completing occluding a red thrombus was developed inside the vessel after 4–5 h in 80% of cases. Smco5 magnet was secured externally to one of the arteries. The constant magnetic field produced by the magnet had no influence on the clot formation. Autologus red cells loaded with ferromagnetic colloid compound and aspirin were administered intravenously, and completely aborted arteriothrombosis on magnetic application side with no deteriorate effect on clot formation in the control artery was recorded. Magnetically responsive ibuprofen-loaded erythrocytes were prepared and characterized in-vitro. The erythrocytes were loaded with ibuprofen and magnetite (ferrofluid) using the press well technique. Various process variables including drug concentration, magnetite concentration sonication of ferrofluids that could affect the loading of the drug and magnetite were optimized. The loaded erythrocytes were characterized for in-vitro drug efflux, haemoglobin release, morphology, osmotic fragility, turbulence shock, in vitro magnetic responsiveness and percent cell recovery. In optimum concentration, erythrocytes could tolerate ibuprofen as no appreciable detrimental effects were noticed on cell morphology, osmotic fragility, turbulence shock, when compared with normal erythrocytes. The drug release profile from the cellular system was observed to follow approximately zero order kinetics. The loaded cells effectively responded to an external magnetic field. Diclofenac sodium loaded magnetic erythrocytes, for active delivery of drug to painful inflamed joints, for possible physical modulation of carrier and contained drug biodistribution is the best-suited example [5, 8–10].

10.7.9   Magnetically modulated device

In recent years, magnetically modulated polymeric controlled drug delivery systems that deliver the drug at increased rate on demand have been developed. These systems consist of polymeric matrix where in drug power is dispersed. The polymeric matrix generally composed of ethylene vinyl acetate copolymer (EVAc) with some magnetic beads. The beads used are either magnetic steel beads composed principally of iron (79%), chromium (17%), carbon (1%), silicon (1%), molybdenum (0.75%) and phosphorous (0.04%) or small samarium cobalt magnets. These systems are formulated by adding approximately 50% of drug polymer mixture. The magnetic tablets are placed in glass vials. An oscillating magnetic field, which is generated by a device that rotates the permanent magnets below the vials, controls release rates [15, 16, 48–51]. The systems can release up to 30 times more drugs when exposed to the magnetic field, and release rates return to normal when the magnetic field is removed. Furthermore, these systems do not cause inflammation in vivo. This was confirmed by lack of edema, cellular infiltrate, or microvascularization as concluded by gross and histological examination of tissues in rabbits [52]. The important factors that control the release rate of magnetic polymeric devices are magnetic field characteristics and the mechanical properties of the polymeric matrix [51]. The release of macromolecules from EVAc systems without magnetic beads, suggest that the molecules with molecular weight greater than 300 cannot penetrate the polymer. The direct incorporation of macromolecules in the polymer macromolecule using cast procedure caused a tortuous and complex series of pores formation within the matrix. The release rate can be determined by factors affecting penetration of water into the polymer. Video recording of the polymeric matrix surface, the magnetic beads showed that the beads actually moved within the matrix in response to the external magnetic field and as a result displace adjustment material containing polymer and drug with it, "Squeezing" out the dissolved drug through the pores. A model for the enhanced release suggested that the major effects originated from the alternate compression and expansion deformation of the pores, causing the fluid within to generate a pustule flow, which alone (no net convection) is able to greatly improve diffusive mass transfer [52].

10.8   Drug delivery and biomedical applications

The most popular application of magnetic carrier technology is bioaffinity chromatography, wastewater treatment, immobilization of enzymes, or other biomolecules, and preparation of immunological assay. It is also used in the delivery of insulin, nitrates as well as in selective β-blockers, in general hormone replacement immunization and chemotherapy.

10.8.1   Controlled release drug delivery

Langer et al. embedded magnetite or iron beads in to a drug filled polymer matrix and then showed that they could activate or increase the release of drug from the polymer by moving a magnet over it or by applying an oscillating magnetic field. The microenvironment within the polymer seemed to have shaken the matrix or produced 'micro cracks' and thus made the influx of liquid, dissolution and efflux of drug possible thereby achieving magnetically controlled drug release. Macromolecules such as peptides have been known to release only at a relatively low rate from a polymer controlled drug delivery system, this low rate of release can be improved by incorporating an electromagnetism triggering vibration mechanism into the polymeric delivery devices with a hemispheric design; a zero-order drug release profile is achieved [53].

10.8.2   Tumour targeting

Clinical cancer therapy is one of the primary applications of this system for the treatment of advanced solid tumors. At the present, preclinical research is investigating the use of magnetic particles encapsulated with different anticancer drugs like mitoxantrone, paclitaxel and cationic magnetic aminodextran microsphere (MADM) which have been synthesized and its potentially towards brain targeting was studied. For example magnetic doxorubin in liposome has significant anticancer effect in nude mice bearing colon cancer [54].

10.8.3   Bio-separation

In bio-separation techniques, magnetic separation is the most promising method. The development of magnetically responsive microsphere has brought an additional driving force to play particle that are bound to magnetic fluids can be used to remove cells and molecules by applying magnetic field and in vivo to concentrate drugs at a anatomical sites with restricted access.

These possibilities from the basis of well established biomedical application in protein and cell separation. Magnetic silica microsphere can be used in the solid phase extraction of genomic DNA from soy-based foodstuffs [55].

10.8.4   Hyperthermia for treatment of cancer

Hyperthermia is the heat treatments of organs or tissues, such that the temperature is increased to 42–46°C to reduce the viability of cancerous cells. It is based on the fact that tumor cells are more sensitive to temperature than normal cells. In hyperthermia it is essential to establish a heat delivery system, such that the tumor cells are heated up or inactivated while the surrounding tissues (normal) are unaffected ferromagnetic magnetite microsphere can also be used in the hypothermic treatment of cancer [56].

10.8.5   Targeting of radioactivity

Magnetic targeting can also be used to deliver the therapeutic radioisotopes. The advantage of these method over external beam therapy is that the dose can be increased, resulting in improved tumor cell eradication, without harm to adjacent normal tissues. Magnetic targeted carriers, which are more magnetically responsive iron carbon particles, have been radio labeled in last couple of years with isotopes such as 188Re, 90Y, 111In, and 125I and are currently undergoing animal trials [39].

10.8.6   Disease diagnosis

The most important application of magnetic particles is as contrast agent for magnetic resonance imaging in diagnosis of diseases. The most commonly used super paramagnetic material is Fe_3O_4 with different coatings such as dextrans, polymers, and silicone. Supramagnetic iron oxide (SPIO) it has been mainly used as a liver-specific contrast agent for intravenous application. It may also be used for detection of metastases in non-enlarged lymph nodes [57].

10.8.7   Gastrointestinal surgery

Magnetic elements have been successfully used in gastrointestinal surgery for tissue fixation. Which form hermetic seal after surgery and passibility of the gastrointestinal tract is maintained and the patient can able to eat immediately after operation. Magnetically guided ferrofluid nanoparticles were used in retinal repair. Magnetically guided interstitial diffusion of the nanoparticles up to 20 mm of the gel over periods of 72 hours was shown to be possible, thus demonstrating that essentially all points on the retinal surfaces are reachable from elsewhere in the ocular interior [58].

10.8.8   Protein immobilization

Magnetic materials were suggested as carriers for protein immobilization. Their property to concentrate near magnetic terminal is used in technological process for selective catalyst removal from the reaction mixture, in immunological studies for separation of cells to which magnetic particles are specifically bound modified targeting *in vivo* into appropriate tissue under guidance if an external magnetic field. A number of methods are available to obtain porous magnetic carriers, containing immobilized matter not only on the surface, but also in the volume of a particle. Normally, these preparations are obtained by granule formation from the suspension of ferromagnetic particles in the solutions or melt of appropriate high molecular weight compound. The drawbacks of the above-mentioned methods include pronounced aggregation of ferromagnetic particles and lead to formation of product with a variety of sizes and magnetic properties [5]. Figure 10.11(A–B) explains the trypsin immobilization onto the amine-functionalized magnetic nanoparticles and process of protein digestion using trypsin-immobilized

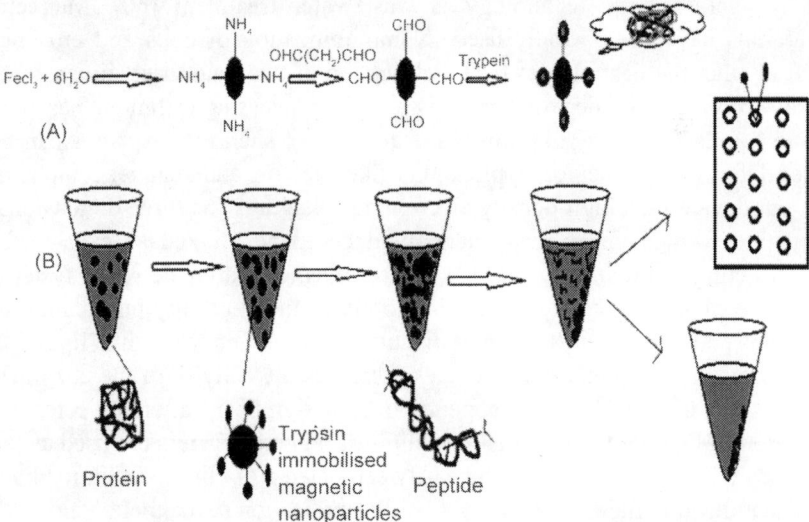

Figure 10.11 Schematic diagrams of trypsin immobilization onto the amine-functionalized magnetic nanoparticles (A), and process of protein digestion using trypsin-immobilized magnetic nanoparticles (B)

magnetic nanoparticles. Three standard proteins, myoglobin (MYO, a protein known to be rather resistant to proteolysis, MW 16 900), cytochrome c (Cyt-C, MW 12 384), and bovine serum albumin (BSA, MW 66 000) were used to test the performance of trypsin-immobilized magnetic nanoparticles [59].

10.8.9   Contraceptive delivery

In this magnetic control system, the drugs and magnetic beads are uniformly dispersed within a polymer material. On exposure to an aquous media, the drug is release in a diffusion controlled fashion. More ever, the rate can be increased or modulated on application of an oscilating external magnetic field. These system may be useful when drug delivery is designed responsive to the changes in steroid excretion during the menstrual cycle [49].

10.8.10   Miscellaneous application

Only the mechanical-physical properties of magnetic particles or ferrofluids can be used for therapy. One example is the immobilization of capillaries under the influence of a magnetic field. In this way, tumors could specifically starve of their blood supply. Another elegant example is the use of magnetic fluids to preventing patients from going blind. The most popular application of magnetic carrier technology is wastewater treatment [60]. Magnetic materials have been widely used in immobilization of cells and enzymes and in other biotechnologies [61]. Fabrication of two- and three-dimensional arrays of magnetic microsphere using a novel processing technique has been done successfully. The two dimensional array of such microspheres can be useful for various device applications like electric and magnetic input or output devices and high-density electronic packaging. The three dimensional structure formed under a magnetic field could be utilized as a basis for constructing photonic crystals [2]. These magnetic drug delivery systems also served as a novel support for immobilized metal affinity purification of proteins and peptides [50]. Immobilization of certain enzymes like lipase on the surface of magnetic microspheres increases the activity of the enzymes. Magnetic carrier technology appears to be a significant alternative for the solution of the biomolecule malformations. Magnetic carriers are used as the support material and they can be easily separated from the reaction medium and stabilized in fluidized bed dryer by the application of magnetic field [40]. Magnetic polymeric microspheres can be use for the adsorption of proteins. It is also having the application in separation from red blood cells (RBC) from the whole blood for photophoresis treatment of white cells [62]. Based on the polyvinyl alcohol affinity medium novel magnetic microspheres can be used in quantitative detection of glycated hemoglobin. The most important application of MMS in medicine is their in-vitro use for the depletion of cancer cells from bone marrow [63]. Table 10.3 describes list of some magnetic modulated products in detail with its application in pharmaceuticals field as given below.

Table 10.3 List of some magnetic modulated products in detail with its application and its recent patent in pharmaceutical field

Manufacturer	Active ingredients	Application	Patent no.
Life Technologies	Supplied sterile, non-pyrogenic suspension containing 4×10^8 Dynabeads®/ml in phosphate buffered saline (PBS), pH 7.4, w0.1% human serum albumin (HSA)	For ex-vivo isolation, activation, and expansion of T-cells in a number of studies, including studies on HIV infection and cancer [Autoimmune diseases]	US 8338168 B2
Merck-Millipore	Human Cytokine/Chemokine Magnetic Bead Panel, HCYTOMAG-60K	Specific cytokines involved in any inflammatory or immune response, it might be necessary to screen panels of cytokines, often requiring some level of automation and/or high throughput. Magnetic Beads can make the process of automation and high throughput screening easier [Inflammatory disease, allergic reactions, irritable bowel disease (IBD), sepsis, and cancer]	US20130011824 A1
Life Technologies	Dynabeads® streptavidin are uniform, superparamagnetic beads of 2.8 μm in diameter with a streptavidin monolayer covalently coupled to the surface. This layer ensures negligible streptavidin leakage while the lack of excess adsorbed streptavidin ensures batch consistency and reproducibility (beads in suspension form)	Dynabeads® M-280 Streptavidin are ideal for numerous applications, including purification of proteins and nucleic acids, protein interaction studies, immunoprecipitation, immunoassays, phage display, biopanning, drug screening and cell isolation	US20120258892 A1
Spherotech	Manufactured by coating a layer of iron oxide and polystyrene onto polystyrene core particle, uniform in size and spherical in shape, Separated using a magnet and resuspended. Available in 2.5% concentration in 3.0 to 3.9 μm size (100 ml)	Used for cell separation, affinity purification, DNA probe assays magnetic particle EIA, etc.	EP1926508 A2

10.9   Future perspectives

Conceptually, magnetic targeting is a very promising approach in the upcoming future however, there are a number of physical, magnetism-related properties which require careful attention. First, the magnetic force, which is defined by its field and field gradient, needs to be large and carefully shaped to fit the target area. For *in vivo* applications, this is not trivial, and collaborations with electrical or biomedical engineers are advisable. Second, the magnetic susceptibility of the MMS needs to be as high as possible. More responsive magnetic materials of defined and homogeneous material properties in a (tissue) stable and defined oxidation state need to be synthesized. Third, the MMS size must be small enough that they do not clog the blood vessels through which they are guided to the target organ. Fourth, altering the surface of MMS with appropriate molecules should always be considered to either decrease the interaction of MMS with tissues or organs, using for example PEG; or to specifically bind to target cell populations using for example antibodies or receptor agonists. Finally, the MMS size must be uniform enough to provide an equal probability of magnetic capture for each MMS and constant drug/ radioactivity content. Beside the magnetic properties, the fate of the particles in the body is an important consideration both for local and systemic short and long-term toxicity. Furthermore, the pharmacokinetic characteristics must be optimized for the specific target organ, taking into account that the normal organ behavior might differ from that of a diseased organ.

10.10   Conclusions

Magnetic seems to serve as a common function of opening a new vista of a multi-barrier of multi-step drug delivery. It has been established that the magnetic drug targeting is an efficient means to localizing toxic or labile pharmaceuticals in a preselective site. Magnetic targeting is a process for choice of delivery of about 45–60% of the new peptide and recombinant proteins at a level of 25–50% localization of injected dose in non-reticuloendothelial target tissues. This means targeting has been exploited achieve adequate drug levels, bioavailability enhancement, localizing the effect of biopharmaceuticals and avoidance of toxic manifestations. Magnetic targeting also offers advantages of magnetic capture and retention to endothelium of microvasculature. In spite of certain drawbacks, such as strong magnetic field requires for the ferrofluid and deposition of magnetite the magnetic micro carriers still play an important role in the selective targeting, and the controlled delivery of various drugs. It is a challenging area for future research in the drug targeting so more researches, long-term toxicity study, and characterization will ensure the improvement of magnetic drug delivery system. The future holds lot of promises in magnetic micro carriers and by further study this will be developed as novel and efficient approach for targeted drug delivery system.

10.11   References

[1] J. Akhtar, R. Chaturvedi, J. Sharma, D. Mittal, P. Pradhan. Magnetized carrier as novel drug delivery system, *Int J Drug Del Tech.* 1 (2009): 28–35.

[2] K.R. Vinod, D. Sridhar, S. Sandhya, D. Banji, T.R. Reddy. Essential of pharmaceutical product development for magnetically modulated drug delivery system, *Int J Pharm Sci Nano.* 4 (2012): 1519–27.

[3] D. Bahadur, J. Giri. Biomaterials and magnetism, *Sadhana.* 28 (2003): 639–56.

[4] S.P. Vyas, R.K. Khar. Targeted & Controlled Drug Delivery. (2004): 459–80.

[5] P. Lokwani, A. Goyal, S. Gupta, R.K. Songara, N. Singh, K.S. Rathode. Pharmaceutical application of magnetic particles in drug delivery system, *Int J of Pharm Res and Dev.* 3 (2011): 147–56.

[6] K.J. Widder, A.E. Senyei, D.F. Ranney. Magnetically responsive microspheres and other carriers for the biophysical targeting of antitumor agents, *Adv Pharmacol Chemother.* 16 (1979): 213–71.

[7] P. Lokwani. Magnetic particles for drug delivery: An overview, *Int J Res Pharm Bio Sci.* 2 (2011): 465–73.

[8] J.M. Gallo, P.K. Gupta, C.T. Hung, D.G. Prrier. Evaluation of drug delivery following the administration of magnetic albumin microspheres containing adriamycin to the rat, *J Pharma Sci.* 78 (1989): 190.

[9] K.J. Widder, A.E. Senyei, D.G. Scarpelli. Magnetic microspheres: a model system for site specific drug delivery in vivo, *Proc. Soc ExpBiol Med.* 58 (1978): 141.

[10] M. Hasegawa, S. Hokkoku. US 18,101,435, 1978.

[11] L.V. Allen, N.G Popovich, H. Ansel. Ansel's pharmaceutical dosage form and drug delivery systems. 8 (2005): 264.

[12] D.F. Ranny, H.H. Huffeker. Annals of the New York Academy of Sciences, Biological Approaches to the Controlled Delivery of Drugs. 507 (1987): 104–19.

[13] Hafeli, W. Scutt, J. Teller, Zborowski. Scientific and clinical applications of magnetic carriers, Springer, USA, (1997) 1–86.

[14] S.P Vyas, A. Malaiya. In vivo characterization of indomethacin magnetic polymethyl methacrylate nanoparticles, *J Microencapsulation.* 6 (1989): 493–99.

[15] H. Kiwada, J. Sato, S. Yananda, Y. Kato. Feasibility of magnetic liposomes as a targeting device for drugs, *Chem Pharm Bull.* 34 (1986): 4253–58.

[16] N.M. Orekhova, R.S. Akchuron, A.A. Belyaev, M.D. Smimor, S.E. Ragimov, A.N. Orekhova. Local prevention of thrombosis in animal arteries by means of magnetic targeting of asprin loaded red cells, *Thrombosis.* 57 (1990): 611–16.

[17] S.P. Was, A. Malaiya. In vivo characterization of indomethacin magnetic polymethyl methacrylate nanoparticles, *J Microencapsul.* 6 (1989): 493–99.

[18] E.E. Hassan, R.C. Parish, J.M. Gallo. Optimized formulation of magnetic chitosan microspheres containing the anticancer agent, oxantrazole, *Pharm Res.* 9 (1992): 390–7.

[19] H. Wen, J. Guo, B. Chang, W. Yang. pH-responsive composite microspheres based on magnetic mesoporous silica nanoparticle for drug delivery, *Eur J Pharm Biopharm.* 84 (2013): 91–8.

[20] R.M. Pabari, T. Sunderland, Z. Ramtoola. Investigation of a novel 3-fluid nozzle spray drying technology for the engineering of multifunctional layered microparticles. *Expert Opin Drug Deliv.* 9 (2012): 1463–74.

[21] J. Zhang, K. Higashi, W. Limwikrant, K. Moribe, K. Yamamoto. Molecular-level characterization of probucol nanocrystal in water by in situ solid-state NMR spectroscopy, *Int J Pharm.* 423 (2012): 571–6.

[22] Y. Liu, B. Zhang, B. Yan. Enabling Anticancer Therapeutics by Nanoparticle Carriers: The Delivery of Paclitaxel, *Int. J. Mol. Sci.* 12 (2011): 4395–4413.

[23] P.V. Finotelli, D. Da Silva, M. Sola-Pennab, A. Malta Rossic, M. Farinad, L. R. Andraded, A. Y. Takeuchie, M.H. Rocha-Leao. Microcapsules of alginate or chitosan containing magnetic nanoparticles for controlled release of insulin, Colloids and Surfaces B: Biointerfaces. 81 (2010): 206–211.

[24] J.Q. Zhang, Z.R. Zhang, H. Yang, Q.Y. Tan, S.R. Qin, X.L. Qiu. Lyophilized Paclitaxel Magnetoliposomes as a Potential Drug Delivery System for Breast Carcinoma via Parenteral Administration: In Vitro and in Vivo Studies, *Pharm. Res.* 22 (2005): 573–583.

[25] V. P. Torchilin. Recent advances with liposomes as pharmaceutical carriers, *Drug discovery.* 4 (2005): 145–160.

[26] P. Pradhana, J. Girib, R. Banerjeea, J. Bellarea, D. Bahadurb. Preparation and characterization of manganese ferrite-based magnetic liposomes for hyperthermia treatment of cancer, *J. Magnetism Magnetic Mat.* 311 (2007): 208–215.

[27] E. Viroonchatapan, M. Ueno, H. Sato, I. Adachi, H. Nagae, K. Tazawa, I. Horikoshi. Preparation and characterization of dextran magnetite-incorporated thermosensitive liposomes: an on-line flow system for quantifying magnetic responsiveness, *Pharm Res.* 12 (1995): 1176–83.

[28] Z. Wanqing, M. Tianqiu, W. Junzheng. Effects of magnetic methotrexate conjugates on the proliferation and ultrastructure of HSC-2 cells, *J Practical Stomat.* 93(2000-01): 42.

[29] M.J. Kim, D.H. Jang, Y.I. Lee, H.S. Jung, H.J Lee, Y.H. Choa. Preparation, characterization, cytotoxicity and drug release behavior of liposome-enveloped paclitaxel/Fe3O4 nanoparticles, *J Nanosci Nanotechnol.* 11(201): 889–93.

[30] W. Yang, R. Xie, X. Pang, X. Ju, L. Chu. Preparation and characterization of dual stimuli-responsive microcapsules with a superparamagnetic porous membrane and thermo-responsive gates, *J. Membrane Sci.* 321(2008): 324–330.

[31] J. Yin, R. Chen, Y. Ji, C. Zhao, G. Zhao, H. Zhang. Adsorption of phenols by magnetic polysulfone microcapsules containing tributyl phosphate, *Chem. Eng. J.* 157 (2010): 466–474.

[32] Q.A. Pankhurst, J. Connolly, S.K. Jones, J. Dobson. Applications of magnetic nanoparticles in biomedicine, J. Phys. D: Appl. Phys. 36 (2003): R167.

[33] C. Alexiou, W. Arnold, P. Hulin, R.J. Klein, H. Renz, F.G. Parak. Magnetic mitoxantrone nanoparticle detection by histology, X-ray and MRI after magnetic tumor targeting. 225 (2001): 187–93.

[34] J.L. Arias, V. Gallardo, S.A. Gomez-Lopera, R.C. Plaza, A.V. Delagdo. Synthesis and characterization of poly(ethyl-2-cyanoacrylate) nanoparticles with a magnetic core, *J Contr Rel.* 77 (2001): 309–21.

[35] V.P. Torchilin. Drug targeting, *Eur J Pharm Sci.* 11 (2000): S81–S91.

[36] M. Babincova, V. Altanerova, M. Lampert, C. Altaner, E. Machova, M. Sramka. Site-specific *in-vivo* targeting of magnetoliposomes using externally applied magnetic field, *Z Naturforsch C.* 55 (2000): 278–81.

[37] S. Bogdansky. Natural polymer as drug delivery systems, *Drugs Pharm Sci.* 45(1990): 231–59.

[38] D.F. Ranney. Magnetically controlled devices and biomodulation, *Drugs Pharm Sci.* 32 (1998): 325–63.

[39] U.O. Hafeli. Magnetically modulated therapeutic systems, *Int Pharm Sci.* 277 (2004): 19–24.

[40] E.B. Denkbas, E. Kilicay, C. Birlikseven, E. Ozturk. Magnetic chitosan microspheres: preparation and Characterization, *Reac Fun Poly.* 50 (2002): 225–32.

[41] J.A. Ritter, A.D. Ebner, K.D. Daniel, L.S. Krystle. Application of high gradient magnetic separation principles to magnetic drug targeting, *J Magnetism Mag Mat.* 280 (2004): 184–201.

[42] S. Bhadra, D. Choubey, G.P. Agrawal. Target oriented microspheres of diclofenac sodium, *Ind Jour Pharm Sci.* 65 (2003): 503–09.

[43] S. Ghassabian, T. Ehtezazi, S.M. Forutan, S.A. Mortazavi. Dexamethasone-loaded magnetic albumin microspheres: preparation and in vitro release, *Int J Pharm.* 130 (1996): 49–55.

[44] J.A. Ritter, A.D. Ebner, K.D. Daniel, A.K. Fahlvik, P. Artursson, P. Edman. Magnetic starch microspheres: Interactions of a microsphere MR contrast medium with macrophages in vitro, *Int J Pharm.* 65 (1990): 249–59.

[45] A.U. Kondoa, H. Fukudab. Preparation of thermo-sensitive magnetic microspheres and their application to bioprocesses colloids and surfaces: Physiochemical and engineering aspects, *Coll Surf A.* 153 (1999): 435–38.

[46] F.G. Shellock, R. Crivelli, R. Venugopalan. Neuromodulation: technology at the neural interface, *Int Neuro Soc.* 11 (2008): 163–70.

[47] R. Asmatulu, R.O. Claus, J.S. Riffle, M. Zalich. Targeting magnetic nanoparticles in high magnetic fields for drug delivery purposes, MRS Spring Meeting. 820 (2004): 63–8.

[48] A.K. Boal. Synthesis and Applications of Magnetic Nanoparticles, *Nanostructure Sci. Tech.* (2004), 1–27.

[49] R.J. Robinson, V.H. Lee. Controlled drug delivery and fundamentals. 2 (1987): 504.

[50] S.R. Rudge, T.L. Kurtz, C.R. Vessely, L.G. Catterall, D.L. Williamson. Preparation characterization and performance of magnetic iron carbon composite microparticles for chemotherapy, *Biomaterials.* 21 (2000): 1411–20.

[51] A.S. Lubbe, C. Bergemann, J. Brock, D.G. McClure. Physiological aspects in magnetic drug targeting, *J Magnetism Mag Mat.* 194 (1999): 149–55.

[52] Hsich, R., Langer, R., Mansodroff, Z., Roseman, J.J. Controlled Release of Bioactive Materials, Edition 1983, Marcel Dekkar, New York, 221.

[53] S. James, J.C. Boylan. Encyclopedia of Pharmaceutical Technology. 2(2002): 825.

[54] C. Pawan, P. Hemchand. Magnetic microsphere: As targeted drug delivery, *J Pharm Res.* (2009): 964–66.

[55] R. Shi, Y. Wang, Y. Hu, L. Chen, Q. Hongwan. Preparation of magnetite-loaded silica microsphere for solid-phase extraction of genomic DNA from soy-based food stuffs, *J Chromt.* 1216 (2009): 6382–86.

[56] M. Kawashita, M. Tanaka, T. Kokubo, Y. Inoue, T. Yao, S. Hamada, T. Shinjo. Preparation of ferromagnetic magnetite microsphere for hyperthermic treatment of cancer, *Biomaterials.* 26 (2005): 2231–33.

[57] W.J. Hsieh, C.J. Liang, J.J. Chieh, S.H. Wang, I.R. Lai, J.H. Chen, F.H. Chang, W.K. Tseng, S.Y. Yang, C.C. Wu, Y.L. Chen. In vivo tumor targeting and imaging with anti-vascular endothelial growth factor antibody-conjugated dextran-coated iron oxide nanoparticles. *Int J Nanomedicine.* 7 (2012): 2833–42.

[58] A. Matthew, M.D. Barish, E. Kent, M.D. Yucel, T. Joseph, M.D. Ferrucci. Magnetic Resonance Cholangiopancreatography, *N Engl J Med.* 341 (1999): 258–64.

[59] Y. Li, X. Xu, C. Deng, P. Yang, X. Zhang. Immobilization of trypsin on superparamagnetic nanoparticles for rapid and effective proteolysis, *J Prot Res.* 6 (2007): 3849–55.

[60] P.E. Podzas, M.E. Daraio, S.E. Jacobo. Chitosan magnetic microsphere for technological application: preparation and characterization, *Physica.* 404 (2009): 2710–12.

[61] Z. Guo, S. Bai, Y. Sun. Preparation and characterization of immobilized lipase on magnetic hydrophobic microspheres, *Enz micro tech.* 32 (2003): 776–82.

[62] J. Chatterjee, Y. Haik, C.J. Chen. Modification and characterization of polysterene based magnetic microspheres and comparison with albumin based magnetic microspheres, *J Magn Mag Mat.* 225 (2001): 21–29.

[63] U.O. Hafeli, R. Ciocan, J.P. Daily. Characterization of magnetic particles and microspheres and their magneto phoretic mobility using a digital microscopy method, *Eur cells mat.* 3 (2002): 24–27.

11

Recent advances in colon-specific drug delivery systems

Suryakanta Swain[1*], Sitty Manohar Babu[1], Chinam Niranjan Patra[2]
and Muddana Eswara Rao[2]

[1]Southern Institute of Medical Sciences, College of Pharmacy, Department of Pharmaceutics, SIMS Group of Institutions, Mangaldas Nagar, Vijyawada Road, Guntur-522 001, Andhra Pradesh, India
[2]Roland Institute of Pharmaceutical Sciences, Department of Pharmaceutics, Khodasinghi, Berhampur-760 010, Ganjam, Odisha, India

11.1   Introduction

The oral route is considered to be most convenient for administration of drugs to patients. Normally drug dissolves in stomach field and intestinal fluid and absorb from these regions of GIT (Patel et al., 2011). Colon was considered as "BLACK BOX" as, most of drugs are absorbed from upper part of GIT tract (Singh et al., 2012). Colonic drug delivery refers to targeted delivery of drugs into the lower gastro intestinal tract (GIT), specifically in the large intestine (i.e., colon). The colon-specific drug delivery system (CDDS) should be capable of protecting the drug in route to the colon, i.e. drug release and absorption should not occur in stomach as well as small intestine, and neither the bioactive agent should be degraded either of the dissolution sites, but only released absorbed once the system reaches the colon (Philip et al., 2010). The GIT can be divided into five regions in terms of drug targeting: the oral cavity, oesophagus, stomach, intestine, and the colon. Each of these regions may be further subdivided for specific targeted drug delivery (Kothawade et al., 2011). Several approaches have been developed for targeted colonic drug delivery. Most of them utilize the physiological properties of the GIT and colon such as pH of GIT, transit time of small intestine, luminal pressure of the colon, and the presence of microbial flora localized in the colon (Sharma et al., 2010).

11.2   Anatomy and physiology of colon

The GI tract is divided into stomach, small intestine and large intestine (Sarasija and Hota, 2000). Human large intestine is about 1.5 m long. The colon is upper 5 ft of the large intestine and mainly situated in the abdomen (Patel et al., 2011). The colon is a cylindrical tube lined by mucosa. Colon is made up of four layers such as serosa, muscularis externa, submucosa and mucosa (Biswal et al., 2013). In terms of size and complexity, the human colon falls between that of carnivores which has no identifiable junction between ileum and colon, and herbivores which have a voluminous cecum. The human cecum is small and there is a rudimentary appendix. The human colon can be divided into three functional areas, the cecum and proximal colon, which act as a fermentation chamber, the transverse colon, the motor patterns of which may hold material in the proximal colon or propel it distally but that may also be an important site for the absorption of water and the rectum, proximal colon acts as a reservoir for fecal material and allows defecation to be delayed until socially convenient. Unlike the small intestine, the colon does not have any villi. However, because of the presence of plicae semilunares, which are crescentic folds, the intestinal surface of the colon is increased to approximately 1300 cm² (Rostad, 1973; Biswal et al., 2013).

11.3   Polymers characteristics for colon-targeted drug delivery

The selection of polymers is an important thing in case of colon-specific drug delivery. The desired properties of colon-targeted drug delivery systems can be achieved by using some polymers either alone or in a combination because it is now recognized that polymers can potentially influence the rate of release and absorption of drugs and play an important role in formulating colon-targeted drug delivery systems. The selected polymers to colon targeting should be able to withstand the pH of the stomach and small intestine. Methacrylic acid esters are most commonly used polymers for colon targeting because they are soluble at above pH 6. The ideal polymer should be able to withstand the lower pH of the stomach and of the proximal part of the small intestine but able to disintegrate at neutral or shortly alkaline pH of the terminal ileum and preferably at ileocecal junction. Biodegradable polymers are also used in CSDD because of its hydrophilic nature and limited swelling characteristic in acidic pH. Linear polysaccharides remains intact in stomach and small intestine but the bacteria of human colon degrades them and thus make them potentially useful in colon-targeted drug delivery systems. Certain polymers used for colon-specific drug delivery according to their pH range such as, Eudragit L is soluble at pH 6 or above and Eudragit S is soluble at pH 7 or above and the combination of these polymers give the desirable release rates (Rajpurohit et al., 2010).

11.4   Drug criteria for selection of colon-specific drug delivery

11.4.1   Drug candidate

Drugs which show poor absorption from the stomach and intestine including proteins and peptides are most suitable for CDDS. The drug used in treatment of IBD (Crohn's disease and ulcerative colitis), irritable bowel syndrome, colon cancer, infectious diseases (amoebiasis and helminthiasis) and diarrhoea are ideal candidates for local colon delivery. Formulations for colonic delivery are also suitable for delivery of drugs, which are polar and or susceptible to chemical and enzymatic degradation in upper GIT (Biswal et al., 2013; Patel et al., 2011).

11.4.2   Drug carrier

The selection of carrier for particular drug candidate depends on the physiochemical nature of the drug as well as the disease for which the system is to be used. The factors like chemical nature, partition coefficient and stability of drug and the type of absorption enhancers chosen influence the carrier selection. The choice of drug carrier also depends on the functional groups of drug molecule. For example, aniline or nitro groups on a drug may be used to link it to another benzene group through an azo bond. In addition to azo-polymers, disulfide bond containing polymers have been utilized as carriers for colon-specific delivery (Table 11.1) (Philip et al., 2010; Sharma et al., 2010).

Table 11.1 Drug carriers along with their suitable polymers used for colon-specific drug delivery system.

Types of drug carriers	Examples	References
Polysaccharide carriers	Chitosan, pectin, chondroitin sulphate, cyclodextrin, dextrans, guar gum, inulin, amylose, sodium alginate	Sharma et al., 2010
Bioadhesive polymers	Polycarbophils, polyurethanes and polyethylene oxidepolypropylene oxide copolymers	Patel et al., 2011
Specific membrane transporters	Integral membrane proteins, Nucleoside transporters, Bile acid transporters	Vinay kumar et al., 2011
Specific enzymes	Bacterial glycosidase	Kothawade et al., 2011

11.5   Factors affecting colon-specific drug delivery

11.5.1   Motility

Studies of colonic motility in vivo usually rely on measurement of changes in muscle electrical activity that may determine contractions. Manometer measure changes in colonic pressure caused by contractions and/or strain gauges measure contractions more directly. All approaches provide useful information but when used separately may not give a complete picture of colonic motor events. Electrical activity may not produce measurable contraction and manometric techniques can only detect contractions that occlude the lumen sufficiently to register as an increase in pressure. In-vitro measurements using strips or segments of colon may suggest mechanisms and patterns of electrical and motor activity, but their role must be assessed in vivo in an intact colon with enteric and autonomic nervous system or central nervous system connection. Since in vivo studies in human involve intubation and often bowel cleansing (sometimes with cathartics that may sensitize the colon), it is difficult to assess whether the same patterns would be seen without the invasive tubes and with a colon full of chemically and mechanically stimulating contents.

11.5.2   pH

In the stomach, pH ranges between 1 and 2 during fasting but increases after eating. There is a pH gradient in the gastrointestinal tract that is in stomach is 1.2. The pH is about 6.5 in the proximal small intestine and about 7.5 in the distal small intestine. From the ileum to the colon, pH declines significantly. However, pH values as low as 5.7 have been measured in the ascending colon in healthy volunteers. The pH in the transverse colon is 6.6 and 7.0 in the descending colon. The change in pH along the gastrointestinal tract has been used as a means for targeted colon drug delivery (Patel et al., 2011; Philip et al., 2010).

11.5.3   Colonic micro flora and their enzymes

Intestinal mucosa contain both phase I and phase II enzymes. Usually, these enzymes are derived from gut micro flora residing in high number in the colon. These enzymes are used to degrade coatings or matrices as well as to break bonds between an inert carrier and an active agent (i.e., release of a drug from a prodrug. The microflora of the colon is in the range of 1011–1012 CFU/mL, consisting mainly of anaerobic bacteria, e.g. bacteroides, bifidobacteria, eubacteria, clostridia, enterococci, enterobacteria, lactobacilli, and ruminococcus, etc. The microflora produces a vast number of enzymes such as reducing enzymes: nitroreductase, azoreducatase, hydrogenase and hydrolytic enzymes: esterase, amidase, sulfatasse, galactosidase, glucoronidase (Patel et al., 2011; Philip et al., 2010).

11.5.4   Colonic transit

The colonic transit may be important in the absorption of some poorly water soluble drugs and sustained release dosage forms. The time taken for the food to pass through the colon is the transit time and in normal subjects it is about 78 hours for expulsion of 50% ingested material, but may range from 18 hours to 144 hours. Like gastric emptying, intestinal transit is influenced by several factors such as food, drugs and diseases (Kothawade et al., 2011; Brahmankar and Jaiswal, 2009).

11.5.5   Blood flow to the GIT

The GIT is extensively supplied with blood capillary network and the lymphatic system. The absorbed drug can thus be taken up by the blood or the lymph. Since blood flow rate to the GIT is 500–1000 times more than the lymph flow, most drugs reach the systemic circulation via blood whereas only a few drugs, especially low molecular weight, lipid soluble compounds are removed by lymphatic system (Brahmankar and Jaiswal, 2009).

11.5.6   Gastric emptying time

The passage of drug from stomach to the small intestine is called gastric emptying and the time required for this process to take place is known as gastric emptying time. Longer the gastric emptying time, lesser the gastric emptying rate. Generally rapid gastric emptying increases bioavailability of a drug. Gastric emptying gets affected by a number of factors like volume, composition & viscosity of meal (Brahmankar and Jaiswal, 2009).

11.5.7   Gastrointestinal disease state

Gastrointestinal diseases such as IBD (inflammatory bowel disease), Crohn's disease, constipation, diarrhoea and gastroenteritis may affect the release and absorption of drug from colon-specific drug delivery system (Brahmankar and Jaiswal, 2009).

11.6   Rational for the development of colon-targeted drug delivery

During early stages of drug development, some new chemical entities (NCE's) present a challenge in efficacy testing due to their instability in gastric fluids and/or irritation in the GIT (Kothawade et al., 2011). To ensure direct treatment at the disease site, and fewer systemic side effects (Patel et al., 2011). Reduces conventional dose and frequency (Sharma et al., 2010). There be certain conditions which demand release of drug after a lag time, i.e, chronopharmacotherapy of diseases which shows circadian rhythms in their pathophysiology (Vinaykumar et al., 2011). Oral delivery of vaccines as the colon is rich in lymphoid tissues (Kothawade et al., 2011).

11.7   Formulation techniques

11.7.1   Granulation

Granules were prepared by using wet granulation method and by using different binder systems. Colonic tablets of metronidazole were prepared by wet granulation method using PVP K30 solution (5%w/v in IPA) as the binder (Chaudhari et al., 2012; Patel et al., 2011). Colonic tablets of nitrofurantoin were prepared by wet granulation method. Lactose was used as diluent and a mixture of talc and magnesium stearate (1:1) was used as lubricant. Guar gum was included in the formulations in various proportions as binder (Singh, 2012). Similarly colonic matrix tablets of mebendazole containing various proportions of guar gum were prepared by wet granulation technique using starch paste as a binder (Biswal et al., 2013). Intragranulation was used for Eudragit polymers in which drug, polymer and MCC were granulated with isopropyl alcohol (IPA) while in the case of extragranulation, drug and MCC were granulated with IPA and then granules were mixed with polymer. For all batches non-aqueous wet granulation method was used (Naikwade, 2008).

11.7.2   Coating

Coating is one of the simplest formulation technologies used for colon-specific delivery. Coating agents used in colon-specific drug delivery are pH sensitive polymers (Kothawade et al., 2011). Coating the peptide capsules with polymers cross linked with azo aromatic group has been found to protect the drug from digestion in the stomach and small intestine. The azo bonds are reduced in the colon and drug is released (Kumari et al., 2013). The coating techniques like fluidized bed-coating technique are used to formulate colon-specific drug delivery system.

11.7.3   Spray drying

High-Amylose cornstarch and Pectin blend microparticles of diclofencac sodium for colon-targeted delivery were prepared by spray drying technique (Choudhury et al., 2012).

11.7.4   Ionotropic gelation

Microspheres are prepared by ionotropic gelatination technique. Here, required amount of guar gum was dispersed in a specified volume of cold water containing the drug and allowed to swell for 2 hours. In another beaker suitable amount of sodium alginate was taken and mixed well with 10 ml of water. The guar gum solution containing the drug was added to sodium alginate solution with stirring to produce a viscous form. After complete mixing 1.0 ml of glutaraldehyde 7 were added to the dispersion, followed by stirring at a constant speed. Then polymer drug solution was added drop wise by using syringe of 22 G in diameter from a height of about 5 cm into a beaker containing 4%w/v solution of calcium chloride with continuous stirring by magnetic stirrer. Then the solution containing the gel formed microspheres was filtered by using Whatman filter paper no-1. The microspheres were allowed to dry at about 30–40°C and stored in well-closed container for further use (Mazumder et al., 2010). A number of techniques such as polymerization, nanoprecipitation, inverse microemulsion can be used to prepare polymeric nanoparticles; however, most of these methods involve the use of organic solvents, heat and vigorous agitation which may be harmful to the peptide and protein drugs. More recently the ionic gelation technique is used as the most favourable method for producing peptide and protein nanoparticles. The nanoparticles prepared by this method have a suitable size and surface charge, spherical morphology as well as a low polydispersity index indicative of a homogenous size distribution. The non-usage of organic solvents, sonication or harsh conditions during preparation reduces the damage to the peptide and proteins and makes this method favorable for the preparation of protein loaded nanoparticles (Kothawade et al., 2011).

11.7.5   Microencapsulation

A new microparticulate system containing budesonide was prepared by microencapsulation for colon-specific delivery (Choudhury et al., 2012).

11.7.6   Emulsification and solvent evaporation

Glutaraldehyde was used as a cross-linking agent and guar gum microspheres were prepared by emulsification method. Core microspheres of alginate were prepared by modified emulsification method in liquid paraffin and by crosslinking with calcium chloride. The core microspheres were coated with Eudragit S-100 by the solvent evaporation technique to prevent drug release in the stomach and small intestine (Biswal et al., 2013).

11.7.7　Quasi-emulsion solvent diffusion

The microsponges can also be prepared by quasi-emulsion solvent diffusion method using the different polymer amounts. To prepare the inner phase, polymer is dissolved in ethyl alcohol. Then, drug can be then added to solution and dissolved under ultrasonication at 35°C. The inner phase is poured into the PVA solution in water (outer phase). After 8 hour of stirring the microsponges were formed due to removal of ethyl alcohol from the system. The microsponges were filtered and dried at 40°C for 12 hours. Microsponges containing flubiprofen and Eudragit RS100 were prepared by quasi-emulsion solvent diffusion method (Biswal et al., 2013).

11.7.8　Entrapment

Flubiprofen was entrapped in to a commercial microsponge-5640 system using entrapment method (Choudhury et al., 2012).

11.7.9　Extrusion spheronization

Extrusion spheronization technique can be used to prepare uniform-size sturdy pellets when it is not possible to obtain mechanically strong granules by other methods. Krogars et al. utilized this technique to produce a commercially viable product of ibuprofen in a matrix base comprising of Eudragit S100 as a binder material, aqueous dispersion of hydroxypropyl methyl cellulose acetate succinate as coating material and citric acid as pH regulating agent.

Multiparticulate dosage form consisting of a hydrophobic core coated with a pH-dependent polymer is proposed for colonic specific delivery of drugs (Fatima et al., 2006).

11.8　Delivery approaches

11.8.1　pH dependent systems

This approach utilizes the existence of pH gradient in the git that increases progressively from the stomach (pH 1.5–3.5) and small intestine (5.5–6.8) to the colon (6.4–7.0). By combining the knowledge of the polymers and their solubility at different pH environments, delivery systems can be designed to deliver drugs at the target site. The most commonly used pH dependent polymers are derivatives of acrylic acid and cellulose (Patel et al., 2011).

11.8.2   Coating of the drug core with pH sensitive polymers

The intact molecule can be delivered to the colon without absorbing at the upper part of the intestine by coating of the drug molecule with the suitable polymers, which degrade only in the colon. The drug core includes tablets, capsules, pellets, granules, microparticles or nanoparticles. The coating of pH-sensitive polymers to the tablets, capsules or pellets provide delayed release and protect the active drug from gastric fluid. The polymers used for colon targeting, however, should be able to withstand the lower pH values of the stomach and of the proximal part of the small intestine and also be able to disintegrate at the neutral of slightly alkaline pH of the terminal ileum and preferably at the ileocecal junction. The majority of enteric and colon targeted delivery systems are based on the coating of tablets or pellets, which are filled into conventional hard gelatin capsules. The problem with this approach is that the intestinal pH may not be stable because it is affected by diet, disease and presence of fatty acids, carbon dioxide, and other fermentation products. Moreover, there is considerable difference in inter- and intraindividual gastrointestinal tract pH, and this causes a major problem in reproducible drug delivery to the large intestine Eudragit-L dissolves at a pH level above 5.6 and is used for enteric coating, where as Eudragit S is used for the colon delivery it dissolves at pH greater than 7.0 (attributable to the presence of higher amounts of esterified groups in relation to carboxylic groups), which results in premature drug release from the system. Problem of premature drug release can be overcome by the use of Eudragit S (Patel et al., 2011).

11.8.3   Embedding in pH-sensitive matrices

The drug molecules are embedded in the polymer matrix. Extrusion spheronization technique can be used to prepare uniform-size sturdy pellets for colon targeted drug delivery when it is not possible to obtain mechanically strong granules by other methods. Excipients had a significant impact on the physical characteristics of the pellets. Eudragit S100 as a pH sensitive matrix base in the pellets increased the pellet size and influenced pellet roundness. Citric acid promoted the pelletization process resulting in a narrower area distribution. However, EudragitS100 could not cause statistically significant delay in the drug release at lower pH (Patel et al., 2011).

11.8.4   Microbially triggered systems

11.8.4.1   Prodrug

Prodrugs can be designed to target specific enzymes or carriers by considering enzyme substrate specificity or carrier substrate specificity in order to overcome various unwanted drug properties. This type of targeted prodrug design requires considerable knowledge related to a particular enzymes or carrier systems including their molecular and functional characteristics. Targeted prodrug design discussed in two categories such as targeting specific enzymes and targeting specific membrane transporters (Vinaykumar et al., 2011).

11.8.4.2   Polysaccharides-based systems

Polysaccharides are the polymers of monosaccharides which retains their integrity because they are resistant to the digestive action of gastrointestinal enzymes. The matrices of polysaccharides are assumed to remain intact in the physiological environment of stomach and small intestine; however, they are acted upon by the bacterial polysaccharidases once they reach in the colon resulting in the degradation of the matrices. Natural polysaccharide polymers have an appeal to the area of drug delivery as they are comprised of polymers having a large number of derivatizable groups, a wide range of molecular weights, biodegradability, varying chemical compositions, a low toxicity yet high stability. They are already approved as pharmaceutical excipient. The number of polysaccharides such as amylose, guar gum, pectin, chitosan, inulin, cyclodextrins, chondroitin sulphate, dextrans and locust bean gum has been investigated for their use in colon targeted drug delivery systems. The selection of a suitable biodegradable polysaccharide is the most important factor in the development of polysaccharide derivatives for colon targeted drug delivery (Vinaykumar et al., 2011).

11.8.4.3   Azo-polymeric approach for drug delivery to colon

Newer approaches are aimed at the use of polymers as drug carriers for drug delivery to the colon. Both synthetic as well as naturally occurring polymers have been used for this purpose. Sub synthetic polymers have been used to form polymeric prodrug with azo linkage between the polymer and drug moiety. These have been evaluated for CDDS. Various azo polymers have also been evaluated as coating materials over drug cores. These have been found to be similarly susceptible to cleavage by the azoreducatase in the large bowel. Coating of peptide capsules with polymers cross linked with azoaromatic group has been found to protect the drug from digestion in the stomach and small intestine. In the colon, the azo bonds are reduced, and the drug is released; e.g. Segmented polynurethanes, Aroatic azo bond containing urethane analogues etc. (Philip et al., 2010).

11.8.5   Combined approaches

11.8.5.1   Combination of pH dependent and microbially triggered CDDS

CODES™ is a unique CDDS technology that was designed to avoid the inherent problems associated with pH or time-dependent systems. CODES™ is combined approach of pH dependent and microbially triggered CDDS. It has been developed by utilizing a unique mechanism involving lactulose, which acts as a trigger form site specific drug release in the colon. The system consists of a traditional tablet core containing lactulose, which is over coated with, acid soluble material (Eudragit E) and then subsequently over coated with an enteric material, Eudragit L as shown in (Fig. 11.1). The premise of the technology is that the enteric coating protects the tablet while it is located in the stomach and then dissolves quickly following gastric emptying. The acid soluble material coating then protects the preparation as it passage through the alkaline pH of the small intestine. Once the tablet arrives in the Colon the bacteria will enzymatically degrade the polysaccharide (lactulose) into organic acid. This lowers the pH surrounding the system sufficient to affect the dissolution of the acid soluble coating and subsequent drug release (Kumar et al., 2012).

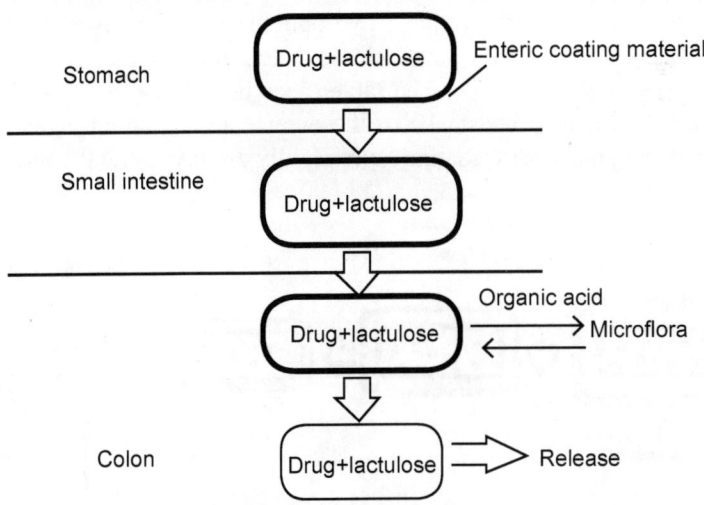

Figure 11.1 Schematics of the conceptual design of CODES

11.8.5.2   Combination of pH and time dependent systems

Appropriate combination of pH sensitivity and time release functions in a dosage form may improve the site specificity of drug delivery to the colon as discussed below:

Recently a novel pH and time controlled nano-particulte colon drug delivery systems was developed by Kshirsagar et al. He used nanoprecipitation method to prepare polymeric nanocapsules (NC) of prednisolone (PD) with pH responsive polymer Eudragit S100. The optimized formulations lead to the preparation of PD-NC with a mean size of 567.87 nm, high encapsulation efficiency of 90.21%. In vitro studies reveal that NC releases the drug after 4.5-h lag time corresponding to time to reach colonic region, and in vivo studies show that NC release drug after 3-h lag time in rat corresponds to arrival in colon (Kshirsagar et al., 2011).

11.8.6   Time-controlled systems

Time-dependent delivery has also been proposed as a means of targeting the colon. Time-dependent system releases their drug load after a pre-programmed time delay. To attain colonic release, the lag time should equate to the time taken for the system to reach the colon. This time is difficult to predict in advance, although a lag time is reported to be relatively constant at three to four hours. Enteric coated time-release press coated (ETP) tablets, are composed of three components, a drug containing core tablet (rapid release function), the press coated swellable hydrophobic polymer layer (Hydroxy propyl cellulose layer (HPC), time release function and an enteric coating layer (acid resistance function). The tablet does not release the drug in the stomach due to the acid resistance of the outer enteric coating layer. After gastric emptying, the enteric coating layer rapidly dissolved and the intestinal

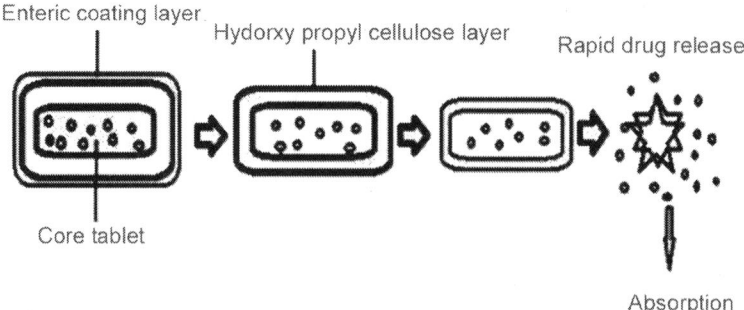

Figure 11.2 Design of enteric coated timed-release press coated tablet

fluid begins to slowly erode the press coated polymer (HPC) layer. When the erosion front reaches the core tablet, rapid drug release occurs since the erosion process takes a long time as there is no drug release period (lag phase) after gastric emptying. The duration of lag phase is controlled either by the weight or composition of the polymer layer (HPC), (Fig. 11.2) (Challa et al., 2011).

11.8.7   Pressure-controlled system

As a result of peristalsis, higher pressures are encountered in the colon than in the small intestine. Takaya et al. developed pressure controlled colon-delivery capsules prepared using ethylcellulose, which is insoluble in water. In such systems, drug release occurs following the disintegration of a water-insoluble polymer capsule because of pressure in the lumen of the colon. The thickness of the ethylcellulose membrane is the most important factor for the disintegration of the formulation. The system also appeared to depend on capsule size and density. Because of reabsorption of water from the colon, the viscosity of luminal content is higher in the colon than in the small intestine. It has therefore been concluded that drug dissolution in the colon could present a problem in relation to colon-specific oral drug delivery systems. In pressure controlled ethylcellulose single unit capsules the drug is in a liquid. Lag times of three to five hours in relation to drug absorption were noted when pressure-controlled capsules were administered to humans (Philip et al., 2010).

11.8.8   Osmotically controlled system

The OROS-CT (Alza corporation) can be used to target the drug locally to the colon for the treatment of disease or to achieve systemic absorption that is otherwise unattainable. The OROS-CT system can be a single osmotic unit or may incorporate as many as 5–6 push-pull units, each 4 mm in diameter, encapsulated within a hard gelatin capsule, each bilayer push pull unit contains an osmotic push layer and a drug layer, both surrounded by a semi permeable membrane. An orifice is drilled through the membrane next to the drug layer (Fig. 11.3). Immediately after the OROS-CT is swallowed, the gelatin capsule containing the push-pull units dissolves. Because of its drug-impermeable enteric coating, each push-pull unit is prevented from absorbing water in the acidic aqueous environment of the stomach, and hence no drug is delivered. As the unit enters the small intestine, the coating dissolves in this higher pH environment (pH > 7), compartment. Swelling of the osmotic push compartment forces drug gel out of the orifice at a rate precisely controlled by the rate of water transport through the semi permeable membrane. For treating ulcerative colitis, each push pull unit is designed with a 3–4 h post gastric delay to prevent drug delivery in the small intestine. Drug release begins when the unit reaches the colon. OROS-CT units can maintain a constant release rate for up to 24 hours in the colon or can deliver drug over a period as short as four hours. Recently, new phase transited systems have come which promise to be a good tool for targeting drugs to the colon. Various in-vitro/in-vivo evaluation techniques have been developed and proposed to test the performance and stability of CDDS. GI pressure is another mechanism that is utilised to initiate the release drug at distal part of GUT (Vinaykumar et al., 2011).

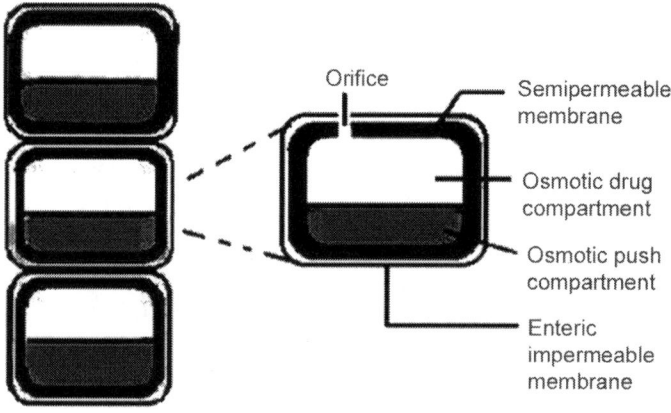

Figure 11.3 Schematic diagrams of different parts of OROS-CT system

11.8.9   Pulsatile drug delivery system

A pulsatile drug release, where the drug is released rapidly after a well-defined lag-time, could be advantageous for many drugs or therapies. Pulsatile release systems can be classified in multiple-pulse and single-pulse systems. A popular class of single-pulse systems is that of rupturable dosage forms. Other systems consist of a drug-containing core, covered by a swelling layer and an outer insoluble, but semi permeable polymer coating or membrane. The lag time prior to the rupture is mainly controlled by: (i) the permeation and mechanical properties of the polymer coating and (ii) the swelling behaviour of the swelling layer. As is frequently found in the living body, many vital functions are regulated by pulsed or transient release of bioactive substances at a specific site and time. Thus it is important to develop new drug delivery systems to achieve pulsed delivery of a certain amount of drugs in order to mimic the function of the living systems, while minimizing undesired side effects. Special attention has been given to the thermally responsive pol (Nisopropylacrylamide) and its derivative hydrogels. Thermal stimuli-regulated pulsed drug release is established through the design of drug delivery devices, hydrogels, and micelles. Therefore, pulsatile drug delivery is one such system that, by delivering drug at the right time, right place and in right amounts, holds good promise of benefit to the patients suffering from chronic problems like arthritis, asthma and hypertension. There are various methods for pulsatile drug delivery system such as capsular system, osmotic system, solubilisation or, erosion of membrane, rupture of membrane (Vinaykumar et al., 2011).

11.8.10   Port system

The Port® system consists of a capsule coated with a semi-permeable membrane. Inside the capsule was an insoluble plug consisting of osmotically active agent and the drug formulation. When this capsule come in contact with the dissolution fluid, the semi permeable membrane allow the entry of water leading to the pressure development inside the capsule and the insoluble plug expelled after a lag time (Fig. 11.4). The dosage form is designed in such a manner that upon ingestion, the first drug release pulse occurs within 1–2 h, followed by period during which no release occurs. Second dose is released in 3–5 h of ingestion. This is again followed by a second no-release interval. Release of third dose occurs within 7–9 h of ingestion. This system avoids the second time dosing (Kothawade et al., 2011).

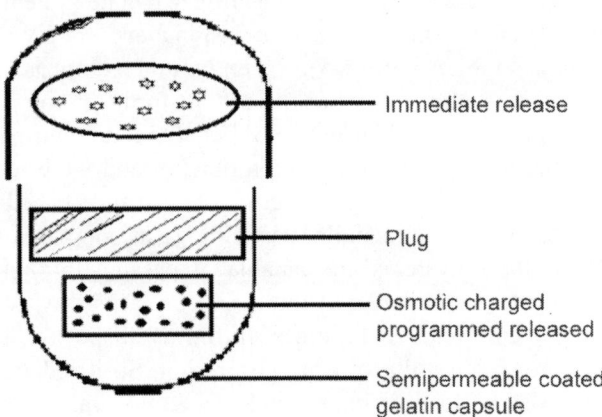

Figure 11.4 Plan of Port® system

11.8.11   Bioadhesive systems

Oral administration of some drugs requires high local concentration in the large intestine for optimum therapeutic effects. Bioadhesion is a process by which a dosage form remains in contact with particular organ for an augmented period of time. This longer residence time of drug would have high local concentration or improved absorption characteristics in case of poorly absorbable drugs. This strategy can be applied for the formulation of colonic drug delivery systems. Various polymers including polycarbophils, polyurethanes and polyethylene oxide polypropylene oxide copolymers have been investigated as materials for Bioadhesive systems. Bioadhesion has been proposed as a means of improving the performance and extending the mean residence time of colonic drug delivery systems (Patel et al., 2011).

11.8.12   Multiparticulate systems

Single unit colon targeted drug delivery system may suffer from the disadvantage of unintentional disintegration of the formulation due to manufacturing deficiency or unusual gastric physiology that may lead to drastically compromised systemic drug bioavailability or loss of local therapeutic action in the colon. Report suggests that drug carrier systems larger than 200 μm possess very low gastric transit time due to physiological condition of the bowel in colitis. And for this reason and considering the selective uptake of micron or submicron particles by cancerous and inflamed cells/tissues a multiparticulate approach is expected to have better pharmacological effect in the colon. Multiparticulate system consists of granule core which is coated with film coating and finally these get converted into matrix tablet (Fig. 11.5). Recently, much emphasis is being laid on the development of multiparticulate dosage forms in comparison to single unit systems because of their potential benefits like, multiparticulate systems enabled the drug to reach the colon quickly and were retained in the ascending colon for a relatively long period of time and hence increased bioavailability. Because of their smaller particle size as compared to single unit dosage forms these systems are capable of passing through the GI tract easily, leading to less inter- and intra subject variability. Moreover, multiparticulate systems tend to be more uniformly dispersed in the GI tract and also ensure more uniform drug absorption. So it reduces risk of systemic toxicity, risk of local irritation and predictable gastric emptying. Multiparticulate system includes micro particulate systems, microspheres, nanoparticles (Patel et al., 2011; Biswal et al., 2013).

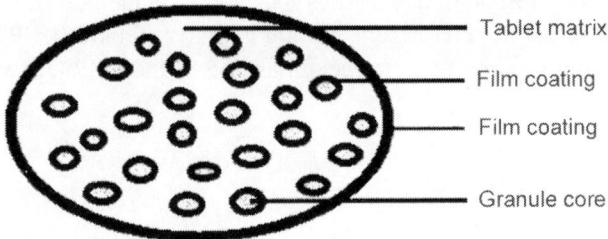

Figure 11.5 Design of multiparticulate system

11.9   Limitations and challenges

One challenge in the development of colon-specific drug delivery system is to establish an appropriate dissolution method in designing in-vitro system. Due to the rationale after a colon delivery system is quite diverse. As, a site for delivery offers a near neutral pH, reduced digestive enzymes activity, a long transit time, and increased responsiveness to absorption enhancers, hence targeting is complicated, with reliability and delivery efficiency. There is less free fluid in the colon than in the small intestine and, hence dissolution could be problematic for poorly water-soluble drugs. In such instances, the drug may need to be delivered in a presolubilized form, or delivery should be directed to the proximal colon, as a fluid gradient exists in the colon with more free water present in the proximal colon than in the distal colon (Vinaykumar et al., 2011; Sharma et al., 2010). In addition, the stability of the drug is also a concern and must be taken into consideration while designing the delivery system. The drug may potentially bind in a nonspecific way to dietary residues, intestinal secretions, mucus or faecal matter. Colonic residence time as commanded by GIT motility. The resident microflora could also affect colonic performances via metabolic degradation of drug. The drug transport across the mucosa and also into the systemic circulation get restricted by lower surface area and relative 'tightness' of the tight junction in the colon (Choudhury et al., 2012).

11.10   Evaluation techniques

Different in vitro and in vivo methods are used to evaluate different carrier systems for their ability to deliver drugs specifically to the colon. The ability of the coats or carriers to remain intact in stomach and small intestine is generally assessed by conducting drug release studies in 0.1N hydrochloric acid for 2 hours followed by phosphate buffer (pH −7.4) for 3 h by using dissolution apparatus. The drug release studies may also be performed by using rat cecal contents. Presently dissolution apparatus such as basket method (USP APP-I), paddle method (USP APP-II), reciprocating cylinder method (USP APP-III) and flow-through cell method are recommended in the USP (Biswal et al., 2013).

11.10.1   *In vitro* evaluation

No standardized evaluation technique is available for evaluation of CDDS because an ideal in vitro model should posses the in vivo conditions of GIT such as pH, volume, stirring, bacteria, enzymes, enzyme activity and other components of food. Generally these conditions are influenced by the diet and physical stress and these factors make it difficult to design a slandered in vitro model. The in-vitro models used for CDDS are as follows (Patel et al., 2011).

11.10.1.1   In-vitro dissolution test

In order to stimulate the pH changes along the gastrointestinal tract, for dissolution media with pH 1.2, pH 6, pH 6.8, and pH 7.2 were sequentially used, referred to as the sequential pH change method. When performing release experiments, the pH 1.2 medium was used for first for 2 h, and then removed, and fresh pH 6 dissolution medium was added. After 1 h, the medium was removed again, and fresh pH 6.8 dissolution medium was and removed after 2 h and finally pH 7.2 was added. Finally the amount of drug released at different time intervals was estimated (Patel et al., 2011).

11.10.1.2   In-vitro enzymatic test

This can be performed by two tests as follows: incubate carrier drug system in fermenter-containing suitable medium for bacteria (Streptococcus faccium or B.ovatus) amount of drug released at different time intervals determined. Drug release study is done in buffer medium containing enzymes (enzyme pectinase, dextranase), or rat/guinea pig/rabbit caecal contents. The amount of drug released at different time intervals during the incubation is estimated to find out the degradation of the carrier under study (Patel et al., 2011).

11.10.2   In-vivo evaluation

When the system design is conceived and prototype formulation with acceptable in vitro characteristics is obtained, in vivo studies are usually conducted to evaluate the site specificity of drug release and to obtain relevant pharmacokinetics information of the delivery system. Although animal models have obvious advantages in assessing colon-specific drug delivery systems, human subjects are increasingly utilized for evaluation of this type of delivery systems with visualization techniques such as γ-scintigraphy imaging.

11.10.2.1   Animal studies

Different animals such as rats, pigs and dogs have been used to evaluate the performance of CSDD system. To closely simulate the human physiological environment of the Colon; the selection of an appropriate animal model for evaluating a colon-specific delivery system depends on its triggering mechanism and system design. For instance, guinea pigs have comparable glycosidase and glucuronidase activities in the colon and similar digestive anatomy and physiology to that of human, so they are more suitable in evaluating glucoside and glucuronate conjugated prodrugs intended for colon delivery. Even though guinea pig is the preferred animal model to investigate the in vivo performance of certain colon-specific delivery systems, it is difficult to administer the delivery system. More often, gastric intubation has to be utilized. Rats were also used to evaluate colon-specific drug delivery systems based on azo-polymers or prodrugs containing azo bonds because the distribution of azoreductase activity in GI tract is similar between rats and human subjects. The in vivo performance of CODES™ was evaluated in beagle dogs using acetaminophen as a model drug and lactulose as the matrix-forming excipient in the core tablet. The ability of intestinal PCDCs to obtain colon-specific delivery was also investigated in beagle dogs. It is well recognized that significant differences exist between human subjects and commonly used laboratory animals in GI tract anatomy and physiology, including GI transit time, pH, distribution of enzyme activity, population of bacteria, etc. Therefore, the data obtained from animal models should be interpreted with caution. In the case of evaluating Colon-specific drug delivery systems, the success of a colon-specific delivery system will be primarily decided by the accomplishment of in vivo drug release in the desired location (i.e. Colon). Therefore, this event can only be ascertained through visualization (Biswal et al., 2013; Choudhury et al., 2012).

11.10.2.2   Studies in human

A variety of techniques like (i) string technique, (ii) endoscopy, (iii) roentgenography were used for monitoring the in vivo behaviour of the oral dosage forms.

11.10.3   String technique

In these studies, a tablet was attached to a piece of string and the subject swallowed the tablet, leaving the free end of the string hanging from his mouth. At various time points, the tablet was withdrawn from the stomach by pulling out the string and physically examining the tablet for signs of disintegration. In some studies, the tablets were recovered by inducing a vomiting reflex.

The presence of foreign object, such as string in the GIT may alter its motility and the physicochemical environment. The psychological stress and anxiety associated with this method may affect the motility of the GI tract (Kumari et al., 2013).

11.10.4   Endoscopy technique

It is an optical technique in which a fibre is used to directly monitor the behaviour of the dosage form after ingestion. This method requires administration of a mild sedative to facilitate the swallowing of the endoscope tube. The sedative itself may alter gastric emptying and GI motility. The physiological factor also contributes to the changes in the motility of the GI tract (Kumari et al., 2013).

11.10.5   Roentgenography

The inclusion of radio-opaque material into the solid dosage form enables it to be visualized by the use of X-rays. By incorporating Barium sulphate into the pharmaceutical dosage forms, it is possible to follow the movement, location and the integrity of the dosage form after oral administration by placing the subject under a fluoroscope and taking a series of X-rays at various time points (Choudhury et al., 2012; Kumari et al., 2013).

11.10.6   Drug delivery index (DDI) and clinical evaluation

DDI is a calculated pharmacokinetic parameter, following single or multiple dose of oral colonic prodrugs. DDI is the relative ratio of RCE (Relative colonic tissue exposure to the drug) to RSC (Relative amount of drug in blood i.e. that is relative systemic exposal to the drug). High drug DDI value indicates better colon drug delivery (Kumar et al., 2012).

DDI can be determine by Eq. (11.1) as below,

$$\text{Drug Delivery Index} = \frac{\text{RCE (Relative colonic tissue exposure to the drug)}}{\text{RSC (Relative amount of drug in blood)}} \quad (11.1)$$

Absorption of drugs from the colon is monitored by colonoscopy and intubation. Currently gamma scintigraphy, high frequency capsules and radio-telemetry are the most preferred techniques employed to evaluate colon drug delivery systems.

11.10.6.1   High frequency capsule

Smooth plastic capsule containing small latex balloon, drug and radiotracer taken orally. Triggering system is high frequency generator. Release of drug and radiotracer triggered by an impulse, the release is monitored in different parts of GIT by radiological localization. It checks the absorption properties of drug in colon (Patel et al., 2011).

11.10.7   Gammascintigraphy

It is the most useful technique which requires the presence of γ-emitting radioactive isotope in the dosage form that can be detected in vivo by an external gamma camera. The dosage form can be radio-labelled using conventional labelling or neutron activation methods. It provides information regarding time of arrival of a colon-specific drug delivery system in the colon, times of transit through the stomach and small intestine, and disintegration. Information about the spreading or dispersion of a formulation and the site at which release from it takes place can also be obtained. It can also provide information about regional permeability in the colon. Information about gastrointestinal transit and the release behaviour of dosage forms can be obtained by combining pharmacokinetic studies and gammascintigraphic studies i.e. pharmacoscintigraphy (Patel et al., 2011).

11.10.8   Radio-telemetry

This technique involves the administration of a capsule that consists of a small pH probe interfaced with a miniature radio transmitter which is capable of sending a signal indicating the pH of the environment to an external antenna attached to the body of the subject (Kumari et al., 2013).

11.11   Pharmaceutical applications

11.11.1   Local action

Corticosteroids such as hydrocortisone and prednisolone are administered via the rectum for the treatment of ulcerative colitis. Irritable bowel syndrome (IBS): abdominal pain or cramping, a bloated feeling, flatulence, diarrhoea or constipation people with IBS may also experience alternating bouts of constipation & diarrhoea, mucus in stool. Drugs for IBS are Dicyclomine, Hyoscine, Propantheline, Cimetropium, Alosetron, Tegaserod (Philip and Philip, 2010).

11.11.2   Systemic action

Chronotherapy, prophylaxis of colon cancer and treatment of nicotine addiction. There be certain conditions which demand release of drug after a lag time, i.e. chronopharmacotherapy of diseases which shows circadian rhythms in their pathophysiology. Diseases such as asthma, angina, hypertension, cardiac arrhythmias, arthritis or inflammation are affected by the circadian biorhythms and these diseases require night-time or early morning onset of drug action. It is therefore highly desirable to have a delayed-release delivery system that can provide nocturnal release of a drug, which in turn may provide considerable relief to the patients while they are resting (Vinaykumar et al., 2011; Singh et al., 2002). The brief description about the drugs used for various colonic disorders as depicted in Table 11.2 and updated patenting systems for colon-specific drug delivery systems is shown in Table 11.3.

Table 11.2 List of drugs used for the different colonic disorders

Disease name	Drugs used	Type of dosage form	Minimum and maximum dose	References
Inflammatory bowel disease	Mesalamine	Eudragit-S coated tablets (dissolves at pH 7)	0.8–2.4 g/day	Fatima et al., 2006
	Mesalamine	Eudragit-L coated tablets (dissolves at pH 6)	1.0–4.0 g/day	Patel et al., 2011
	Budesonide	Eudragit-L coated beads	9 mg/day	
Colorectal cancer	5-Flourouracil	Guar gum matrix tablet	0.83 mg/kg	Philip; Philip B, 2010
Arthritis	NSAIDs like Fenoprofen calcium	Coated tablets of Fenoprofen calcium	250 mg/day	Potu A et al., 2011
Amoebiasis	Metronidazole	Guar gum matrix tablet	50 mg/day	Sharma et al., 2011

Table 11.3 Updated patent literature for colon-specific drug delivery system.

Title of patent	Purpose of invention	Major claims	Patent no. and approval year
Colonic delivery of adsorbents	Particulate delivery system including encapsulated adsorbent and method of preparation	The method of claim 1, wherein the adsorbent is selected from the group consisting of activated charcoal, clays, talc, silica, and resins	US8,106,000 B2 and 2012

Contd..

Contd..

Title of patent	Purpose of invention	Major claims	Patent no. and approval year
Colonic drug delivery formulation	Delayed release formulation with a core comprising a drug and a delayed release coating	A formulation as claimed in any of the preceding claims, wherein the thickness of the coating as measured by the theoretical weight gain (TWG) of the coated formulation is from 5% to 10%.	EP 2 018159B1 and 2012
Drug for treatment of colon cancer	Invention of a method for treatment, prevention of hyperproliferative disease	The method according to claim 1, in which hyperproliferative disease is cancer	US2011/0257270A1 and 2011
Multiparticulate pharmaceutical formulation for colon absorption	Development of multiparticulate formulation for colon targeting	The composition of claim 1, wherein the pharmaceutically active agent is an opoid.	US2012/0328707 A1 and 2012
Indigestible polymer: starch acetate-based film coatings for colon targeting	A colon-targeted delivery dosage form for controlled release of an active ingredient	The colon-targeted delivery dosage form according to claim 1, wherein the polymer mixture comprises a plasticizer	US2013/0078289 A1 and 2013
Colon disease targets and uses thereof	A method for diagnosing and detecting diseases associated with colon	A composition comprising the protein of claim 1 and a pharmaceutically acceptable carrier	US2013/0111613 A1 and 2013
Colon and pancrease cancer peptidomimetics	Use of Colon and pancrease cancer peptidomimetics in diagnostic therapeutic methods	A fusion protein comprising the polypeptide of any one of claim 1, 2 and 3	WO2012/040617 A2 and 2012
Compositions and methods for the controlled repopulation of gram-negative bacteria in the colon	Development of method for controlled repopulation of the colon of a subject who has undergone selective digestive contamination	The E.coli strain or method according to claim 9, wherein the amyloglycoside is neomycin	WO2012/085241 A1 and 2012

11.12   Conclusions

The colonic region of the GIT has become an increasingly important site for drug delivery and absorption. CDDS offers considerable therapeutic benefits to patients in terms of both local and systemic treatment. The release of drug load in colon region is depended on pH of GIT, gastro intestinal transit time and microbial flora and their enzymes to degrade coated polymers and breaking bonds between carrier molecule and drug molecule. The preferred CDDS is that should release maximum drug load in colon region. Successful colonic delivery could be achieved by protecting the drug from absorption and /or the environment of the upper GIT and then be abruptly released into the proximal colon, which is considered the optimum site for colon targeted delivery of drugs. The various strategies for targeting orally administered drugs to the colon include coating with pH-sensitive polymers, formulation of timed released systems, exploitation of carriers that are degraded specifically by colonic bacteria, bioadhesive systems. Among different approaches the pH dependent system is less suitable than others due to the large inter and intra subject variation in the gastro intestinal pH, but gives better results with combination of time-dependent system, microbially activated system and others. Different polymers are used to prepare CDDS by various approaches and are evaluated for their efficiency and safety.

11.13   Current and future scope

Currently, there are several modified release solid formulation technologies available for colonic delivery. These technologies rely on GI pH, transit times, enterobacteria and luminal pressure for site-specific delivery. Each of these technologies represents a unique system in terms of design but has certain shortcomings, which are often related to degree of site-specificity, toxicity, cost and ease of scale up/manufacturing. It appears that microbially controlled systems based on natural polymers have the greatest potential for colonic delivery, particularly in terms of site-specificity and safety. In this regard, formulations that employ a film coating system based on the combination of a polysaccharide and a suitable film forming polymer represents a significant technological advancement. Further developments in this area require means to develop the co-processing of the polymeric blend of a polysaccharide(s) and a film forming material while maintaining the propensity of the composition to microbial degradation in the colon. Recent reports indicate interest in colon as a site where poorly absorbed drug molecules may have improved bioavailability. The distal colon is considered to have less hostile environment as well as enzyme activity compared to stomach and small intestine. The development of a dosage form that improves the oral absorption of peptide and protein drugs whose bioavailability is very low because of instability in the GI tract (due to pH or enzymatic degradation) is one of the greatest challenges for oral peptide delivery in the pharmaceutical field. Colon targeted multiparticulate systems like microspheres and nanoparticles can provide a platform for spatial delivery of candidates like peptides, proteins, oligonucleotides and vaccines. However, drug release is not the end point of oral delivery. The bioavailability of protein drugs delivered at the colon site needs to be addressed. The use of drug absorption enhancers into the drug delivery systems is likely to enhance therapeutic efficacy. Studies on drug absorption by the intestinal system have focused on drug transporters that mediate drug influx and efflux and agents which can enhance drug absorption. The colon segment is designed by nature mainly to expel metabolism products rather than to absorb nutrients. Therefore, more research that is focused on the specificity of drug uptake at the colon site is necessary. Such studies will be significant in advancing the cause of colon targeted delivery of therapeutics in future.

11.14   References

1. Biswal P.K., Kumar A. and Bhadouriya A.S. Design and evolution of colon specific drug delivery system. *Int. J. Pharm. Chem. Biolog. Sci.,* 2013; 3(1): 150–167.

2. Brahmankar D.M. and Jaiswal S.B. Biopharmaceutics and Pharmacokinetics-A Treatise. 2009; 2nd edn., 67–78.

3. Challa T., Vynala V., Allam K.V. Colon specific drug delivery systems: a review on primary and novel approaches. *Int. J. Pharm. Sci. Rev. Res.* 2011; 7(2): 171–181.

4. Chaudhari P.S., Slunkhe K.S., Amrutkar P.P., Chaudhari S.V., Oswal R.J. Formulation and development of colon specific drug delivery using dextrin. *Int J Pharm Bio Sci.* 2012; 3(1): 269–276.

5. Choudhury P.K., Panigrahi T.K., Murthy P.N., Tripathy N.K., Behera S., Panigrahi R. Novel Approaches and Developments in Colon Specific Drug Delivery Systems-A Review. *Webmed Cent Pharm Sci.* 2012; 3(2): 1–20. WMC003114.

6. Fatima L., Asghar A., Chandran S. Multiparticulate Formulation Approach to Colon Specific Drug Delivery: Current Perspectives. *J Pharm Pharmaceut Sci.* 2006; 9(3): 327–338.

7. Kothawade P.D., Gangurde H.H., Surawase R.K., Wagh M.A., Tamizharasi S. Conventional and novel approaches for colon specific drug delivery: A review. *J Sci. Tech.* 2011; 6(2): 33–56.

8. Kshirsagar S.J., Bhalekar M.R., Patel J.N., Mohapatra S.K., Shewale N.S. Preparation and characterization of nanocapsules for colon-targeted drug delivery system. *Pharm Dev Technol.* In press 2011.

9. Kumar S.P., Prathibha D., Parthibarajan R., Reichal C.R. Novel colon specific drug delivery system: a review. *Int J Pharm Sci.* 2012; 4(1): 22–30.

10. Kumari A., Syal P., Kumar S. Colon targeted drug delivery system: a review. Novel Sci. *Int. J. Pharm. Sci.* 2013; 2(1–2): 9–12.

11. Vinaykumar K.V., Sivakumar T., Tamizhmani T., Sundar Rajan T., Sarath Chandran I. Colon targeting drug delivery system: A review on recent approaches. *Int J Pharm Biomed Sci.* 2011, 2(1): 11–19.

12. Mazumder R., Nath L.K., Haque A., Maity T., Choudhury P.K., Shrestha B., Chakraborty M., Pal R.N. Formulation and in vitro evaluation of natural polymers based microspheres for colonic drug delivery. *Int J Pharm Pharmaceu Sci.* 2010; 2(1): 211–219.

13. Naikwade S.R., Kulkarni P.P., Jathar S.R., Bajaj A.N. Development of time and pH dependent controlled release colon specific delivery of tinidazole. DARU. 2008; 16(3): 119–127.

14. Patel A., Bhatt N., Patel K.R., Patel N.M., Patel M.R. Colon targeted drug delivery system: a review system. JPSBR. 2011; 1(1): 37–49.

15. Patel G.N., Patel R.B., Patel H.R. Formulation and in-vitro evaluation of microbially triggered colon specific drug delivery using sesbania gum. *J Sci. Tech.* 2011; 2(6): 33–45.

16. Philip A.K., Philip B. Colon Targeted Drug Delivery Systems: A Review on Primary and Novel Approaches. *Oman Med. J.* 2010; 25(2): 70–78.

17. Potu A., Pasunooti S., Reddy P., Reddy V., Burra S. Formulation and evaluation of fenoprofen calcium compressed coated tablets for colon specific drug delivery. *Asi. J. Pharm. Clin. Res.* 2011; 4(2): 88–95.

18. Rajpurohit H., Sharma P., Sharma S. and Bhandari A. Polymers for Colon Targeted Drug Delivery. *Ind J Pharm Sci.* 2010; 72(6): 689–696.

19. Rostad H. Acta Physiol, Scand. 1973; 89:91.

20. Sarasija S, Hota A. Colon-specific drug delivery systems. *Ind J. Pharm. Sci.* 2000; 62(1): 1–8.

21. Sharma A., Jain K Amit. Colon targeted drug delivery using different approaches. IJPSR. 2010; 1(1): 60–66.

22. Sharma M., Joshi B., Bansal M., Goswami M. The colon specific delivery system: The local drug targeting. *Int. Res. J. Pharm.* 2011; 2(12): 103–107.

23. Singh B.N., Kim K.H., In: Swarbrick J., Boylan J.C. (Eds.), (2002). Encyclopedia of Pharmaceutical Technology. New York, Marcel Dekker, Inc, 886–909.

24. Singh K.I., Sharma D., Singh J., Sharma A. Colon specific drug delivery system: review on novel approaches. *Int. J. Nat. Product Sci.* 2012; 1: 232.

25. Singh R. Formulation and evaluation of colon targeted drug delivery. *Int. J. Pharm. Life Sci.* 2012; 3(12): 2265–2268.

26. Sekhri, R. Compositions and methods for the controlled repopulation of gram-negative bacteria in the colon. WO 085241 A1, June 28, 2012.

27. Teskin, R.L. Colon and pancrease cancer peptidomimetics. WO 040617 A2, March 29, 2012.

28. Yeounjin K., Tao H.E., Steve R. Colon disease targets and uses thereof. US 0111613 A1, May 2, 2013.

29. Juargen S., Youness K., Guerin D., Jerome K., Daniel W. Indigestible polymer: starch acetae-based film coatings for colon targeting. US 0078289 A1, March 28, 2013.

30. Hermann L.H., Schindellegi. Multiparticulate pharmaceutical formulation for colon absorption. US 0328707 A1, Dec 27, 2012.

31. Klopman G., Sarasota F.L., Chakravati S.K., Beachwood O.H. Drug for treatment of colon cancer. US 0257270A1, Oct 20, 2011.

32. Huguet H.C., Fattal E., Andremont A., Tsapis N. Colonic delivery of adsorbents. US 8,106,000 B2, Jan 31, 2012.

33. Abdul, W., Valentine, C. Colonic drug delivery formulation. EP 2 018159B1, Jun 27, 2012.

Fast dissolving films as drug delivery system

Satya Prakash Singh[1], **Chinam Niranjan Patra**[2], **Suryakanta Swain**[2] **and**
Anup Kumar Sirbaiya[1]

[1]Faculty of Pharmacy, Integral University, Lucknow-226026, India.
[2]Department of Pharmaceutics, Roland Institute of Pharmaceutical Sciences,
Berhampur, Odisha-760010, India.

12.1   Introduction

The oral route is one of the most preferred routes of drug administration as it is more convenient, cost effective, and ease of administration lead to high level of patient compliance. The oral route is problematic because of the swallowing difficulty for paediatric and geriatric patients who have fear of choking. Patient convenience and compliance oriented research has resulted in bringing out safer and newer drug delivery systems. Recently, fast dissolving drug delivery systems have started gaining popularity and acceptance as one such example with increased consumer choice, for the reason of rapid disintegration or dissolution, self-administration even without water or chewing. Fast dissolving drug delivery systems were first invented in the late 1970s as to overcome swallowing difficulties associated with tablets and capsules for paediatric and geriatric patients. Buccal drug delivery has lately become an important route of drug administration. Various bioadhesive mucosal dosage forms have been developed, which includes adhesive tablets, gels, ointments, patches, and more recently the use of polymeric films for buccal delivery, also known as mouth-dissolving films.

The surface of buccal cavity comprises of stratified squamous epithelium which is essentially separated from the underlying tissue of lamina propria and submucosa by an undulating basement membrane [1]. It is interesting to note that the permeability of buccal mucosa is approximately 4–4,000 times greater than that of the skin, but less than that of the intestine [2]. Hence, the buccal delivery serves as an excellent platform for absorption of molecules that have poor dermal penetration [3]. The primary barrier to permeability in otiral mucosa is the result of intercellular material derived from the so-called 'membrane coating granules' present at the uppermost 200 µm layer [4]. These dosage forms have a shelf life of 2–3 years, depending on the active pharmaceutical ingredient but are extremely sensitive to environmental moisture [5].

An ideal fast dissolving delivery system should have the following properties: high stability, transportability, ease of handling and administration, no special packaging material or processing requirements, no water necessary for application, and a pleasant taste. Therefore, they are very suitable for pediatric and geriatric patients; bedridden patients; or patients suffering from dysphagia, Parkinson's disease, mucositis, or vomiting. This novel drug delivery system can also be beneficial for meeting current needs of the industry. Fast dissolving films (FDF) were initially introduced in the market as breath fresheners and personal care products such as dental care strips and soap strips.

However, these dosage forms are introduced in the United States and European pharmaceutical markets for therapeutic benefits. The first of the kind of oral strips (OS) were developed by the major pharmaceutical company Pfizer who named it as Listerine® pocket packs™ and were used for mouth freshening. Chloraseptic® relief strips were the first therapeutic oral thin films (OTF) which contains benzocaine and were used for the treatment of sore throat [6]. Formulation of fast dissolving buccal film involves material such as strip-forming polymers, plasticizers, active pharmaceutical ingredients, sweetening agents, saliva stimulating agents, flavouring agents, colouring agents, stabilizing and thickening agents, permeation enhancers, and superdisintegrants. All the excipients used in the formulation of fast dissolving film should be approved for use in oral pharmaceutical dosage forms as per regulatory perspectives [7].

12.2   Fast dissolving films

Fast dissolving films, a new drug delivery system for the oral delivery of the drugs, was developed based on the technology of the transdermal patch. This delivery system consists of a very thin oral strip, which is simply placed on the patient's tongue or any oral mucosal tissue, instantly wet by saliva the film rapidly hydrates and adheres onto the site of application. It then rapidly disintegrates and dissolves to release the medication for oromucosal absorption or with formula modifications, will maintain the quick-dissolving aspects allow for gastrointestinal absorption to be achieved when swallowed. In contrast to other existing, rapid dissolving dosage forms, which consist of liophylisates, the rapid films can be produced with a manufacturing process that is competitive with the manufacturing costs of conventional tablets [8–10].

12.2.1   Special features of fast dissolving films

- Thin elegant film
- Available in various size and shapes
- Unobstructive
- Excellent mucoadhesion
- Fast disintegration
- Rapid release

The fast dissolving films has also a clear advantage over the oral dissolving tablets (ODTs):

- ODTs are sometimes difficult to carry, store and handle (fragility and friability).
- Many ODTs are prepared by using the expensive lyophillisation process.

A large number of drugs can be formulated as mouth dissolving films. Innovative products may increase the therapeutic possibilities in the following indications.

- Pediatrics (antitussives, expectorants, antiasthamatics)
- Geriatrics (antiepileptic, expectorants)
- Gastrointestinal diseases
- Nausea (e.g. due to cytostatic therapy)
- Pain (e.g. migraine)
- CNS (e.g. antiparkinsonism therapy) [11]

12.2.2   Classification of oral films

There are three different subtypes of oral films such as flash release, mucoadhesive melt-away wafer and mucoadhesive sustained-release wafers. Types of oral films and their properties are described in Table 12.1.

Table 12.1 Different types of oral films and their properties

Property/sub type	Flash release	Mucoadhesive melt-away wafer	Mucoadhesive sustained release wafer
Area (cm^2)	2–8	2–7	2–4
Thickness [μm]	20–70	50–500	50–250
Structure	Film: single layer	Single or multilayer	Multilayer system
Excipients	Soluble, highly hydrophilic polymers	Soluble, hydrophilic polymers	Low/non-soluble polymers
Drug phase	Solid solution	Solid solution or suspended drug particles	Suspension and/or solid solution
Application	Tongue (upper palate)	Gingival or buccal Region	Gingival, (other region in the oral cavity)
Dissolution	Maximum 60 seconds	Disintegration in a few minutes, forming gel	Maximum 8–10 hours

12.2.3   Composition of the fast dissolving oral film system

Mouth dissolving film is a thin film with an area of 5–20 cm^2 containing an active ingredient. The immediate dissolution, in water or saliva respectively, is reached through a special matrix from water-soluble polymers. Drugs can be incorporated upto a single dose of 15 mg. Formulation considerations (plasticizers etc.) have been reported as important factors affecting mechanical properties of the films, such as shifting the glass transition temperature to lower temperature.

Components of mouth dissolving film includes: (Table 12.2)

Table 12.2 A typical composition contains the following components

Agents	Concentration
(i) Drug	1–25%
(ii) Water soluble polymer	40–50%
(iii) Plasticizers	0–20%
(iv) Fillers, colours, flavours etc.	0–40%

12.2.3.1   Active pharmaceutical agents

Active pharmaceutical substance can be from any class of pharmaceutically active substances & that can be administered orally or through the buccal mucosa. For the effective formulation, dose of drug should be in mgs (less than 20 mg/day). Various categories of drugs such as antiemetic, neuroleptics, cardiovascular agents, analgesics, antiallergic, antiepileptic, anxiolytics, sedatives, hypnotics, diuretics, antiparkinsonism agents, anti-bacterial agents and drugs used for erectile dysfunction, antialzheimers, expectorants and anitussive can be used as active pharmaceutical agent for fast dissolving formulations [12–19]. The ideal characteristics of a drug to be selected are as follows:

• The drug should have pleasant taste.
• The drug to be incorporated should have low dose generally less than 30 mg.
• The drugs with smaller and moderate molecular weight should be preferable.
• The drug has should be stable and soluble in water as well as in saliva.
• It should be partially unionized at the pH of oral cavity.
• It should have the ability to permeate oral mucosal tissue

Water-soluble polymers

Water-soluble polymers are used as film formers. The use of film forming polymers in dissolvable films has attracted considerable attention in medical and nutraceutical application. The water-soluble polymers achieve rapid disintegration, good mouthfeel and mechanical properties to the films. The disintegration rate of the polymers is decreased by increasing the molecular weight of polymer film bases. Some of the water soluble polymers used as film former are HPMC (hydroxy propyl methyl cellulose) E-3 and K-3, Methyl cellulose A-3, A- 6 and A-15, Pullulan, Carboxy methyl cellulose cekol 30, Polyvinyl pyrollidone PVP K-90, Pectin, Gelatin, Sodium alginate, Hydroxy propyl cellulose, Polyvinyl alcohol (PVA), Maltodextrins and Eudragit-RD10. . Polymerized rosin is a novel film-forming polymer [19, 20].

Plasticizers

Formulation considerations (plasticizer, etc.) have been reported as important factors affecting mechanical properties of films. The mechanical properties such as tensile strength and elongation to the films have also been improved by the addition of plasticizers. Variation in their concentration may affect these properties. The commonly used plasticizers are glycerol, dibutylpthallate, and polyethylene glycols, etc.

Surfactants

Surfactants are used as solublising or wetting or dispersing agent so that the film is getting dissolved within seconds and release active agent immediately. Some of the commonly used are sodium lauryl sulfate, benzalkonium chloride, bezthonium chloride, tweens etc. One of the most important surfactant is polaxamer 407 that is used as solubilizing, wetting and dispersing agent.

Flavor

Any flavor can be added, such as intense mints, sour fruit flavors or sweet confectionery flavors.

Color

A full range of colors is available, including FD&C colors, EU colours, natural colours and custom Pantone-matched colours.

*Some saliva stimulating agents may also be added to enhance the disintegration and to get rapid release. Some of these agents are citric acid, tartaric acid, malic acid, ascorbic acid and succinic acid. List of commercially available FDFs as depicted in Table 12.3.

Table 12.3 List of commercially available FDFs

Product	Active drug	Dose strength (mg)	Application	Company
Triaminic	Dextromethorphan HBr	5/7.5	Seasonal allergy	Novartis
Triaminic	Diphenhydramine HCl	12.5	Thin strip for long acting cough	Novartis
Theraflu	Dextromethorphan HBr	10/20	For long acting cough	Novartis
Gas-X	Simethicone	62.5	Gas-X thin strip anti gas	Novartis
Sudafed PE	Phenylephrine HCl	10	Decongestant oral strips	Pfizer
Benadryl	Diphenhydramine HCl	12.5	Antihistaminic oral strips	Pfizer
Chloraseptic	Benzocaine: Menthol	3/3	Chloraseptic relief strips	Prestige
Suppress	Dextromethorphan	2.5	Suppress cough strips	InnoZen
Suppress	Menthol	2.5	Suppress herbal cough relief Strips	InnoZen
Orazel	Menthol/Pectin	2/30	Cough and cold relief strips	Del
Listerine	Cool mint	–	Antiseptic mouthwash	Pfizer
Little Colds	Pectin	–	Sore throat strips	Prestige brands
Eclipse	Sugar-free mints	–	Chewing gum, breath mint	Wringley's
Donepzil	Donepzil HCL	5/10	In Alzheimer's disease	Labtec GmbH
Ondansetron	Ondensteron	4/8	Antiemetic, helps in nausea and vomiting	Labtec GmbH

12.2.4   Classifications of fast dissolve technology

For ease of description, fast-dissolve technologies can be divided into three broad groups:

    (i)   The lyophilized systems

   (ii)   Compressed tablet-based systems

  (iii)   Thin film strips

(i) The lyophilized systems

The technology around these systems involves taking a suspension or solution of drug with other structural excipients and, through the use of a mould or blister pack, forming tablet-shaped units. The units or tablets are then frozen and lyophilized in the pack or mould. The resulting units have a very high porosity, which allows rapid water or saliva penetration and very rapid disintegration. Dose-handling capability for these systems differs depending on whether the active ingredients are soluble or insoluble drugs, with the dose capability being slightly lower for the former than for some tablet based systems.

(ii) Compressed tablet-based systems

This system is produced using standard tablet technology by direct compression of excipients. Depending on the method of manufacture, the tablet technologies have different levels of hardness and friability. These results in varying disintegration performance and packaging needs, which can range from standard HDPE bottles or blisters through to more specialist pack designs for product protection The speed of disintegration for fast-dissolve tablets compared with a standard tablet is achieved by formulating using water soluble excipients, or superdisintegrant or effervescent components, to allow rapid penetration of water into the core of the tablet.

The one exception to this approach for tablets is Biovail's Fuisz technology. It uses the proprietary Shear form system to produce drug-loaded candy floss, which is then used for tableting with other excipients. These systems can theoretically accommodate relatively high doses of drug material, including taste-masked coated particles. The potential disadvantage is that they take longer to disintegrate than the thin-film or lyophilized dosage forms. The loose compression tablet approach has increasingly been used by some technology houses, branded companies and generic pharmaceutical companies, for in-house development of line extension and generic fast-dissolve dosage forms.

(iii) Thin film strips

Thin film strips, also called oral wafers in the related literature, are a group of flat films which are administered into the oral cavity. Although thin film systems, the third class, have been in existence for a number of years, they have recently become the new area of interest in fast dissolve pharmaceutical drug delivery. Dissolvable oral thin films (OTFs) or oral strip (OS) evolved over the past few years from the confection and oral care markets in the form of breath strips and became a novel and widely accepted form by consumers for delivering vitamins and personal care products.

Companies with experience in the formulation of polymer coatings containing active pharmaceutical ingredients (APIs) for transdermal drug delivery capitalized on the opportunity to transition this technology to OTF formats.

Today, OTFs are a proven and accepted technology for the systemic delivery of APIs for over-the-counter (OTC) medications and are in the early-to mid development stages for prescription drugs. This is largely as a result of the success of the consumer breath freshener products such as Listerine Pocket Paks in the US consumer market. Such systems use a variety of hydrophilic polymers to produce a 50–200 mm film of material. This film can reportedly incorporate soluble, insoluble or taste-masked drug substances. The film is manufactured as a large sheet and then cut into individual dosage units for packaging in a range of pharmaceutically acceptable formats (Fig. 12.1).

Figure 12.1 Thin film strip and buccal film strip

12.2.5   Methodology employed for fast dissolving formulations

1. Melt granulation
2. Phase transition process
3. Sublimation
4. Three-dimensional Printing (3DP)
5. Mass Extrusion
6. Spray Drying
7. Cotton Candy Process
    a) Floss Blend
    b) Floss Processing
    c) Floss Chopping and Conditioning
    d) Blending and Compression

8. Tablet Molding
9. Lyophilization or Freeze-Drying
10. Direct Compression
    a) Superdisintegrants
    b) Sugar Based Excipients
11. Nanonization

1. Melt granulation

Melt granulation technique is a process by which pharmaceutical powders are efficiently agglomerated by a meltable binder. The advantage of this technique compared to a conventional granulation is that no water or organic solvents is needed. Because there is no drying step, the process is less time consuming and uses less energy than wet granulation. It is a useful technique to enhance the dissolution rate of poorly water-soluble drugs, such as griseofulvin. This approach to prepare FDT with sufficient mechanical integrity, involves the use of a hydrophilic waxy binder (Superpolystate©, PEG–6–stearate). Superpolystate© is a waxy material with a melting point of 33–37°C and a HLB value of 9. So it will not only act as a binder and increase the physical resistance of tablets but will also help the disintegration of the tablets as it melts in the mouth and solublises rapidly leaving no residues [21].

2. Phase transition process

It is concluded that a combination of low and high melting point sugar alcohols, as well as a phase transition in the manufacturing process, are important for making FDTs without any special apparatus. FDT were produced by compressing powder containing erythritol (melting point: 122°C) and xylitol (melting point: 93–95°C), and then heating at about 93°C for 15 min. After heating, the median pore size of the tablets was increased and tablet hardness was also increased. The increase of tablet hardness with heating and storage did not depend on the crystal state of the lower melting point sugar alcohol [22].

3. Sublimation

In this method a subliming material like camphor, is removed by sublimat ion from compressed tablets and high porosity is achieved due to the formation of many pores where camphor particles previously existed in the compressed tablets prior to sublimation of the camphor [23]. A high porosity was achieved due to the formation of many pores where camphor particles previously existed in the compressed mannitol tablets prior to sublimation of the camphor. These compressed tablets which have high porosity (approximately 30%) rapidly dissolved within 15 seconds in saliva. Granules containing nimusulide, camphor, crospovidone, and lactose were prepared by wet granulation technique. Camphor was sublimed from the dried granules by vacuum exposure [24]. Conventional methods like dry granulation, wet granulat ion and direct compression with highly soluble excipients, super disintegrants and/or effervescent systems can also be used.

4. Three-dimensional printing (3DP)

Three-dimensional printing (3DP) is a rapid prototyping (RP) technology. Prototyping involves constructing specific layers that uses powder processing and liquid binding materials. A novel fast dissolving drug delivery device (DDD) with loose powders in it was fabricated using the three-dimensional printing (3DP) process. Based on computer-aided design models, the DDD containing the drug acetaminophen were prepared automatically by 3DP system [25]. It was found that rapidly disintegrating oral tablets with proper hardness can be prepared using TAG. The rapid disintegration of the TAG tablets seemed due to the rapid water penetration into the tablet resulting from the large pore size and large overall pore volume [26].

5. Mass extrusion

This technology involves softening of the active blend using the solvent mixture of water soluble polyethylene glycol and methanol and expulsion of softened mass through the extruder or syringe to get a cylindrical shaped extrude which are finally cut into even segments using heated blade to form tablets. This process can also be used to coat granules of bitter drugs to mask their taste [15, 27].

6. Spray drying

In this technique, gelatin can be used as a supporting agent and as a matrix, mannitol as a bulking agent and sodium starch glycolate or crosscarmellose or crospovidone are used as superdisintegrants. Tablets manufactured from the spray-dried powder have been reported to disintegrate in less than 20 seconds in aqueous medium. The formulation contained bulking agent like mannitol and lactose, a superdisintegrant like sodium starch glycolate and croscarmellose sodium and acidic ingredient (citric acid) and/or alkaline ingredients (e.g. sodium bicarbonate). This spray-dried powder, which compressed into tablets showed rapid disintegration and enhanced dissolution. Maximum drug release and minimum disintegration time were observed with Kollidon CL excipient base as compared to tablets prepared by direct compression, showing the superiority of the spray dried excipient base technique over direct compression technique [28].

7. Cotton candy process

The FLASHDOSE® is a MDDDS manufactured using Shearform™ technology in associate ion with Ceform TI™ technology to eliminate the bitter taste of the medicament [29, 30]. The Shearform technology is employed in the preparation of a matrix known as 'floss', made from a combination of excipients, either alone or with drugs. The floss is a fibrous material similar to cotton-candy fibers, commonly made of saccharides such as sucrose, dextrose, lactose and fructose at temperatures ranging between 180°F and 266°F [31]. However, other polysaccharides such as polymaltodextrins and polydextrose can be transformed into fibers at 30–40% lower temperature than sucrose. This modification permits the safe incorporation of thermolabile drugs into the formulation tablets manufactured by this process are highly porous in nature and offer very pleasant mouthfeel due to fast solubilization of sugars in presence of saliva. The manufacturing process can be divided into four steps as detailed below [32].

(a) Floss blend

In this step, 80% sucrose in combination with mannitol/dextrose and 1% surfactant is blended to form the floss mix. The surfactant acts as a crystallization enhancer in maintaining the structural integrity of the floss fibers. It also helps in the conversion of amorphous sugar into crystalline form from an outer portion of amorphous sugar mass and subsequently converting the remaining portion of the mass to complete crystalline structure. This process helps to retain the dispersed drug in the matrix, thereby minimizing migration out of the mixture [33].

(b) Floss processing

The floss formation machine uses flash heat and flash flow processes to produce matrix from the carrier material. The machine is similar to that used in 'cotton-candy' formation which consists of a spinning head and heating elements. In the flash heat process, the heat induces an internal flow condition of the carrier material. This is followed by its exit through the spinning head (2000–3600 rpm) that flings the floss under centrifugal force and draws into long and thin floss fibers, which are usually amorphous in nature [34–36].

(c) Floss chopping and conditioning

This step involves the conversion of fibers into smaller particles in a high shear mixer granulator. The conditioning is performed by partial crystallization through an ethanol treatment (1%) which is sprayed onto the floss and subsequently evaporated to impart improved flow and cohesive properties to the floss [31].

(d) Blending and compression

Finally, the chopped and conditioned floss fibers are blended with the drug along with other required excipients and compressed into tablets. In order to improve the mechanical strength of the tablets, a curing step is also carried out which involves the exposure of the dosage forms to elevated temperature and humidity conditions, (40°C and 85% RH for 15 min). This is expected to cause crystallization of the floss material that results in binding and bridging to improve the structural strength of the dosage form [37].

8. Tablet molding

Molding process is of two types, i.e. solvent method and heat method. Solvent method involves moistening the powder blend with a hydro alcoholic solvent followed by compression at low pressures in molded plates to form a wetted mass (compression molding). The solvent is then removed by air-drying. The tablets manufactured in this manner are less compact than compressed tablets and posses a porous structure that hastens dissolution. The heat molding process involves preparation of a suspension that contains a drug, agar and sugar (e.g. mannitol or lactose) and pouring the suspension in the blister packaging wells, solidifying the agar at the room temperature to form a jelly and drying at 30°C under vacuum. The mechanical strength of molded tablets is a matter of great concern. Binding agents, which increase the mechanical strength of the tablets, need to be incorporated. Taste masking is an added problem to this technology. The taste masked drug particles were prepared by spray congealing a molten mixture of hydrogenated cottonseed oil, sodium carbonate, lecithin, polyethylene glycol and an active ingredient into a lactose-based tablet triturate form. Compared to the lyophillization technique, tablets produced by the molding technique are easier to scale up for industrial manufacture [38].

9. Lyophilization or freeze-drying

Freeze drying is the process in which water is sublimed from the product after it is frozen. This technique creates an amorphous porous structure that can dissolve rapidly. A typical procedure involved in the manufacturing of ODT using this technique is mentioned here. The active drug is dissolved or dispersed in an aqueous solution of a carrier/polymer. The mixture is done by weight and poured in the walls of the preformed blister packs. The trays holding the blister packs are passed through liquid nitrogen freezing tunnel to freeze the drug solution or dispersion. Then the frozen blister packs are placed in refrigerated cabinets to continue the freeze-drying. After freeze drying the aluminum foil backing is applied on a blister-sealing machine. Finally the blisters are packaged and shipped. The freeze-drying technique has demonstrated improved absorption and increase in bioavailability. The major disadvantages of lyophillization technique are that it is expensive and time consuming; fragility makes conventional packaging unsuitable for these products and poor stability under stressed conditions [39].

10. Direct compression

Direct compression represents the simplest and most cost effective tablet manufacturing technique. This technique can now be applied to preparation of ODT because of the availability of improved excipients especially superdisintegrants and sugar based excipients.

(a) Superdisintegrants

In many orally disintegrating tablet technologies based on direct compression, the addition of super disintegrants principally affects the rate of disintegration and hence the dissolution. The presence of other formulation ingredients such as water-soluble excipients and effervescent agents further hastens the process of disintegration. For the success of fast dissolving tablet, the tablet having quick dissolving property which is achieved by using the super disintegrants.

(b) Sugar-based excipients

This is another approach to manufacture ODT by direct compression. The use of sugar based excipients especially bulking agents like dextrose, fructose, isomalt, lactilol, maltilol, maltose, mannitol, sorbitol, starch hydrolysate, polydextrose and xylitol, which display high aqueous solubility and sweetness, and hence impart taste masking property and a pleasing mouthfeel. Mizumito et al. have classified sugar-based excipients into two types on the basis of molding and dissolution rate.

11. Nanonization

A recently developed nanomelt technology involves reduction in the particle size of drug to nanosize by milling the drug using a proprietary wet-milling technique. The nanocrystals of the drug are stabilized against agglomeration by surface adsorption on selected stabilizers, which are then incorporated into MDTs. This technique is especially advantageous for poor water-soluble drugs. Other advantages of this technology include fast disintegration/ dissolution of nanoparticles leading to increased absorption and hence higher bioavailability and reduction in dose, cost-effective manufacturing process, conventional packaging due to exceptional durability and wide range of doses (up to 200 mg of drug per unit) [15].

12.2.6   Advantages of oral strips

- Availability of larger surface area that leads to rapid disintegrating and dissolution in the oral cavity.

- The disadvantage of most ODT is that they are fragile and brittle, which warrants special package for protection during storage and transportation. Since the films are flexible they are not as fragile as most of the ODTs. Hence, there is ease of transportation and during consumer handling and storage.

- As compared to drops or syrup formulations, precision in the administered dose is ensured from each of the strips.

- No need of water has led to better acceptability amongst the dysphagic patients. The difficulty encountered in swallowing tablets or capsules is circumvented. The large surface area available in the strip dosage form allows rapid wetting in the moist buccal environment. The dosage form can be consumed at anyplace and anytime as per convenience of the individual.

- The oral or buccal mucosa being highly vascularized, drugs can be absorbed directly and can enter the systemic circulation without undergoing first-pass hepatic metabolism. This advantage can be exploited in preparing products with improved oral bioavailability of molecules that undergo first pass effect.

- Since the first pass effect can be avoided, there can be reduction in the dose which can lead to reduction in side effects associated with the molecule.

- Patients suffering from dysphagia, repeated emesis, motion sickness, and mental disorders prefer this dosage form, as they are unable to swallow large quantity of water [40–42].

12.2.7   Disadvantage of oral strip

The disadvantage of OS is that high dose cannot be incorporated into the strip. However, research has proven that the concentration level of active can be improved up to 50% per dose weight. Novartis Consumer Health's Gas-X® thin strip has a loading of 62.5 mg of simethicone per strip.

There remain a number of technical limitations with the use of film strips. The volume of the dosage unit is clearly proportional to the size of the dose, which means these extremely thin dosage forms are best suited to lower-dose products. As an example of this, Labtec claim that the Rapid Film technology can accommodate dose of up to 30 mg. This clearly limits the range of compatible drug products. The other technical challenge with these dosage forms is achieving dose uniformity.

12.2.8   Application of oral strip in drug delivery

Oral mucosal delivery via Buccal (Fig. 12.1), sublingual, and mucosal route by use of OTFs could become a preferential delivery method for therapies in which rapid absorption is desired, including those used to manage pain, allergies, sleep difficulties, and central nervous system disorders.

Dissolvable oral thin films (OTFs) evolved over the past few years from the confection and oral care markets in the form of breath strips and became a novel and widely accepted form by consumers for delivering vitamins and personal care products.

Topical applications

The use of dissolvable films may be feasible in the delivery of active agents such as analgesicsor antimicrobial ingredients for wound care and other applications.

Gastro retentive dosage systems

Dissolvable films are being considered in dosage forms for which water-soluble and poorly soluble molecules of various molecular weights are contained in a film format. Dissolution of the films could be triggered by the pH or enzyme secretions of the gastrointestinal tract, and could potentially be used to treat gastrointestinal disorders.

Diagnostic devices

Dissolvable films may be loaded with sensitive reagents to allow controlled release when exposed to a biological fluid or to create isolation barriers for separating multiple reagents to enable a timed reaction within a diagnostic device [43–45].

12.2.9   Manufacturing methods

One or combination of the following process can be used to manufacture the mouth dissolving films.

1. Solvent casting
2. Semisolid casting
3. Hot melt extrusion
4. Solid dispersion extrusion
5. Rolling

1. Solvent casting method

In solvent casting method water-soluble polymers are dissolved in water and the drug along with other excipients is dissolved in suitable solvent then both the solutions are mixed and stirred and finally casted in to the Petri plate and dried (Fig. 12.1).

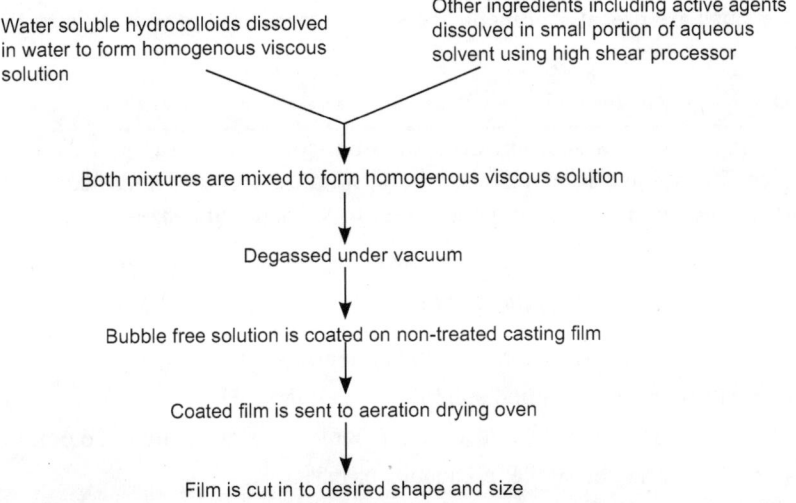

Figure 12.2 Method of preparation of films

2. Semisolid casting

In semisolid casting method firstly a solution of water-soluble film-forming polymer is prepared. The resulting solution is added to a solution of acid insoluble polymer (e.g. cellulose acetate phthalate, cellulose acetate butyrate), which was prepared in ammonium or sodium hydroxide. Then appropriate amount of plasticizer is added so that a gel mass is obtained. Finally the gel mass is casted in to the films or ribbons using heat controlled drums. The thickness of the film is about 0.015–0.05 inches. The ratio of the acid insoluble polymer to film forming polymer should be 1:4. Both mixtures are mixed to form homogenous viscous solution. Degassed under vacuum Bubble free solution is coated on non-treated casting film coated film is sent to aeration drying oven film is cut into desired shape and size [46–49].

3. Hot melt extrusion

In hot melt extrusion method firstly the drug is mixed with carriers in solid form. Then the extruder having heaters melts the mixture. Finally the melt is shaped in to films by the dies. There are certain benefits of hot melt extrusion.

- Fewer operation units
- Better content uniformity
- An anhydrous process

4. Solid dispersion extrusion

In this method immiscible components are extrude with drug and then solid dispersions are prepared. Finally the solid dispersions are shaped in to films by means of dies.

5. Rolling method

In rolling method a solution or suspension-containing drug is rolled on a carrier. The solvent is mainly water and mixture of water and alcohol. The film is dried on the rollers and cut into desired shapes and sizes.

12.2.10   Characteristics of compounds suitable for this system

- Compounds should have good aqueous solubility.
- They must have good solubility at salivary pH.
- They must have low dose so that could be incorporated into oral film.
- They must be stable in aqueous or basic pH.
- Compounds with problem of first pass metabolism are suitable for this system.
- Compounds with gastric irritating property are suitable for this system.

12.2.11   Quality control tests

Thickness

It can be measured by micrometer screw gauge at different strategic locations. This is essential to ascertain uniformity in the thickness of the film as this is directly related to the accuracy of dose in the strip.

Dryness test/tack tests

About eight stages of film drying process have been identified i.e. set-to-touch, dust-free, tack-free (surface dry), dry-to-touch, dry-hard, dry-through (dry-to-handle), dry-to-recoat and dry print-free. Although these tests are primarily used for paint films most of the studies can be adapted intricately to evaluate pharmaceutical OS as well. The details of evaluation of these parameters can be checked elsewhere.

Tack is the tenacity with which the strip adheres to an accessory (a piece of paper) that has been pressed into contact with the strip.

Tensile strength

Tensile strength is the maximum stress applied to a point at which the strip specimen breaks. It is calculated by the applied load at rupture divided by the cross-sectional area of the strip as given in the equation below:

$$\text{Tensile strength} = \frac{\text{Load at failure} \times 100}{\text{Strip thickness} \times \text{Strip width}}$$

Percent elongation

When stress is applied, a strip sample stretches and this is referred to as strain. Strain is basically the deformation of strip divided by original dimension of the sample. Generally elongation of strip increases as the plasticizer content increases.

Tear resistance

Tear resistance of plastic film or sheeting is a complex function of its ultimate resistance to rupture. Basically very low rate of loading 51 mm (2 inch)/min is employed and is designed to measure the force to initiate tearing. The maximum stress or force (that is generally found near the onset of tearing) required to tear the specimen is recorded as the tear resistance value in Newton (or pounds-force).

Folding endurance

Folding endurance is determined by repeated folding of the strip at the same place till the strip breaks. The number of times the film is folded without breaking is computed as the folding endurance value.

Disintegration time

The disintegration time limit of 30 seconds or less for orally disintegrating tablets described in CDER Center for Drug Evaluation and Research (FDA) guidance can be applied to fast dissolving oral strips. Although, no official guidance is available for oral fast disintegrating films strips, this may be used as a qualitative guideline for quality control test or at development stage. Pharmacopoeial disintegrating test apparatus may be used for this study. Typical disintegration time for strips is 5–30 seconds.

Dissolution test

Dissolution testing can be performed using the standard basket or paddle apparatus described in any of the pharmacopoeia. The dissolution medium will essentially be selected as per the sink conditions and highest dose of the API. Many times the dissolution test can be difficult due to tendency of the strip to float onto the dissolution medium when the paddle apparatus is employed.

Assay/drug content and content uniformity

This is determined by any standard assay method described for the particular API in any of the standard pharmacopoeia. Content uniformity is determined by estimating the API content in individual strip. Limit of content uniformity is 85–115 percent [50–52].

12.3   Conclusions

The development of fast-dissolving films also provides an opportunity for a line extension in the marketplace; a wide range of drugs (e.g., antiemetic, neuroleptics, cardiovascular agents, analgesics, antiallergic, antiepileptic, anxiolytics, sedatives, hypnotics, diuretics, antiparkinsonism agents, anti-bacterial agents and drugs used for erectile dysfunction, antialzheimers, expectorants and anitussive) can be considered candidates for this dosage form. Pharmaceutical marketing is another reason for the increase in available fast dissolving/disintegrating products. As a drug entity nears the end of its patent life, it is common for other pharmaceutical manufacturers to develop a given drug entity in a new and improved dosage form. A new dosage form allows a manufacturer to extend market exclusivity, while offering its patient population a more convenient dosage form or dosing regimen. In this regard, fast dissolving/ disintegrating tablet formulations are similar to many sustained release formulations that are now commonly available. An extension of market exclusivity, which can be provided by a fast-dissolving/disintegrating dosage form, leads to increased revenue, while also targeting underserved and under-treated patient populations. Although the cost to manufacture these specialized dosage forms exceeds that of traditional tablets.

12.4   References

1. Siddiqui M.D., Garg G., Sharma P. A short review on "A Novel Approach in Oral Fast Dissolving Drug Delivery System and their Patents" *Adv Biol Res.* 2011, 5: 291–303.

2. Galey W.R., Lonsdale H.K., Nacht S. The in vitro permeability of skin and buccal mucosa to selected drugs and tritiated water. *J Invest Dermatol.* 1976, 67: 713–7.

3. Malke M., Shidhaye S., Kadam V.J. Formulation and evaluation of oxcarbazepine fast dissolve tablets. *Indian J Pharm Sci.* 2007; 69: 211–4.

4. Mishra R., Amin A. Formulation and characterization of rapidly dissolving films of cetirizine hydrochloride using pullulan as a film forming agent. *Indian J Pharm Educ Res.* 2011; 45: 71–7.

5. Mahajan A., Chabra N., Aggarwal G. Formulation and characterization of fast dissolving buccal films: A review. Sch Res Libr Der Pharm Lett. 2011; 3:152–65.

6. Arya A., Chandra A., Sharma V., Pathak K. Fast dissolving oral films: An innovative drug delivery system and dosage form. *Int J Chem Tech Res.* 2010; 2: 576–83.

7. Chemical Market Reporter. Fuisz sign deal for drug delivery. *Chem Mark Report.* 1998; 253: 17.

8. Bhyan B., Jangra S., Kaur M., Singh H. Orally fast dissolving films: Innovations in formulation and technology. *Int J Pharm Sci Rev Res.* 2011; 9: 50–6.

9. Bhura N., Sanghivi K., Patel U., Parmar B. A review on fast dissolving film. *Int J Res Bio Sci.* 2012; 3: 66–9.

10. Fulzele S.V., Satturwar P.M., Dorle A.K. Polymerized rosin: Novel film forming polymer for drug delivery. *Int J Pharm.* 2002; 249: 175–84.

11. Barnhart SD, Sloboda MS. The future of dissolvable films. *Drug Deliv Technol.* 2007; 7: 34–7.

12. Hariharn M., Bogue A. Orally dissolving film strips (ODFS): The final evolution of orally dissolving dosage forms. *Drug Deliv Technol.* 2009; 9: 24–9.

13. Nagar P., Chauhan I., Yasir M. Insight into polymers: Film formers in mouth dissolving films. *Drug Invent Today.* 2011; 3: 280–9.

14. Dixit R.P., Puthli S.P. Oral strip technology: Overview and future potential. *J Control Release.* 2009; 139: 94–107.

15. Saurabh R., Malviya R., Sharma P.K. Trends in buccal film: Formulation characteristics, recent studies and patents. *Eur J Appl Sci.* 2011; 3: 93–101.

16. Gauri S., Kumar G. Fast dissolving drug delivery and its technologies. *Pharm Innov.* 2012; 1: 34–9.

17. Deshmane S.V., Joshi U.M., Channwar M.A., Biyani K.R., Chandewar A.V. Design and characterization of carbopol-HPMC-ethyl cellulose based buccal compact containing propranolol HCl. *Indian J Pharm Educ Res.* 2010; 44: 67–78.

18. Khairnar A., Jain P., Bhaviskar D., Jain D. Development of mucoadhesive buccal patches containing aceclofenac: *In vitro* evaluation. *Int J Pharm Sci.* 2009; 1: 91–5.

19. Shinde A.J., Garala K.C., More H.N. Development and characterization of transdermal therapeutics system of tramadol hydrochloride. *Asian J Pharm.* 2008; 2: 265–9.

20. Dong Y., Kulkarni R., Behme R.J., Kotiyan P.N. Effect of the melt granulation technique on the dissolution characteristics of griseofulvin. *International Journal of Pharmaceutics* 2007; 329 (1, suppl 2): 7280.

21. Abdelbary G., Prinderre P., Eouani C., Joachim J., Reynier J.P., Piccerelle P. The preparation of orally disintegrating tablets using a hydrophilic waxy binder. *International Journal of Pharmaceutics* 2004; 278(2): 423–33.

22. Kuno Y., Kojima M., Ando S., Nakagami H. Evaluation of rapidly disintegrating tablets manufactured by phase transition of sugar alcohols. *Journal of Controlled Release* 2005; 105(1–2): 16–22.

23. Koizumi K., Watanabe Y., Morita K., Utoguchi N., Matsumoto M. New method of preparing high-porosity rapidly saliva soluble compressed tablets using mannitol with camphor, a subliming material. *International Journal of Pharmaceutics* 1997; 152(1): 127–31.

24. Gohel M., Patel M., Amin A., Agrawal R., Dave R., Bariya N. Formulation design and optimization of mouth dissolve tablets of nimesulide using vacuum drying technique. *AAPS PharmSciTech* 2004; 5(3): 10–15.

25. Yu D.G., Shen X.X., Han J., Zhu L.M., Branford-White C., Li X.Y., Yang X.L. Oral Fast-Dissolving DDD Fabricated Using 3DP, Bioinformatics and Biomedical Engineering. The 2nd International Conference 2008; 1602–605.

26. Ito A., Sugihara M. Development of oral dosage forms for elderly patients: use of agar as base of rapidly disintegrating oral tablets. *Chem Pharm. Bull.* 1996; 44(11): 2132–136.

27. Bhaskaran S., Narmada G.V. Rapid dissolving tablets a novel dosage form. *Indian Pharmacist.* 2002; 1: 9–12.

28. Mishra D.N., Bimodal M., Singh S.K., Vijaya Kumar S.G. Spray dried excipient base: a novel technique for the formulation of orally disintegrating tablets. *Chem Pharm Bull* 2006; 54(1): 99–102.

29. Myers G.L., Battist G.E., Fuisz R.C. Process and apparatus for making rapidly dissolving dosage units and product there from. PCT Patent WO 95/34293-A1; 1995.

30. Cherukuri S.R., Myers G.L., Battist G.E., Fuisz R.C. Quickly dispersing comestible unit and product. PCT Patent WO 95/34290-A1;1995.

31. Cherukuri S.R., Myers G.L., Battist G.E., Fuisz R.C. Process for forming quickly dispersing comestible unit and product there from. US Patent 5587172; 1996.

32. Fuisz R. Ulcer prevention method using a melt-spun hydrogel. US Patent 5622717; 1997.

33. Fuisz R., Cherukuri S.R. Process and apparatus for making tablets and tablets made there from. US Patent 5654003; 1997.

34. Myers G.L., Battist G.E., Fuisz R.C. Delivery of controlled-release systems. US Patent 5567439; 1996.

35. Myers G.L., Battist G.E., Fuisz RC. Apparatus for making rapidly dissolving dosage units.US Patent 5871781; 1999.

36. Cherukuri S.R., Fuisz R. Process and apparatus for making tablets and tablets made there from. US Patent 5654003; 1997.

37. Myers G.L., Battist G.E., Fuisz R.C. Ulcer prevention method using a melt-spun hydrogel. US Patent 5622719; 1997.

38. Bhowmik D., Chiranjib B., Pankaj K., Chandira R.M. Fast Dissolving Tablet: An Overview. *Journal of Chemical and Pharmaceutical Research* 2009; 1(1): 16377.

39. Elan Corporation, plc. Orally Disintegrating Tablets (ODT) - Nanomelt™. http://www.elan.com/EDT/nanocrystal%5Ftechnology/orally_disintegrating_tablet.asp

40. Mizumoto T., Tamura T., Kawai H., Kajiyama A. Formulation Design of TasteMasked Particles, Including Famotidine, for an Oral Fast-Disintegrating Dosage Form. *Chem. Pharm. Bull.* 2008; 56(4): 530–35.

41. Sugimoto M., Narisawa S., Matsubara K., Yoshino H. Effect of Formulated Ingredients on Rapidly Disintegrating Oral Tablets Prepared by the Crystalline Transit ion Method. *Chem. Pharm. Bull.* 2006; 54(2): 175–80.

42. Watanabe Y., Zama Y., Matsumoto M. New compressed tablet rapidly disintegrating in saliva in the mouth using crystalline cellulose and a disintegrants. *Biol. Pharm. Bull.* 1995; 18(9): 1308–10.

43. Sakellariou P., Rowe R.C. Interactions in cellulose derivative films for oral drug delivery. *Prog Polym Sci.* 1995; 20: 889–942.

44. Browhn G.L. Formation of film from polymer dispersions. *J Polym Sci.* 1956; 22: 423–34.

45. Brown D. Orally disintegrating tablets-taste over speed. *Drug Deliv Technol.* 2003; 3: 58–61.

46. Zerbe H., Guo J. Water soluble films for oral administration with instant wettability. US Patent 5948430, Sep 7, 1999.

47. Tapolsky G., Osborne D. Bioerodable film for delivery of pharmaceutical compounds to mucosal surface. US Patent, 6159498, Dec 12, 2000.

48. Lori D. Fast dissolving orally consumable films containing sweetners. US Patent 2003/0211136 Nov 13, 2003.

49. Friend D.R., Levine A.W., Ziegler K.L., Manna E. Fast dissolving films for oral administration of drugs. US Patent 2004/0208931 A1. 2004.

50. Fadden D.J., Kulkarni N., Sorg A.F. Fast dissolving oral consumable film containing modified starch for improved heat and moisture resistance. US Patent 2004/0247648 May 3, 2003.

51. Leung S.S., Leone R.S., Kumar L.D., Kulkarni N., Sorg A.F. Fast dissolving orally consumable film. US Patent 7025983, Apr 11, 2006.

52. Kupper R., Smothers M. Dissolving thin film xanthone supplement. US Patent 7182964 B2, Feb 27, 2007.